P9-CME-056

DO-IT-YOURSELF PLUMBING

A POPULAR SCIENCE BOOK

DO-IT-YOURSELF
PLUMBING

MAX ALTH

Drawings by

Carl J. De Groote

POPULAR SCIENCE

HARPER & ROW

New York, Evanston, San Francisco, London

ACKNOWLEDGMENT

The author wishes to acknowledge his appreciation to his editor, Henry Gross, for his help, dedication and patience in the preparation of this book.

Published by Book Division, Times Mirror Magazines, Inc.

Library of Congress Catalog Card Number: 74-27320
SBN: 0-06-010122-9

Designed by Jeff Fitschen

Manufactured in the United States of America

To

Char,
Misch,
Syme,
Sal,
Arabella
and
Mendel

Contents

Introduction

When you call for immediate plumbing assistance today, you are often answered by a machine that offers to record your cry for help as soon as it beeps. A few hours later a young lady phones to inform you that next Tuesday is the soonest you can have an appointment.

The repair, when it is finally made, is expensive. This is to be expected, for every workman is worthy of his hire. Yet when we see the bill we realize that we have not hired an individual, but a corporation, and that our biggest mistake last year was buying stock in the phone company instead of in Plumbers Inc. Of course, it is unrealistic to expect plumbers to remain one-man operations, forever working out of the back of a truck. And it may be unfair to expect them to rush out in the middle of the night like firemen.

What really hurts, though, is when you are standing knee deep in water next to the plumber who has finally arrived and watch him cure the trouble with a turn of a valve handle or a flick of the pipe wrench. It hurts because you know you could have done the same thing—without the wait, without the flood damage and without the bill. All you needed was knowledge.

Plumbing is a vast body of knowledge composed of little bits and pieces. This is what makes plumbing seem so complicated—all the different kinds of pipes, valves, faucets, fittings and fixtures, each requiring simple, yet somewhat different, treatment to assemble, join and repair. When you realize that a plumbing system is put together much like an Erector set, you will have no hesitancy in tackling the problems that arise.

The first chapter of this book provides the insight you need into how a plumbing system works. This is followed by chapters on the tools you need to do your own plumbing and the parts and materials that compose a plumbing system. You don't have to master all of this information; just review it and refer back when necessary.

Chapters 5 through 8 deal with repairs. Here is where to look when you have a specific plumbing problem—when you want to repair a faucet, valve, toilet or leaking pipe, or unclog a drain.

However, there is more to plumbing than just repairs. An additional room or wing on a house needs water and heat. Old fixtures need to be replaced with new ones. A new house needs a complete plumbing system. The principles underlying all this work are explained in three chapters on designing a system. Here you learn how to calculate the size of pipes, and how to run drainpipes in conformity with local codes. Instructions in designing a system are followed by practical guidance in installing pipes and fixtures in a new house or in an addition, and in adding fixtures and appliances in your present dwelling.

As a home is not a home without central heating—and in some instances, a septic tank—I have devoted chapters to maintaining and repairing these vital parts of the plumbing installation.

Throughout this book I have been guided by the conviction that how-to instructions alone are excellent for quickly conveying information, but when the device or project on hand is not covered by the how-to instructions, there is nowhere to turn. There is no way of figuring the "thing" out. On the other hand, if you learn *how* things work—the principles involved—you can generally solve most of the problems you encounter. At the very least, you will have an inkling of which way to go.

It is my hope that when you have read this book you will be a plumber, albeit a beginning plumber, but with sufficient knowledge to successfully master any repair or construction you undertake.

DO-IT-YOURSELF PLUMBING

1

Your Home Plumbing System

As used in this book, the term "plumbing" includes everything in a house that has to do with heat, hot and cold water and gas. The electrical side of the equipment and controls does not fall within the scope of plumbing and is not covered in this book beyond mention.

Taken as a unit, a modern home certainly has a maze of piping, valves, gauges and controls. But there is no need to ever consider the entire system of plumbing as a unit because there are actually several systems, some interconnected, some not. In any event, interconnected systems can always be separated by existing controls for the purpose of repair, alteration and additions. In all, the modern home may contain the following systems:

Cold water	Heating
Hot water	Gas
Drainage	Outdoor

THE COLD-WATER SYSTEM

If you live in a city or suburb, and most of us do, your cold water comes from a large pipe called a main, which usually runs down the length of your street. In some localities, it may run through your backyard and those of your neighbors.

The municipality or a water company working under the direction of the municipality supplies clean water under pressure to the main. When you open a hot- or cold-water faucet in your home, it is the pressure on the water in the main that drives the water out of your faucet.

Should a pipe in your home break, or should one of your faucets leak, there are any number of valves you can close by simply turning them to shut off the flow of water. If you do not close a valve in the pipeline, water will continue to flow.

As the accompanying diagram shows, there are three valves (sometimes two) in the pipeline—called the water-service line—coming into your home. One is called the curb valve, and it is located just your side of the street curb, under a hinged metal cover sometimes marked "Water." This valve is underground and can only be operated by means of a long-handled wrench used by

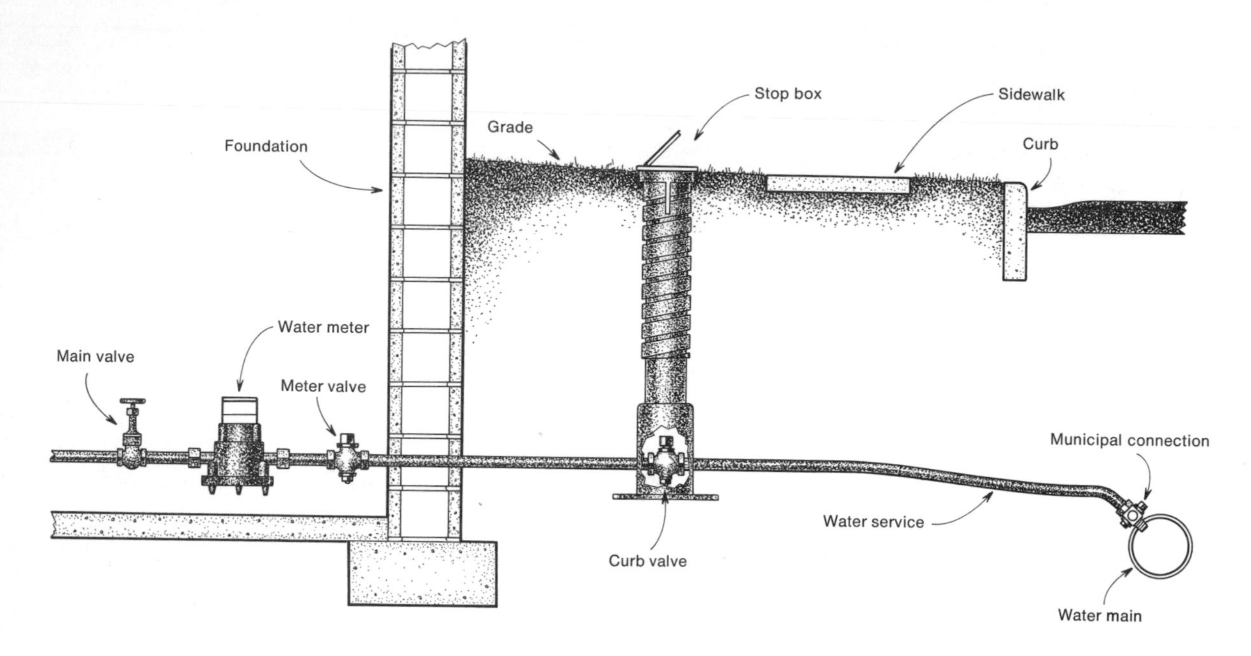

Standard water-service installation. The city maintains water under constant pressure in its water main. From there the water flows through valves and the water meter into the house. Access to the curb valve is through the stop box.

the water company. The second main valve, the meter valve, is located just inside your basement or cellar wall where the pipe enters. Closing this valve shuts off the flow of water into the house, and permits the water meter to be repaired. Whether or not there is a meter stop valve in your water-service line, there is a main valve on the other side of the water meter. This is the valve you use to shut off the water flow. This does not remove water already in the pipes, however, for every hot- and cold-water pipe is filled with water. Thus shutting the main valve will stop water from flowing from an open faucet or break in a pipe on the top floor, but not from pipes and faucets lower down. When these are opened, all the water contained in the piping at a higher level will flow out. In some instances, siphoning may take place and water from a lower level may flow out.

The curb valve is located under this hinged cover. Find your curb valve and keep it clear of grass so it is accessible to the water company when it has to be closed in a hurry.

When a pipeline has to be opened for a repair or alteration, the system must be drained of water in addition to shutting off the incoming flow. To speed draining, faucets above the desired level are opened to admit air.

It is important you locate the main valve and make certain it is kept accessible. This is the valve you run to when something is wrong and you don't have time to locate a secondary valve near the trouble spot.

If you are unfamiliar with valves, it is suggested you try shutting off the water just to see how it works and to make certain you know in which direction it turns. If the valve has not been used for a long time, and this is usual, it sometimes sticks and it is difficult for a beginner in a hurry to know which is the correct direction to force the reluctant handle.

THE WATER METER

The cold water entering the house from the service line passes through the main valve

Typical residential water-meter installation. In this house, the main valve was installed on the inside, in *front* of the water meter. Closing this valve stops the flow of water into the house. The cap is lifted to read the meter.

and then the water meter before continuing on. The water meter is present for one purpose only, to measure the quantity of water consumed. Outside of it going tap, tap, tap when the water is turned on anywhere in the house, the meter does nothing but sit there and add up the cubic feet of water

This meter face reads 77,834 and a fraction cubic feet. The figure is derived by adding the readings of all the dials. Each reading is expressed in a round number. Thus, starting with the dial marked 100,000, the column would look like this:

70,000
7,000
800
30
4
77,834
cubic feet

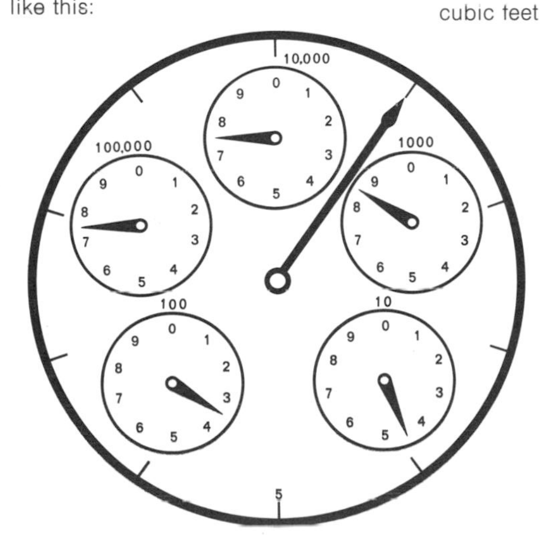

This meter reads 96,860 and a fraction. Note that the pointers on this meter do not all rotate in the same direction. The one-cubic-foot dial is used for tests.

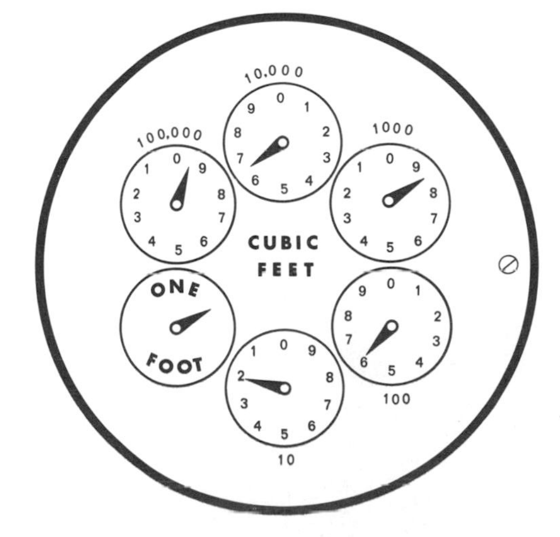

used. If you wish, the meter can be used to calculate the water cost of filling a pool or watering a lawn. The meter can also be used to detect if there is leakage when the water is shut off. If the hands on the meter move, you have a hidden leak.

The accompanying drawing illustrates two water meters and how they may be read. The "1 cubic foot" dial is the one to watch. Once around is 1 cubic foot of water, which equals 7.481 gallons. If there is any movement of the indicator after everything has been closed, you have a leak. It is quite possible to have a leak in a pipe in a wall or a partially open valve in the basement and believe a damp cellar is the result of rain water seepage. It is worth checking.

If you have any doubts about the accuracy of your water meter, call the water works. They will check it without charge. If the meter leaks, call them too. It is their meter and their responsibility. Do not touch it.

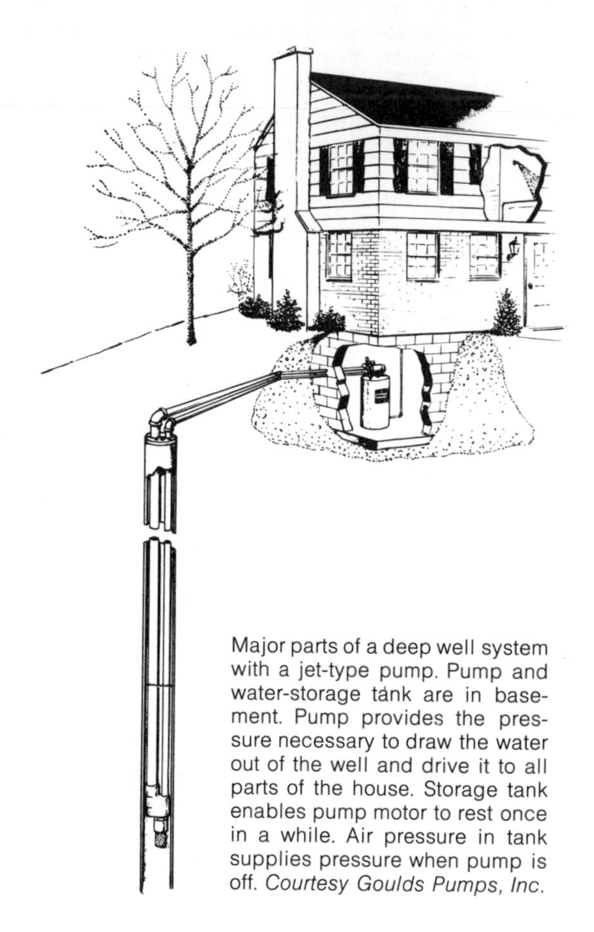

Major parts of a deep well system with a jet-type pump. Pump and water-storage tank are in basement. Pump provides the pressure necessary to draw the water out of the well and drive it to all parts of the house. Storage tank enables pump motor to rest once in a while. Air pressure in tank supplies pressure when pump is off. *Courtesy Goulds Pumps, Inc.*

WELL SYSTEMS

There are any number of different well systems. Essentially, a well system is a means of extracting water from a local well and supplying it to the home under pressure. This is usually accomplished by a wind or electrically driven pump.

If pressure is developed by a pump alone, shutting this off relieves the system of pressure, just as closing the water-service valve in the basement does. Water still remains in the house pipelines and must be drained if any of the lines are to be opened for repairs.

When there is a water-storage tank, the connection from the tank to the house must be closed to prevent water from the tank from flowing after the pump has been shut down. If the tank is pressurized, the air within it must be released. If it is not released, the pressure will empty the tank even though the pump is turned off.

From the user's point of view, it makes no difference how the water in the pipes has gotten there. It flows the same way, private pump system or municipal pump system.

THE HOT-WATER SYSTEM

The hot water that comes out of the faucet on the kitchen sink is called domestic hot water to distinguish it from the water used in a hot-water heating system. Under no circumstance are these two systems ever connected so that the water from one flows into the other.

The equipment used to heat cold water for domestic use varies in design, size and type. Depending on the volume of water wanted, the cost of various fuels and their availability, the domestic hot-water heating equipment may be oil, gas or electrically fired. Some older installations even depend on coal. In many homes the hot water is heated in the furnace that supplies the house with heat. But no matter what means are used to heat the water, the basic piping is always just about the same.

Cold water is led into the heating device via one pipeline. Hot water comes out of the device via a second pipeline. Except for a slight loss of pressure due to the heating apparatus and associate fittings, hot-water pressure is normally the same as cold-water pressure. Closing the main valve, or any valve in the cold-water pipeline leading to the hot-water heater, stops the flow of water through the heater.

The hot-water system may therefore be considered as starting at the domestic hot-water heater and branching out to the various faucets from there. Generally hot- and cold-water pipes to paired faucets are the same size and are run parallel to one another. To identify the hot-water pipe, simply touch it; it is hot. Should there be little or no difference in temperature, open the hot-water tap and let the hot water run a while. The pipe carrying the hot water will quickly grow warm.

Hot-water temperature and hot-water quantity are only partially related. It is quite possible for a domestic hot-water system to be in such condition and adjustment that when a hot-water tap is opened one gets a burst of boiling hot water and then nothing but lukewarm water. It is also possible for another system to put out a continuous stream of moderately warm water.

Water temperature depends on the setting of the temperature-control device on the domestic hot-water heater and the temperature of the water coming in through the service pipe. Domestic hot water is therefore always a bit hotter in the summer than the winter.

Therefore it is always possible, excepting for defective equipment, to adjust the domestic hot-water heater to produce very hot water. The quantity of hot water produced, meaning how much hot water can be drawn before it runs lukewarm, depends on the capacity of the device, and its condition, which is almost directly related to its age.

The following will explain how the various factors are interrelated.

There are two different types of domestic hot-water heating methods or systems. One type consists of a tank and a heating means which can be gas or electric. The other type does not have a tank. It is called tankless.

When the temperature of the water in a tank heater drops below a preset temperature, the heating device is automatically turned on. When all the water in the tank comes up to another preset temperature the heating device is turned off. The immediate availability of hot water in a tank heater is about 70 percent of the hot water in the tank. When a hot-water tap is opened, hot water flows out of the tank and cold enters. The time necessary for the heating device to raise the temperature of a cold tank of water to its hot temperature is called its "recovery time." As can be realized, this is as important as tank size. Recovery time is an integral function of tank design—coil size, heater BTU, etc. It cannot be changed, and as the tank ages and corrosion and mineral deposits form on the heat exchanger, recovery time grows longer. Water temperature, however, does not drop appreciably. Therefore, one cannot evaluate a tank system by water temperature alone.

The tankless system is not connected to

a tank. It is merely a heat exchanger comprising a coil of copper pipe. In one design, called external, the heat exchanger is wrapped around a heating main, which is connected to the boiler in the furnace used to heat the house. The boiler may be steam or hot water.

In the other tankless heater designed for producing domestic hot water, the heat exchanger coil is immersed in the water inside the furnace boiler.

The external design, sometimes called takeoff or even Taco (after the name of one manufacturer), is not as efficient as the im-

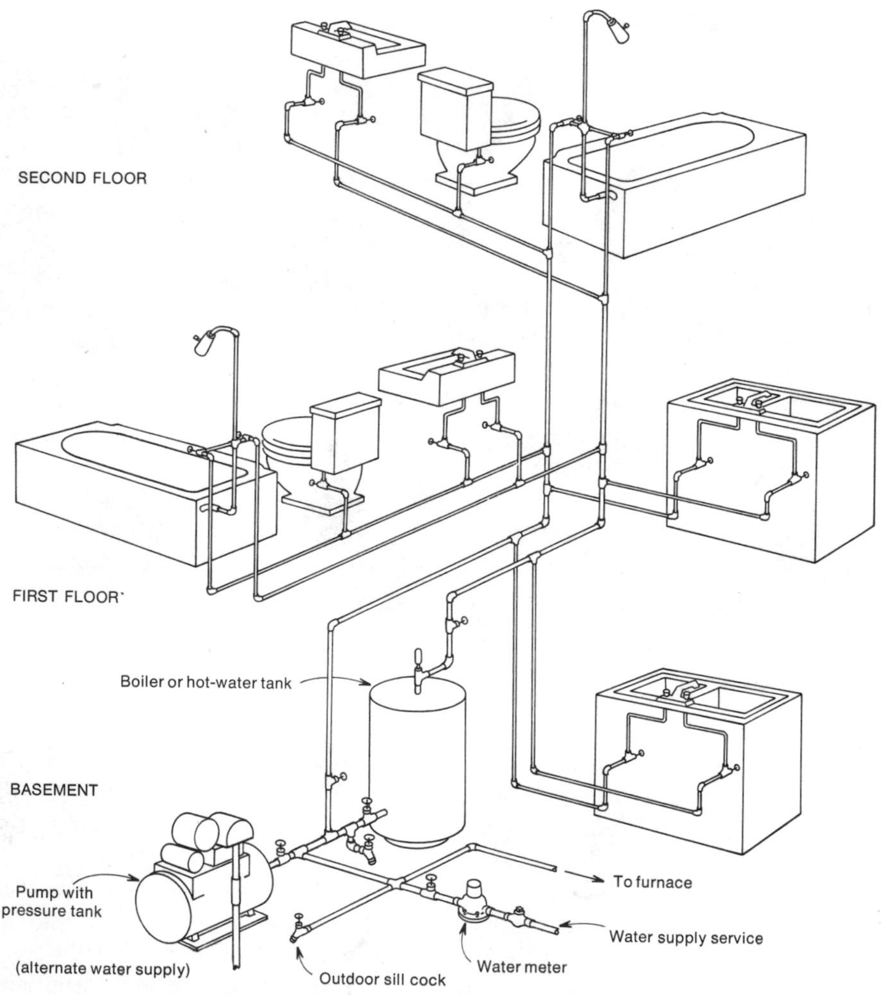

Hot- and cold-water system in a typical two-story house. Water originates at either service pipe or well pump and is piped directly to all the cold-water faucets and outlets and the domestic hot-water heater. From the heater, the hot water goes to all the hot-water faucets. In addition there is a cold-water pipe running to the furnace.

mersed design. However, it has the advantage of being easier to repair and to replace.

The hot water coming out of a tankless heater is mixed with cold water. This is accomplished by a mixing valve which connects the cold-water pipe entering and the hot-water pipe leaving the heat exchanger. When a tankless heater is new it will produce its specified flow of hot water endlessly. As it ages and its efficiency drops, the mixing valve is adjusted to permit more hot water and less cold to flow into the hot-water main. The heater is unable to keep up with its specified flow rate. The result is that the first surge of water is too hot, the balance is too cool. When the heat exchanger becomes so inefficient that the mixing valve must be closed entirely to cold water, continued use of the tankless heater becomes dangerous. The first few gallons of water coming out will be at scalding temperature.

Tankless heaters have another peculiarity that is worth mentioning. When the heat in a home heated by a hot-water heat system is turned on, there is a tendency for the domestic hot water to suffer a drop in temperature. When the heat in a home heated by steam is turned on there is a tendency for the domestic hot water to increase in temperature. Both these responses are increased by turning the mixing valve to admit less cold water. As the valve is always adjusted this way in response to system aging and decreasing efficiency, changes in domestic hot-water temperature following the calling for heat by the room thermostat is always most noticeable in old systems.

DRAINAGE SYSTEM

When you wash your hands or flush the toilet, the used water goes down a pipe and is led out of the house to the municipal sewer line, or to your septic tank. The system of pipes that do this job are collectively called the drainage system, and also the discharge system. The pipes carrying dirty water from the sink are sometimes called waste pipes, and pipes carrying human excreta are called soil pipes. The pipes external to the house that are used to carry rain water off are known as the storm pipes or the storm sewer pipes.

There are three parts to a drainage system: the pipe that carries the used water away, the vent pipe that admits air to the drainpipe, and the trap which prevents sewer gas from entering the house. The accompanying drawing shows the parts of the system.

The putrefaction of the waste materials discharged into city sewers and septic tanks give rise to various gases. All are odorous, some are poisonous and some, like methane, are even explosive. The gas is prevented from entering the house by the plug of water in the trap. When you open a tap and let water run down the drain, fresh water replaces the water present, so this water doesn't of itself cause noxious gases.

If you examine the drawing of the drainage system you will see that every sink, tub and basin drains through a trap. In some instances several drains share one trap. In other instances, each fixture drains into a trap. There is a trap built into the toilet bowl.

If you examine the same drawing a little further you will see that each trap is connected to a vent, which is a pipe leading to fresh air. The vents serve three purposes: (1) They prevent siphoning from emptying the trap. If all the drains were filled with water at the same time, or there was a heavy flow of water in two interconnected drains, a siphon effect could readily develop which would empty one or more traps. With the sewer side of the trap open to the air, this

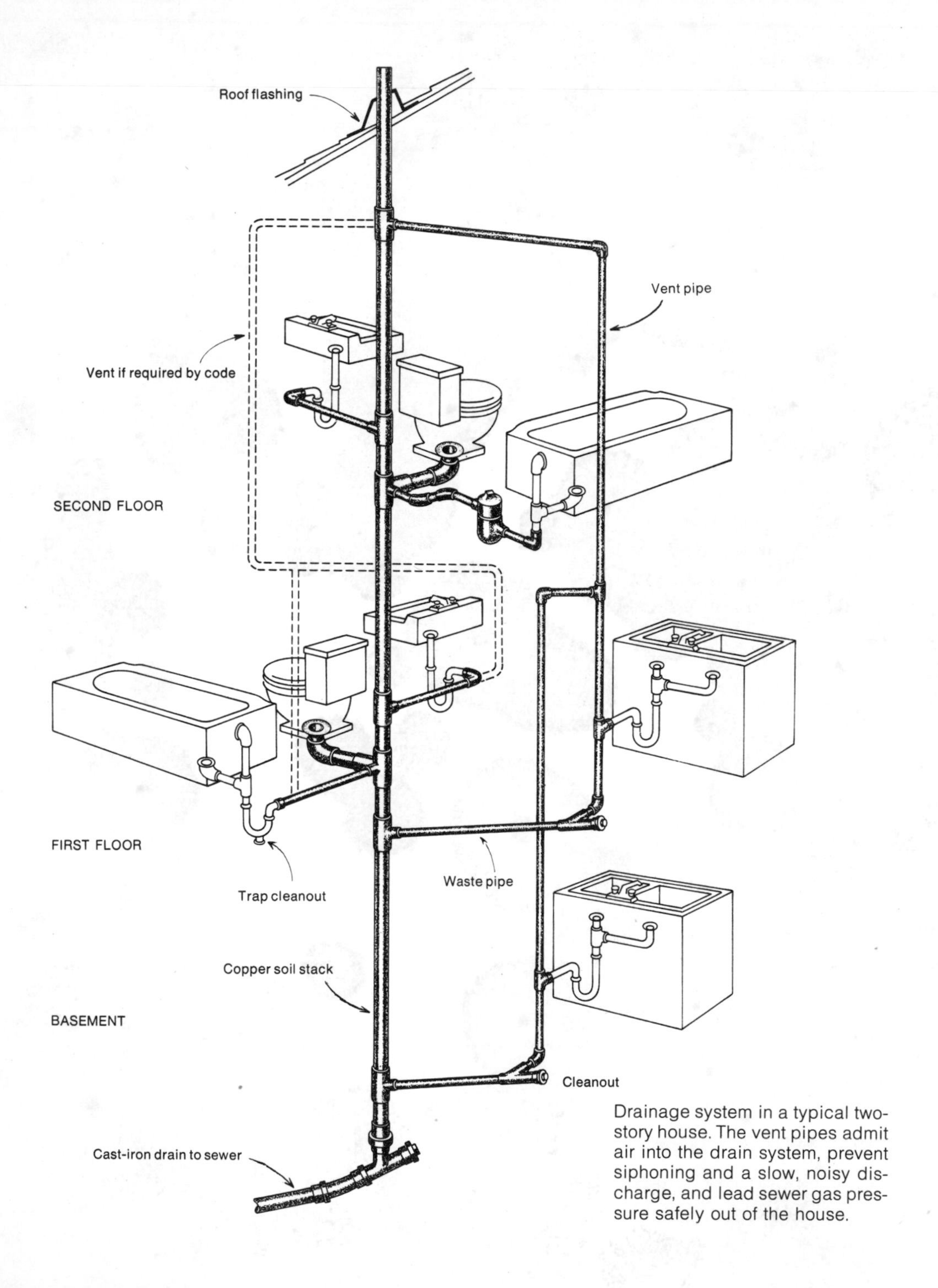

Drainage system in a typical two-story house. The vent pipes admit air into the drain system, prevent siphoning and a slow, noisy discharge, and lead sewer gas pressure safely out of the house.

cannot happen. (2) They lead whatever sewer gas pressure develops safely out of the house. Should a strong positive gas pressure develop in the sewer, the gas could conceivably drive the water in the trap into the house, and the gas could follow. (3) The presence of fresh air circulating through the drain and sewer lines reduces corrosion and the growth of slime.

We have gone to the trouble of explaining why traps are desirable and necessary in the plumbing system because traps tend to collect all kinds of muck and are the first place stoppages occur. Many homeowners tend to view traps as bureaucratic blunders visited upon them by the plumbing department. They are not. They are necessities.

Whether or not a portion of your home's drain system becomes clogged depends on the design of the system and the care exercised by your family.

Obviously nonsoluble objects that fall or are thrown into the drain by children will cause stoppage. What is not as commonly realized is that ordinary kitchen wastes can also cause stoppages if dumped into the drain in sufficient quantities and for sufficiently long periods of time.

Should your drain system clog periodically despite the precautions listed, there may be some small object in the line that effectively reduces its diameter. Or, the drain lines have been poorly designed or incorrectly installed. In the latter case there is nothing for it but to redouble the preventive maintenance, or rebuild the defective portion of the drain system.

HOT-WATER HEAT

Hot-water heat is the familiar term used to describe the system of heating a home by means of hot water. There is a boiler in which the water is heated, pipes that lead the heated water from the boiler through the radiators and back to the boiler and various controls and valves.

In older installations the flow of water from the boiler through the radiators was powered by convection. The water in the boiler, being lighter in weight than the cold water in the radiators, moved upwards. The cold water, being heavier, moved down and into the boiler. In the modern installation an electrically driven pump forces the water to circulate through the system. This permits the use of considerably smaller diameter pipes and radiators, increases overall efficiency and makes for much more rapid heating. As soon as the water in the boiler becomes hot, the circulator drives the hot water through the radiators, which quickly become hot.

A hot-water heating installation is a closed system. Water heated in the boiler moves to the radiators, where the water gives up heat and becomes cooler. From the radiators the water moves back to the boiler, where it is reheated. In this way the heat of the furnace is transported to the various rooms in the house. Normally, no water is lost.

STEAM HEAT

A steam-heat system has a boiler in which water is boiled to generate the steam, which is led to one or more radiators where the steam gives up its heat and condenses back to water. The simpler, less efficient systems have but one pipe leading from the boiler or a steam main to each radiator. The main, the pipe to the radiator, and the radiator itself are pitched down towards the boiler. The condensate, the water that forms in the radiator, runs downhill and back into the

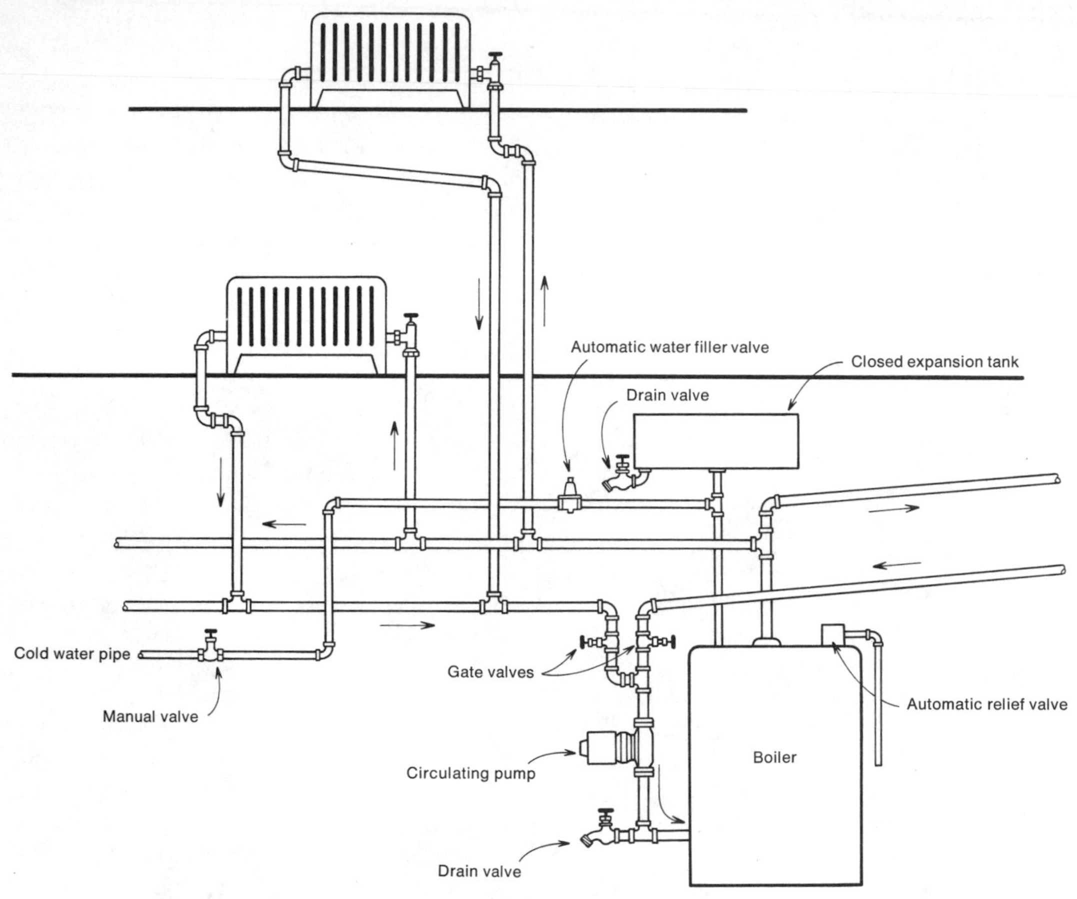

Basic parts of a forced-flow hot-water space-heating system. A circulating pump forces water through the system. Hot water passes through the pipe that emerges from the top of the boiler through the radiators and back to the bottom of the boiler through the pump. The gate valves are used to shut off or vary the flow of hot water through the radiators. The drain valve is used to drain the entire system for repairs. The automatic relief valve prevents damage from excess pressure. The expansion tank permits the water to expand and contract with temperature changes.

boiler. One pipe carries both steam and returning water. This setup is called the single-pipe system.

When a second pipe is connected to each radiator to collect and return the water to the boiler the setup is called a two-pipe system and all the return pipes and radiators are pitched to make the water flow down the return pipe.

It is easy to recognize the single-pipe steam radiator. No hot water system ever has only one pipe. But the two-pipe system requires more careful examination for recognition. The hot-water radiators are equipped with a small air-bleeding valve. It is no more than a short stem. Steam radiators are always equipped with a shiny air vent that looks like the top of the U.S. Capitol building.

A steam boiler may have the same general configuration as a hot-water heat boiler. However, the steam boiler always has a water sight glass on its side. The level of the water in the boiler is at the level of the water in the glass. The sight glass is always provided because it is dangerous to operate a steam boiler with little or no water in the tank—it will explode should the safety valve be corroded fast. This is almost impossible with hot-water heat because all the water in

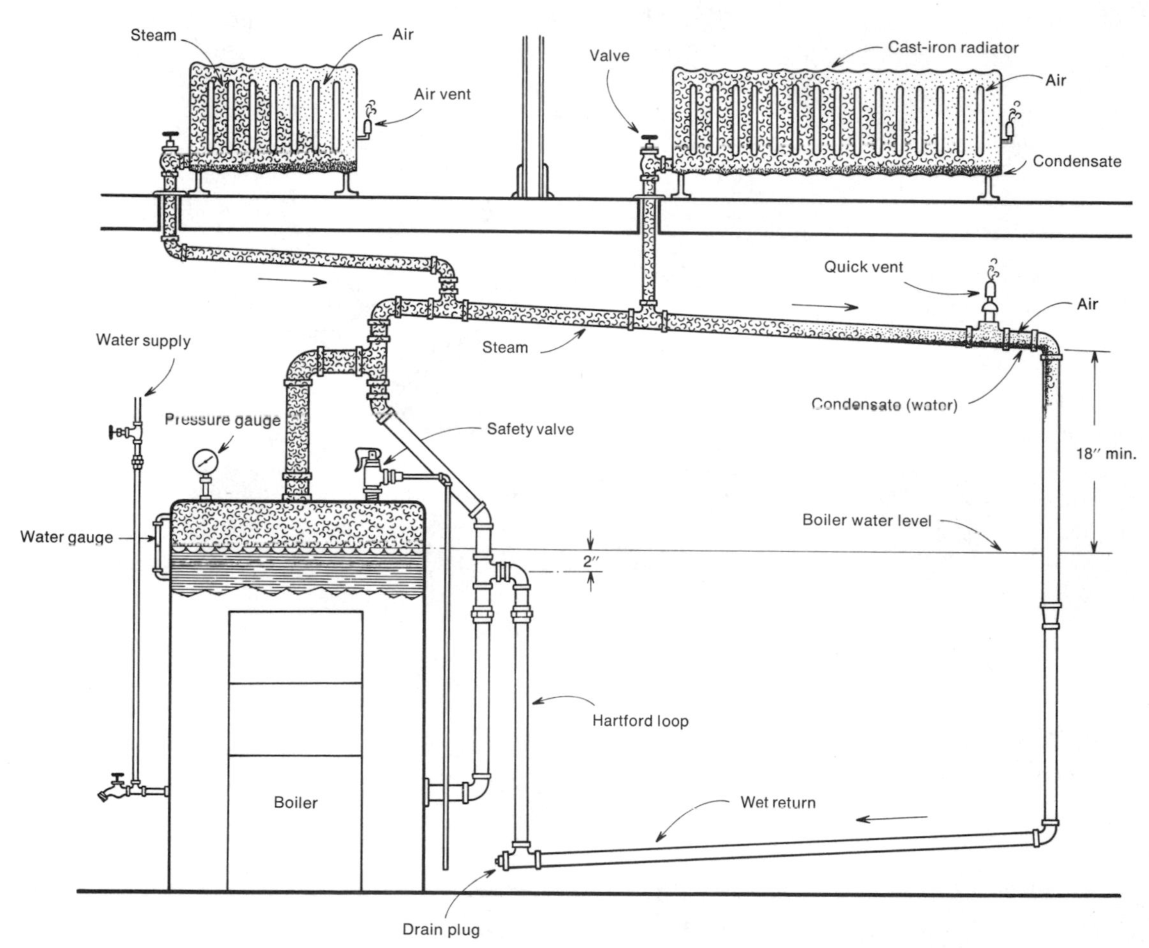

A standard steam-heat system for a small building. Flow of steam through each radiator may be controlled by valve on radiator. Safety valve prevents system damage from excess pressure. Water gauge must be checked frequently to be sure level is at about midpoint.

all the radiators must be removed before the water in the boiler can become dangerously low.

In operation, steam generated in the boiler enters the steam main and continues on into all the radiators, pushing the air present out through the air vents on the radiators. In the radiators the steam cools, returns to its water form and runs by gravity back down into the boiler. Once the boiler is up to boiling temperature (actually higher), there is always plenty of steam because water expands approximately 1,000 times in volume when it changes to steam.

The accompanying illustration shows the major components of a typical single-pipe steam-heat system.

As some of the steam generated is lost through the air vents, there is always a quantity of fresh water entering the boiler. And, when the heat is turned off and all the steam returns to its water phase, air enters the steam lines. Together, the air in the fresh water and the air entering through the vents work to create rust and scale inside the steam boiler and pipes. To remove this rust and scale, it is customary to drain a steam boiler once a month or so during the winter. This is accomplished by means of a provided valve. The valve is opened and the water is permitted to flow into a bucket until the stream of emerging water flows clear. Then the valve is closed.

At the same time, the homeowner is advised to watch the water level in the sight glass. In normal operation it will drop. The automatic cold-water feed mechanism will open and water will be admitted to the boiler until the water level is correct. Usually this is about mid-point on the sight glass.

Should the automatic feed quit, should the water level fail to return to its midpoint level, admit water to the boiler by means of the hand feed valve. At this time steps should be taken to repair or replace the automatic cold-water feed mechanism. *A steam boiler should never be operated by hand water feed.* Should you forget, the boiler will run dry and may explode if the safety valve doesn't let go. If it does let go and the thermostat continues calling for heat the furnace will be damaged.

For the sake of safety, homeowners having steam-heat systems should make it a practice to inspect the sight glass regularly throughout the year and drain the blow-down once a month. (The blow-down is the valve under the automatic water feed. It is opened and the water permitted to drain until it runs clear.)

2

Tools for Plumbing

Few tools are required for the general run of home plumbing and heating repairs, and not too many more are necessary for altering plumbing and installing new fixtures. One reason is that much of plumbing consists of screwing and unscrewing parts and fittings. Another is that a large selection of cut and threaded pipe is available in many localities. And still another is that where and when it isn't feasible to use precut and threaded pipe, one can have pipe cut to sketch at a plumbing supply house, and can rent the expensive tools needed for cutting and threading pipe.

MINIMUM TOOLS

While it is impossible to forecast the plumbing troubles that may strike, the following list includes tools most often used. These will handle the jobs a homeowner is likely to encounter and willing to tackle. Many may already be present in your toolbox.

MOST FREQUENTLY USED TOOLS

- Closet auger
- Force pump
- 10-inch standard screwdriver
- 6-inch standard screwdriver
- 10-inch Phillips screwdriver
- 6-inch Phillips screwdriver
- Crescent wrench
- Pipe wrench
- Gas pliers
- Valve-seat removal tool
- Valve-seat refacing tool

Closet auger. Used for clearing stoppages in the trap that is part of the toilet itself. Primary consideration in its selection is that it be sturdy. The "snake" portion, the flexible cable, should be close to ⅜ inch thick. If the cable portion of the tool isn't thick enough to be stiff, the tool is almost useless.

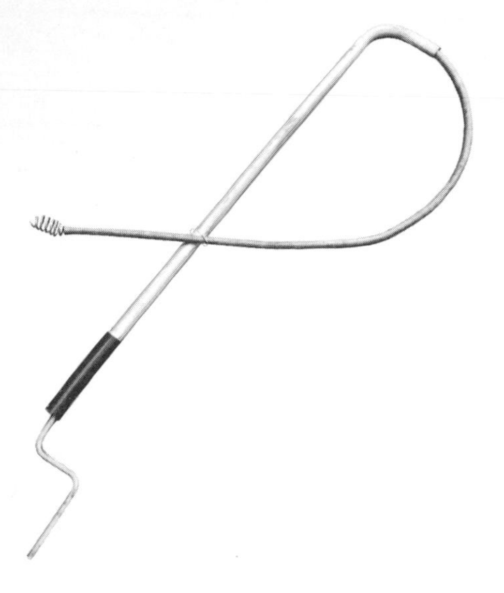

Closet Auger

Force pump. Used to force an obstruction through a clogged pipe or fitting. The obstruction causes water to back up in the pipe. The flexible end of the force pump, or plumber's friend as it is sometime called, is repeatedly forced down against the incompressible water. The pressure on the pump is thus transmitted to the obstruction.

Force Pump

There are two force pump designs on the market today. One is a hemisphere. The other is a hemisphere with a flexible tube extension. This can be folded inward to duplicate the simple hemisphere or can be extended to fit into the drain in a toilet bowl or sink. The second design is usually of heavier rubber and far more effective.

Screwdrivers. There isn't much to choose from in screwdriver design, but their size in relation to the screw head is very important. If you don't use a driver of the proper size it will chew up the head of the screw and make it impossible to turn.

If you "strip" the screw head, you must drill a hole through the length of the screw and remove it with a screw remover, a special hardened-steel screw with a left-hand thread.

Crescent wrench. Used for the large nuts encountered on faucets and joints. The 10-inch wrench suggested as one of the "must have" tools will not take every flat-sided fitting encountered, but it will take most of them. When you buy one, get one of good quality. The cheap wrenches give under strain and will not hold.

Crescent Wrench

Stillson Wrenches

Pipe wrench. A form of stillson wrench, an adjustable wrench with one jaw on a spring-loaded pivot. To use the wrench properly, adjust the jaws to lightly contact the sides of the pipe and then pull on the handle in the direction pointing towards the jaws.

The pipe wrench is made for pipe. It is normally not used for flat-sided fittings and nuts. The pipe wrench takes a healthy bite into the pipe. The teeth on the wrench—the sharp ridges running across its mouth—dig into the pipe. When pressure is applied, these teeth may chew some of the pipe's surface off and almost always leave deep marks. A pipe wrench therefore should never be used on polished fittings or polished pipe without protecting the metal with several layers of tape or a rag.

When the pipe-wrench handle is pulled in the correct direction, the pivoted jaw moves to compress the object between the jaw. Therefore the harder the wrench is pulled the more firmly it holds onto the pipe. The pipe wrench has a more powerful grip than any of the other tools and it can be used to loosen nuts, bolts, and fittings that have been rounded by slipping wrenches and pliers.

However, because the moving jaw clamps on the object it holds, a pipe wrench will "flatten" to some extent any soft object that resists it. Therefore a pipe wrench should not be used on soft pipe or fittings, such as copper valves made to be "sweated" (soldered) in place, nor on flat-sided fittings, such as valves and unions.

Gas pliers. Used for unscrewing small-diameter pipe after it has been loosened, for working with small fittings and for other odd tasks. Gas pliers are not suited for tightening or loosening that requires a great deal of pressure. When used on a tight nut or fitting they will often strip the corner edges off, making the part more difficult to hold.

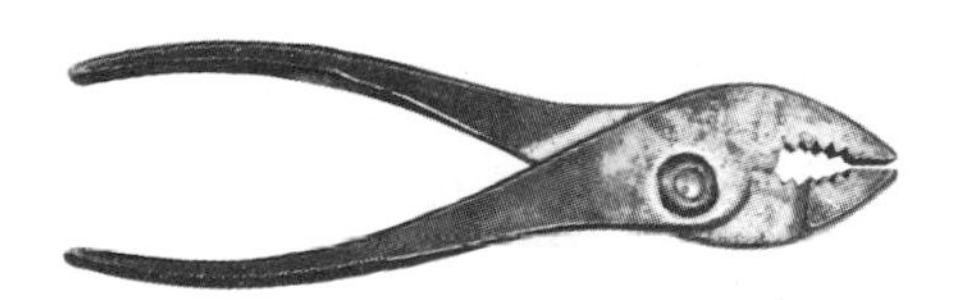

Standard Gas Pliers

Channel-Lock Pliers

Table 1

Wrench Size and Pipe Diameters

Overall length of wrench, jaws open	6″	8″	10″	14″	18″	24″	36″	48″
Pipe diameter wrench will fit properly	1/8–1/2″	1/8–3/4″	1/8–1″	1/4–1 1/2″	1/4–2″	1/4–2 1/2″	1/2–3 1/2″	1–5″

The ordinary 6- or 8-inch gas pliers are limited to grasping pipe and fittings no more than 1½-inches in diameter. For larger objects a pair of Channelock-type pliers are advisable. Channelock is a trade name for pliers with a number of channels cut across the side of one half of the tool. These engage a projection on the other half of the tool and permit the pliers to be set to a very wide opening without slipping apart under load. A somewhat similar pair of pliers can be purchased from other manufacturers. They are useful for working with large nuts and fittings that do not require high torques (turning pressure).

Two valve-seat removal tools and a small sampling of the many valve seats in use today. The straight tool is easier to use although you need a wrench to turn it.

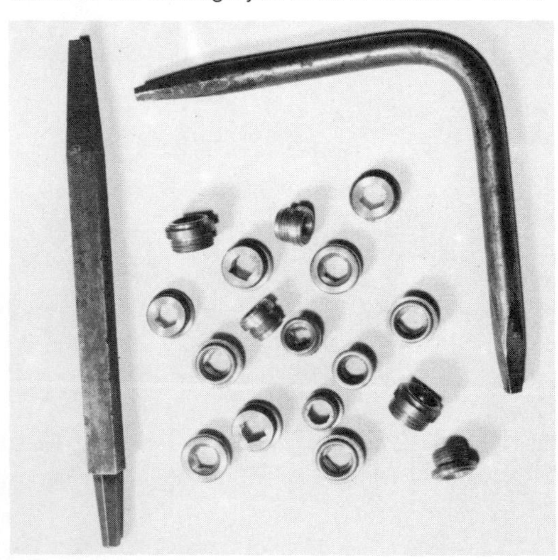

Valve-seat removal tool. Two kinds are available: one a straight bar about 8 inches long; the other, a 10-inch bar bent in the middle to form a right angle. One end of the tool fits valve seats with square holes; the other end fits valve seats with octagonal holes. Both are used the same way, but the straight tool requires a wrench. To remove a seat, the valve (or faucet) is dismantled —that is, the cap nut and spindle are removed. The valve seat is now visible. The square or octagonal end of the tool is inserted into the center of the valve seat. The fit must be snug. If it is not, it is possible the square plug is in an octagonal hole, vice versa, or the tool end is too long. It hits the bottom of the valve before its sides make solid contact with the inside edges of the valve seat. If this is the case, shorten the tool by cutting a small piece off the end. (This is the simplest, least expensive solution to the problem.)

Valve-seat refacing tool. Used to reface valve seats. The inexpensive valve-seat refacer sold in five and dime stores and some hardware shops is a sad joke. It looks like the letter "T." Don't waste your money on it.

The commercial valve-seat refacer costs a lot more, is much easier to use and once in a while does a perfect job. Perfect in this case means there is no drip whatsoever after the faucet is reassembled. However, it is always better to replace the valve seat than to reface it. The refacing tool should only be

Professional valve-seat refacing tool with three extra cutters. Cone screws into lower half of faucet or valve.

Basin wrench is used to loosen or tighten one of the nuts holding a faucet assembly in place. The wrench can reach into spaces too narrow for any other type of wrench, such as the underside of a wash basin.

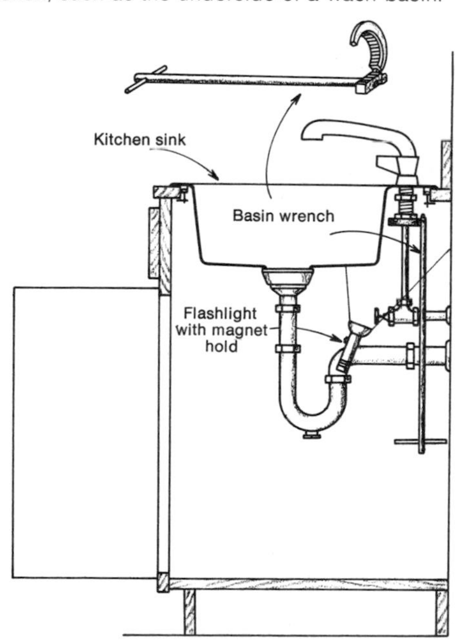

used when there is no removable seat or when it is impossible to remove.

ADDITIONAL TOOLS

Most of the tools just described should be in every homeowner's workshop. The following tools include a few you may require for routine repairs and others you'll need only for special work or extensive improvements on the plumbing system.

Basin wrench. Used to reach up beneath a basin or sink and work on the nuts holding the faucet or fixture in place. The task can be done without this tool, but lack of space makes it very difficult. To use a basin wrench, it is best to lie down beneath the basin, face up, and direct a flashlight on the nut.

Adjustable spud wrench. Used to turn large, flat-sided nuts such as are found holding drain pipes in place beneath a kitchen sink or lavatory. They are also used for turning brass and copper valves and flat-sided fittings which can be damaged by a pipe wrench.

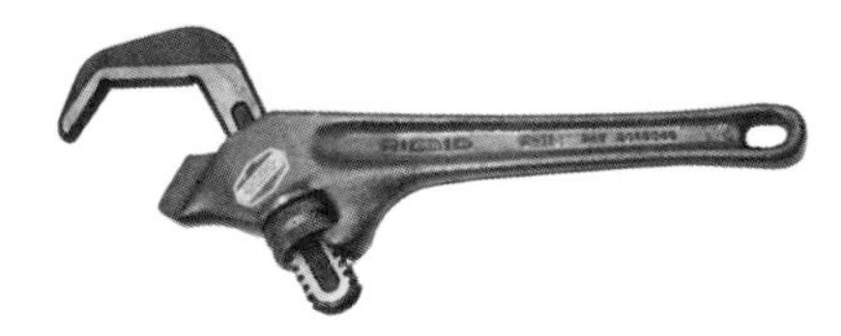

Adjustable Spud Wrench

There are many forms of adjustable spud wrenches. Essentially the wrench is a form of monkey wrench which means it has two adjustable, parallel gripping surface that have no teeth. One form of spud wrench looks like a monkey wrench. The other looks somewhat like an open hand.

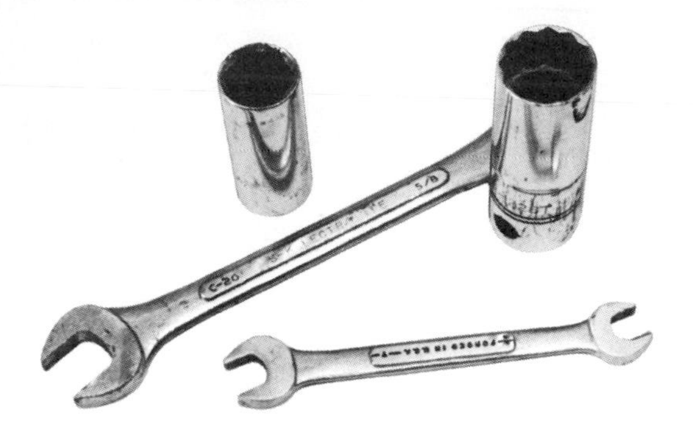

Open-end and Socket Wrenches

Strap Wrench

Open-end wrench. Used when it is difficult or impossible to slip a socket wrench over a nut or fitting, as for example the nut portion of a plumbing union. Generally the two openings on a single wrench are of different sizes, to accommodate nuts of different sizes. The ends of the wrench are usually angled to give maximum play.

Socket wrench. Used to slip over a valve or faucet stem and turn the packing nut and upper half of the valve body when the valve is within a wall and cannot be reached by other means. Automotive-type socket wrenches, two or more inches in length are used with a pipe wrench. The wrench is a metal tube with a serrated inner surface that slips over the nut.

Large pipe wrench. If you are planning to do any threaded pipe work, you will need a second pipe wrench. One wrench will suffice if you are screwing on a fitting and the pipe remains firm. If it doesn't, if you work on two pieces of pipe, you need a second wrench to hold the second piece of pipe or fitting. For most homes a 24-inch wrench which will handle pipe to 2½ inches in diameter is large enough.

Strap wrench. Used for turning plated pipe you do not want to scratch and thin-wall pipe that would be crushed by a pipe wrench. It is also useful for getting into tight corners no other wrench can reach.

Chain wrench. Used in place of a pipe wrench. The chain wrench will reach pipes and fittings that are inaccessible to the pipe wrench because of its bulk. The chain wrench is almost much lighter in weight for any given pipe size.

Chain Wrench

Table 2

Choosing a Hacksaw Blade

Teeth per inch	16/inch (coarse)	24/inch (medium)	32/inch (fine)
Application	Material ¼–1″ pipe, angle iron, bars, heavy plate	Material ⅛–¼″ thin-wall pipe, tubing, thin rod, etc.	Material less than ⅛″ thick. Sheet metal, leaders, gutters

Hacksaw. Used for cutting metal. Blades should be mounted in the saw frame with their teeth pointing *forward,* away from the handle. Choice of blade—coarse, medium, fine—depends on the thickness of the material to be sawn. Thick material requires blades with few teeth. Thin material requires blades with many teeth. The hardness of the metal to be cut is not a factor in blade selection.

In using the hacksaw, apply very little pressure during the cutting stroke, that is, when the blade is moving forward over the work. Use no pressure at all on the return stroke. Do not lift the blade, but allow it to slide over the work. When cutting steel and iron it is advisable to use a cutting lubricant. A mixture of light oil and kerosene can be used, or oil alone if no kerosene or no commercial cutting liquid is available. The lubricant extends the useful life of the hacksaw blade and speeds cutting, especially through thick sections.

Files. A round file, sometimes called a rattail file, is a milled steel rod, tapered at one end and formed at the other into a thin point designed to accept a handle. The round file is used for filing the inside edge of a pipe to remove burrs.

A flat file is generally used to remove burrs from the outside surface of a pipe and to smoothen the end of a pipe.

Both files are normally held with their handle ends towards the user. Light pressure is applied on the forward stroke, no pressure on the backstroke. This is because the hundreds of little cutting edges all point forward.

Large snake. Used for clearing obstructions in pipes which cannot be cleared by the closet auger, force pump or chemicals.

As the snake must be both flexible to go around turns and stiff enough not to coil up on itself, a snake that is much less than half inch in diameter is almost useless. The type made of steel wire wound in a tight spring is best. The tube with the flat cross section is not nearly as good, as it folds up upon itself when you push too hard and once kinked retains its kinks.

A large snake fitted into a rotating drum. When you turn the handle, the snake spins as it is fed into the drain. Note U-shaped cutter at tip.

Wire-wound snakes a half inch in diameter are effective in lengths up to 30 feet. If you have to probe farther you need to use a thicker, heavier snake.

For really long runs of 75 feet or more you must use a "rod," which consists of a series of flexible rods screwed end to end. The rods are more than an inch thick. When you use a rod, the process is called "rodding."

Propane torch. Used for soldering. The modern, bottle-powered propane torch consists of a nozzle and valve that screw onto a replaceable bottle of liquid gas. The amount of heat produced by the torch and the shape of the flame can be varied by using different nozzles. However, the straight nozzle about a half inch in diameter is satisfactory for most of the work encountered in plumbing.

Don't use a gasoline blowtorch. It is inconvenient and very dangerous. If you still have one, junk it.

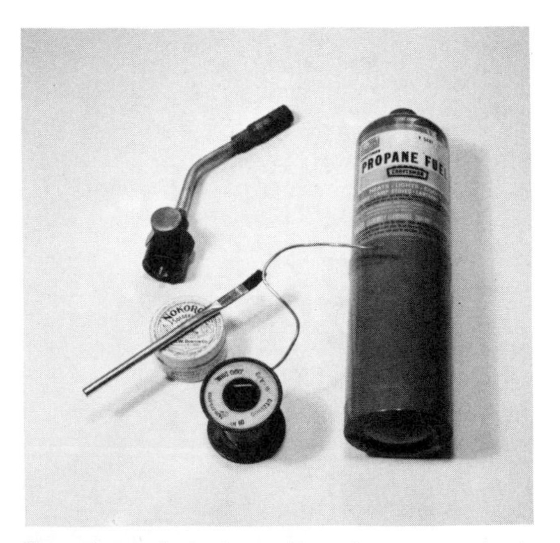

The valve and nozzle section of a propane torch screws onto the tank section. The two parts should be disassembled for safety when not in use, as shown. Can of soldering flux, brush and wire solder are at left.

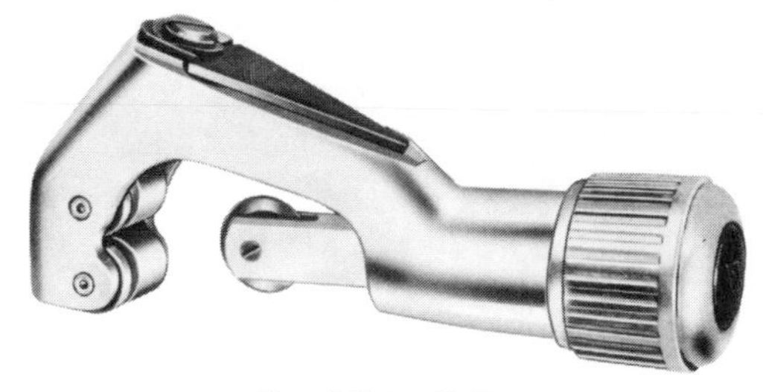

Small Tube Cutter

Tube cutter. Used for cutting copper and plating tubing. The better types come with a foldaway pipe reamer. The only important choice to make in selecting a tubing cutter is that it will handle the tube size you plan to cut.

Tube bender. This is used to bend tubing without the possibility of flattening it. Select a size that just slips over the tube to be bent. If the bender's diameter is too large, the tube may kink.

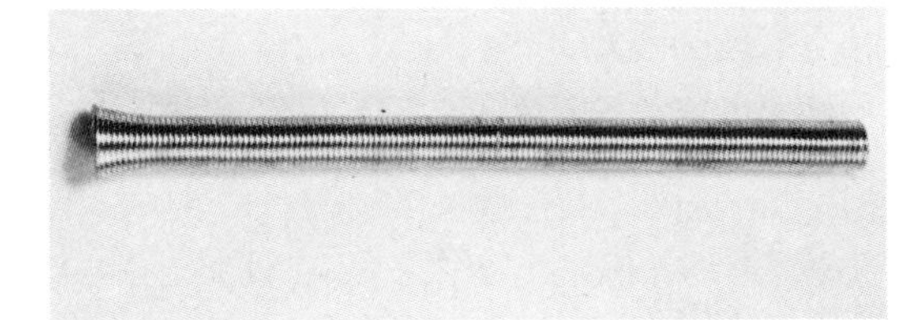

Tube Bender

Flaring tool. Used for making flared joints in copper tubing. There are two basic designs. The better one has a small depression near the top of the clamp. With this design you bring the top of the tube flush with the surface of the clamp. With the other type you have to measure the distance the tube end projects beyond the clamp in order to make a perfect joint.

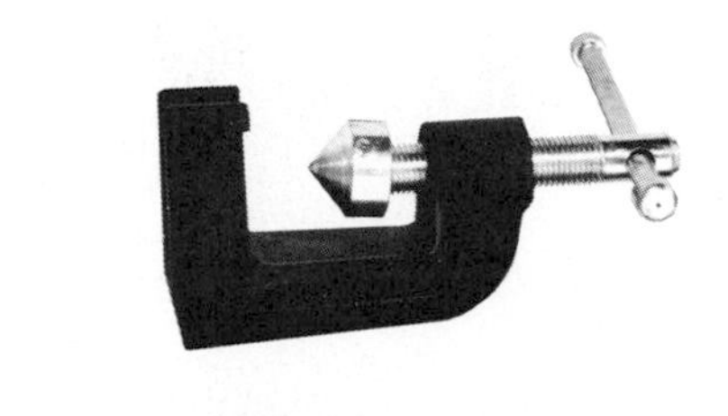

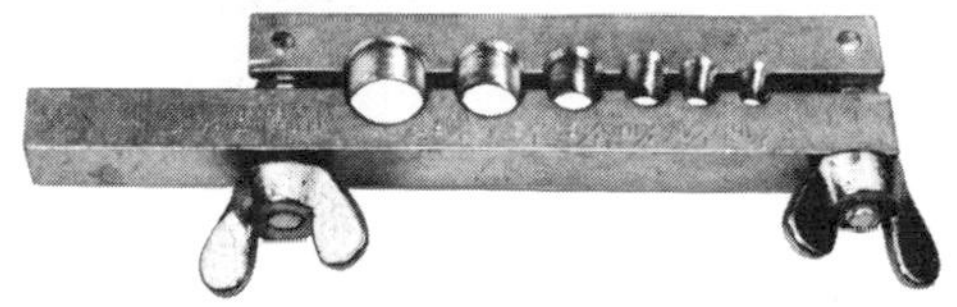

Flaring Tool

Wheeled pipe cutter. Used for cutting thick-walled iron, copper, brass, and steel pipe. It is similar to the tubing cutter, but is larger and stronger.

For hand cutting the cutter is turned or rotated around the pipe by hand. For power cutting the pipe is rotated by machine while the cutter is held immobile by a rod extending from the machine.

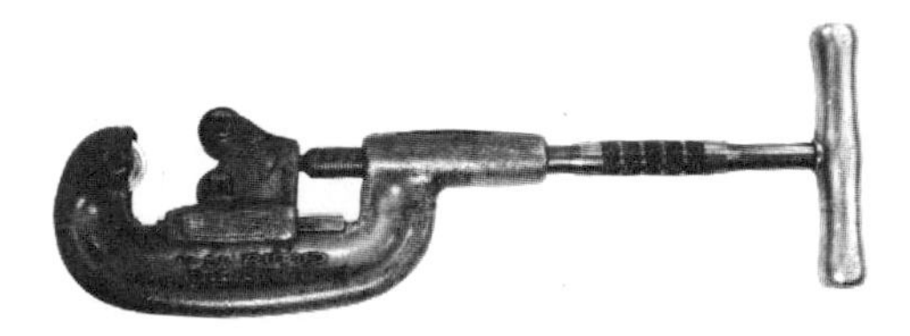

Wheeled Pipe Cutter

Pipe vise. Used to hold pipe. It differs from an ordinary vise in that the jaws go around the pipe and prevent it from rotating. An ordinary vise can be converted to holding pipe by fitting the jaws with properly shaped pieces of hardwood or soft metal (aluminum).

Pipe vises are available in any number of designs. Some have their own stands, some clamp to the back of a truck or workbench. Some have a quick-release mechanism and some have to be laboriously opened and closed by turning a screw. If you are planning to cut, thread and join pipe, select any suitable vise that opens wide enough for the pipe size you are going to work with. If you plan to do any pipe bending, get a large vise with its own stand that has provisions for bending pipe.

Quick-release pipe vise on its own stand. The half-round channels on the right side of the stand are for bending pipe. There is a hole of the proper size next to each channel.

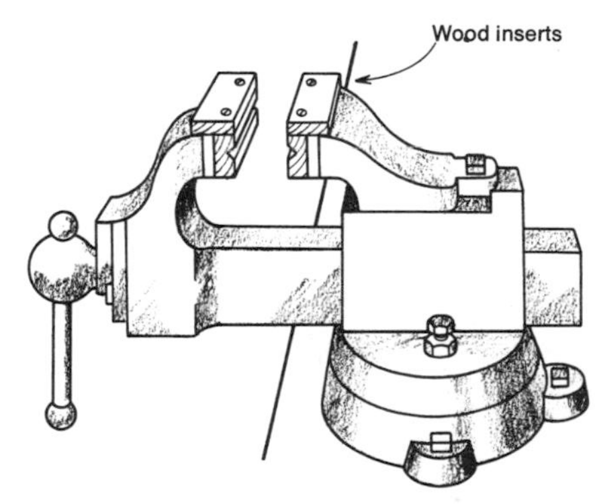

You can convert a standard bench vise to a pipe vise by adding wood inserts as shown. Alternately, you can sometimes just hook the pipe under the vise jaws, if the pipe diameter is the correct size.

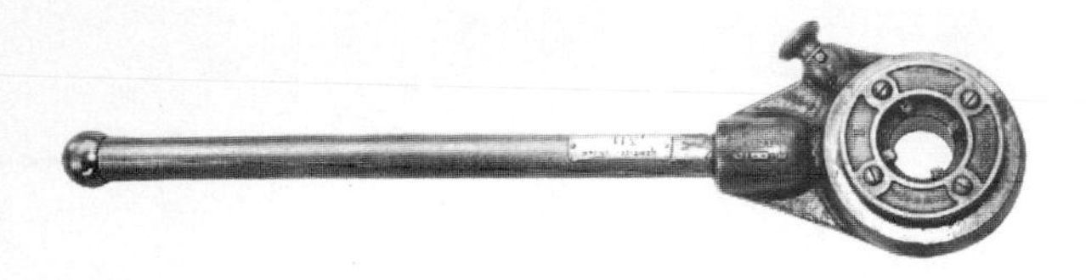

Ratchet Stock and Die

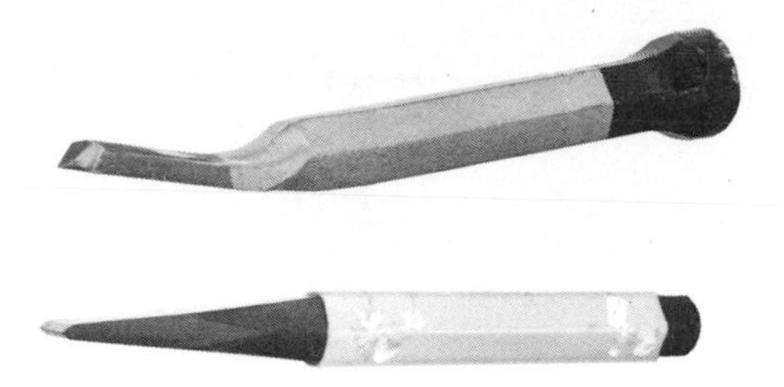

Caulking Iron and Cold Chisel

Pipe stock and dies. Used for cutting external thread on pipe. Modern stocks can accept a number of different-sized dies for pipe of different diameters. You can purchase a range of die sizes or just the ones you need.

There are two stock designs. The two-handled type and the ratchet type. The latter is more expensive but somewhat faster and easier to use. Note that dies are generally not interchangeable. The die that fits one stock usually will not fit another type or make.

Pipe reamer. Used to remove the burrs formed on the inside edge of a pipe end after it is cut. The tubing cutters usually have a reamer attached. Pipe cutters do not. There are reamers that fit into a carpenter's brace and there are reamers that have their own handles. The simpler types are good enough if you don't plan to do much pipe cutting.

Cold chisel. Used for cutting cast-iron pipe and vitrified clay pipe. Any cold chisel will do; one about ¾ inch thick is about right.

Caulking iron. This is a blunt-ended chisel that is used for driving lead into a lead-caulked pipe joint. It is only required when you make lead-caulked joints and when you tighten old joints. These chisels are manufactured in a large variety of shapes and sizes to accommodate the various conditions and joints encountered in the plumbing trade. One is all you will ever need if you are not going to make lead caulked joints. If you are, you will need to select chisels to suit each particular joint and condition.

Taps. These are cutting tools used for making threads inside pipes and holes in metal. A tap looks like a machine screw with four slots down its side. In use, the tap is inserted part way into the hole. The end of the tap is thinner than its body. The tap or the metal surrounding the hole is revolved. The tap is screwed into the hole, its threads cutting threads in the softer metal.

Ratchet Pipe Reamer

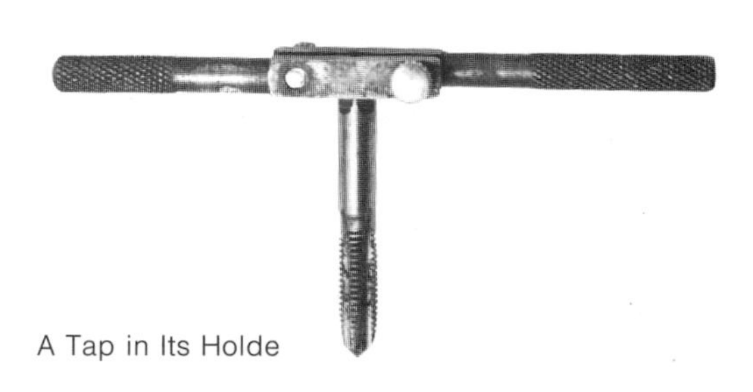

A Tap in Its Holde

3

Pipes and Fittings

If you plan to do your own plumbing repairs, you will encounter a wide assortment of pipes and fittings. You should be sufficiently familiar with these components of the system to be able to tell the difference between, say, a copper and a galvanized steel pipe, and to know why a certain fitting is used in a particular place. A knowledge of pipes and fittings will help you to determine when the wrong ones are used (perhaps causing a leak), and to update your system with new ones that were unavailable when the system was installed.

GALVANIZED PIPE

Galvanized steel pipe is steel pipe that has been coated on the inside and outside with a layer of zinc. Note, however, that the threaded portion of the pipe is not zinc coated and that rusting of the exposed thread is normal. Pipe with zinc-plated threads (hot-dipped after threading) can be obtained but is rarely if ever used for home plumbing.

Of all the kinds of pipe presently installed, galvanized is by far the most common. The pipe has many advantages. It is strong and needs no protection whether installed within walls or exposed. Galvanized is corrosion resistant. Although its life expectancy when buried is fifteen to twenty-five years, fifty years without disintegration is not uncommon. More often a buried main has to be exhumed because of sediment and interior mineral deposits long before its iron walls have given way.

Examples of galvanized steel pipe, threaded and unthreaded, and the fittings used to connect them.

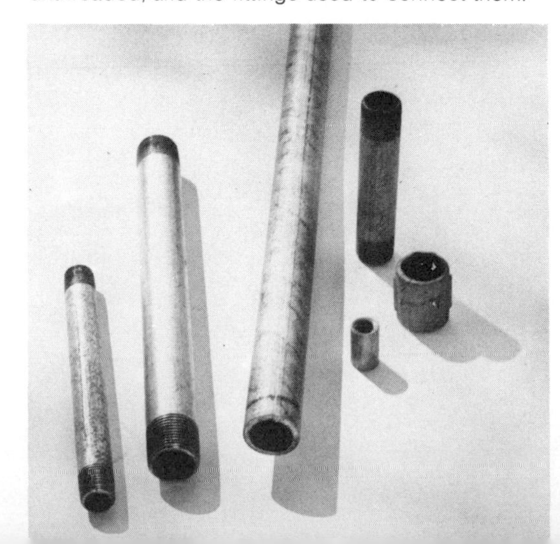

SOLDERABLE GALVANIZED STEEL

This is a new series of pipes and accompanying fittings having the stiffness and strength of conventional galvanized pipe, but which can be joined by soldering. The advantage is that joining is much faster, easier and simpler than possible with screw joints. The disadvantage, at present at least, is that base costs are higher and, as it is a fairly new material, few plumbing supply houses carry it. (As it is, they complain they have more pipe and fittings right now than they care to stock.)

BLACK PIPE

Black steel pipe is galvanized pipe without its zinc coating. The advantage is cost and somewhat less internal friction. Generally, black pipe is not used for water and waste lines, but only for heating lines leading from the furnace to steam and hot-water radiators. Pipe sizes and fittings are the same as those manufactured for galvanized work but are not zinc coated. In addition, there is a series of fittings, valves, etc., used only for steam and hot-water work. Note that pipe diameters and threads are exactly alike, and

Table 3

GALVANIZED AND BLACK PIPE

Nominal Pipe Size (Inches)	Outside Diameter (Inches)	Wall Thickness (Inches)	Inside Diameter (Inches)	Weight of Pipe (Lbs. per Ft.)	Weight of Water (Lbs. per Ft. of Pipe)	Threads per inch
1/8	0.40	.06	.26	.24	.02	27
		.09	.21	.31	.01	
1/4	0.54	.08	.36	.42	.04	18
		.11	.30	.53	.03	
3/8	0.67	.09	.49	.56	.08	18
		.12	.42	.73	.06	
1/2	0.84	.10	.62	.85	.13	14
		.14	.54	1.08	.10	
3/4	1.05	.11	.82	1.13	.23	14
		.15	.74	1.47	.18	
				1.94	.12	
1	1.31	.13	1.04	1.67	.37	11 1/2
		.17	.95	2.17	.31	
1 1/4	1.66	1.40	1.38	2.27	.64	11 1/2
		.19	1.27	2.99	.55	
1 1/2	1.90	.14	1.61	2.71	.88	11 1/2
		.20	1.50	3.63	.76	
2	2.37	.15	2.06	3.65	1.45	11 1/2
		.21	1.93	5.02	1.28	
2 1/2	2.87	.20	2.46	5.7	2.07	8
		.27	2.32	7.6	1.87	
3	3.50	.21	3.06	7.5	3.20	8
		.30	2.90	10.2	2.86	
3 1/2	4.00	.22	3.54	9.1	4.29	8
		.31	3.36	12.5	3.84	
4	4.50	.23	4.02	10.7	5.50	8
		.33	3.82	14.9	4.98	
5	5.56	.25	5.04	14.6	8.67	8
		.37	4.81	20.7	7.88	

Courtesy Crane Company

Table 4

THREADED PIPE FITTINGS

Size	A	B	C	D	E	F	G	J	K	Z
1/8	11/16	1/2	1	11/16	13/16			15/16	1	1/4
1/4	13/16	3/4	1 3/16	5/8	15/16			1 1/16	1	3/8
3/8	15/16	13/16	1 7/16	11/16	1 1/16	2 1/8	1 7/16	1 3/16	1 1/8	3/8
1/2	1 1/8	7/8	1 5/8	13/16	1 3/16	2 7/16	1 11/16	1 5/16	1 1/4	1/2
3/4	1 5/16	1	1 7/8	15/16	1 5/16	2 13/16	2 1/16	1 1/2	1 7/16	9/16
1	1 1/2	1 1/8	2 1/8	1 1/16	1 1/2	3 3/8	2 7/16	1 11/16	1 11/16	11/16
1 1/4	1 3/4	1 5/16	2 7/16	1 1/4	1 11/16	4 1/16	2 15/16	1 15/16	2 1/16	11/16
1 1/2	1 15/16	1 7/16	2 11/16	1 3/8	1 7/8	4 1/2	3 5/16	2 1/8	2 5/16	11/16
2	2 1/4	1 11/16	3 1/4	1 11/16	2 1/4	5 7/16	4	2 1/2	2 13/16	3/4
2 1/2	2 11/16	1 15/16	3 13/16	1 7/8	2 9/16	6 1/4	4 11/16	2 7/8	3 1/4	15/16
3	3 1/8	2 3/16	4 1/2	2 1/8	3	7 1/4	5 9/16	3 3/16	3 11/16	1

RETURN BENDS

	Close Pattern			Open Pattern	
Size	M	N	Size	M	N
1/2	1	1 3/4	1/2	1 1/2	1 7/8
3/4	1 1/4	2 3/16	3/4	2	2 1/4
1	1 1/2	2 1/2	1	2 1/2	2 5/8
1 1/4	1 3/4	2 13/16	1 1/4	3	3 3/16
1 1/2	2 3/16	3 3/16	1 1/2	3 1/2	3 5/8
2	2 5/8	3 7/8	2	4	4 3/8

Shapes and dimensions of common threaded pipe fittings. Dimensions of these fittings are given in table above.

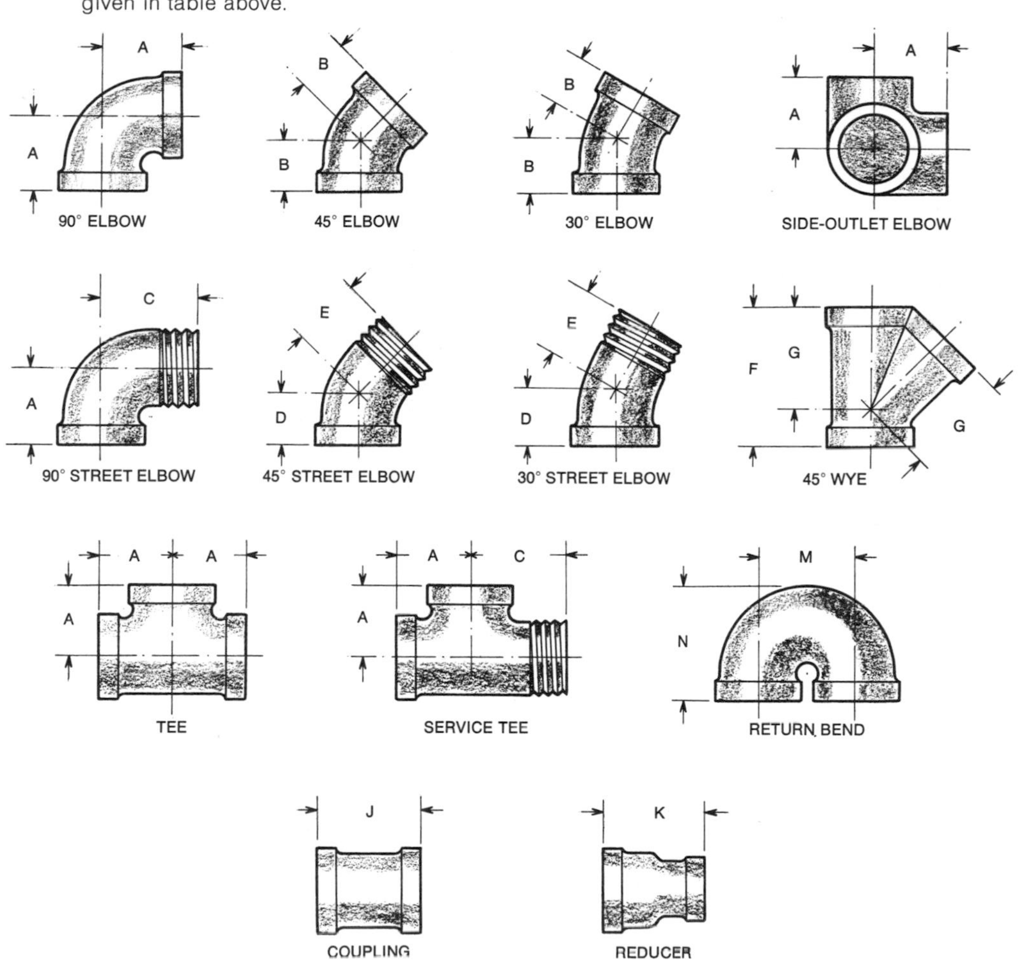

that in a pinch, bare, nonzinc-coated black pipe and fittings *can* be connected to galvanized pipelines, but because of accelerated corrosion, it is not advisable.

Threaded fittings. Galvanized and black iron fittings are available in threaded form. They are widely used and easy to work with. Fittings in both materials come in the same range of sizes. The shapes and dimensions of frequently used threaded pipe fittings are provided in Table 4 and its illustration.

COPPER TUBING

Copper tubing is available in two general types, hard and soft, which further subdivide into wall thicknesses as well as lengths and diameters. Hard (tempered) copper tubing is rigid. It comes in 10- and 20- foot lengths. Some supply houses will cut it for you into any length over 1 foot. Flexible copper tubing comes in coils that are usually 30, 60 or 100 feet long.

The hard, rigid tubing is chosen for exposed runs where appearance is important and where physical damage is possible. It is also used in homes that are to be closed for the winter. The rigid tubing can be lightly pitched to aid drainage. The soft tubing is used when a lot of turns and bends are necessary. It is difficult to run in a neat, perfectly straight line. It retains its original coil shape to some extent, so that without coaxing, it looks somewhat like a snake when first uncoiled.

All copper tubing and fittings may be used for drainage lines, but the type designated DWV (drain, waste, vent) is especially made for the job. It is lighter in weight, costs less than standard copper large-diameter piping. Even so, the fittings are expensive, so use as few as possible. DWV pipe should never be in direct contact with the earth.

Copper tubing has many advantages over galvanized and other types of piping. It is considerably smoother than galvanized. As a result, water encounters far less friction when passing through copper pipe than galvanized or black. Because of the reduced friction, it is common to use one size smaller copper tubing than galvanized for an identical run. For example, if the line calls for ¾-inch galvanized, ½-inch copper can be used instead without any loss of water flow.

When overall ease of handling, bending, joining are considered, copper tubing is far

Installing hard copper tubing in a house. The difference between hard and soft tubing is shown in photo below. Hard tubing is always straight; soft tubing is shipped in coils and always retains the curve.

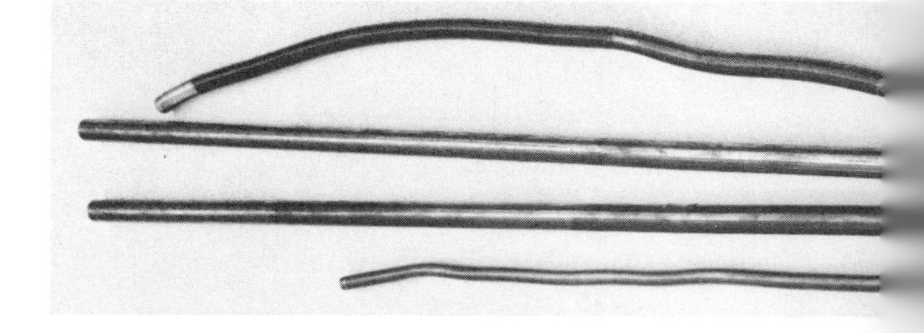

Table 5

COPPER TUBING

Designation	Color	Diameters	Temper	Joining	Wall Thickness	Lengths
K	Green	¼" to 12"	Hard, soft	Flare, compression, solder	Heavy	10′ and 20′ 30′, 60′, 100′
L	Blue	¼" to 12"	Hard, soft	Flare, compression, solder	Medium	10′ and 20′ 30′, 60′, 100′
M	Red	⅜" to 12"	Hard	Solder	Light	30′, 60′, 100′
DWF (Drain, Waste, Vent)	Yellow	All	Hard	Solder, compression	Light, medium	20′

Copper drainpipe (DWV) being installed in a new house. The plumber is setting up a run with a Wye fitting, which he'll sweat (solder) in place when he's sure the fit is right.

and away easier and faster to install than galvanized piping, screw thread or soldered. Beginning plumbers find copper easier to work with than galvanized, even when they have never soldered pipe joints.

Besides its higher cost, copper tubing has one drawback when compared to galvanized, its softness. This means that extra care and extra materials are necessary for satisfactory, long-lasting installations. The soft tubing must be protected from damage, either by placing it where it cannot be reached or shielding it with a stronger metal. However, this is rarely much of a problem and is generally an insignificant portion of the overall job.

Copper sweat fittings. Fittings that are to be soldered to copper pipe are called sweat fittings. They are available in a large variety of sizes and shapes. Some are joined by sweating alone. Others are made with a

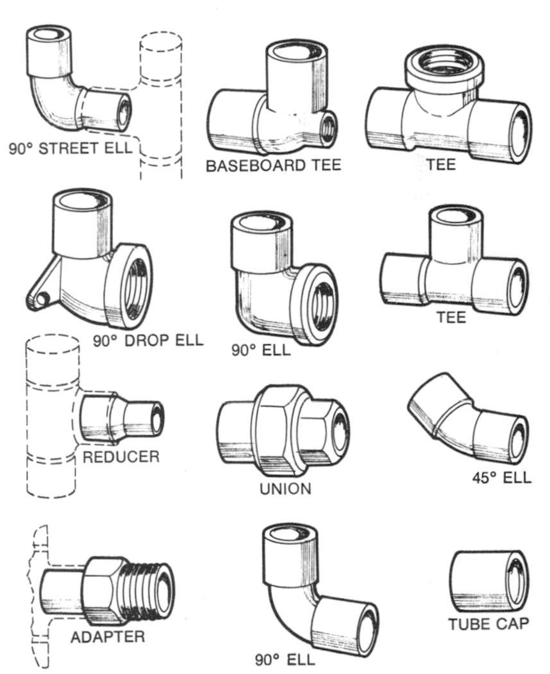

Various types of sweat fittings used with copper water-supply tubing. Some have threads for making dual connections.

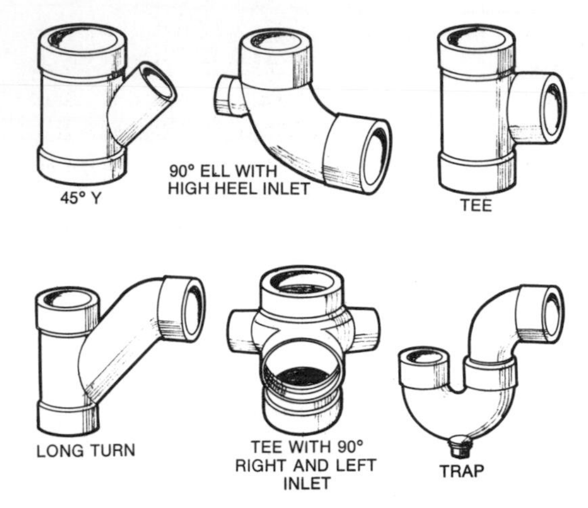

Copper drainage fittings are also made to be sweated. They come in a variety of shapes; these are only a few examples.

sweat joint at one end and screw threads, male or female at the other end. A few of the many types available are illustrated.

BRASS PIPE

Once considered the ultimate in piping material, brass pipe was a main feature of many a real-estate salesman's pitch. The all-brick house with all-brass plumbing was supposed to last forever, but despite its high cost, brass has not proved its worth. Brass has been found peculiarly susceptible to corrosion when used for hot-water lines, especially so when the water temperature is very high.

Brass pipe and fittings come in the same sizes as galvanized, and one can be directly replaced with the other. *However, as a galvanic action is set up when iron and brass join in the presence of water, the two metals should not be used together except for temporary repairs.* And when replacing brass with brass always use the same "kind" of brass, red or yellow. (The red has more copper.)

LEAD PIPE

There are some historians who hold lead to blame for Rome's downfall. The Romans, they say, used so much lead in their water

Thin-wall brass pipe used for drains sometimes has preformed lip (right) for connecting to sink strainer. Threaded and nonthreaded types are also shown.

Table 6

BRASS PIPE

Pipe Size, in.	Outside Diameter, in.	Wall Thickness, in.	Weight per Foot 67% Copper, lb. "Yellow"	Weight per Foot 85% Copper, lb. "Red"
¼	0.540	0.083	0.437	0.450
½	0.840	0.103	0.911	0.938
¾	1.050	0.114	1.240	1.270
1	1.315	0.127	1.740	1.790
1¼	1.660	0.146	2.560	2.630
1½	1.900	0.150	3.040	3.130
2	2.375	0.157	4.020	4.140
2½	2.875	0.188	5.830	6.000

pipes and dishes that they unknowingly suffered acute lead poisoning. True or false, lead is not used to conduct water to the home today. Some of the older homes still have lead goosenecks in their cellars, but they are not used on new installations. And, it would be a wise move to replace those lead sections with copper.

Lead pipe is used for making closet bends, which are the sections of pipe connecting the soil stack to the floor flange on which the toilet bowl rests. The use of lead at this point, rather than a nonflexible material, permits the toilet and the floor flange upon which it rests to be shifted to correct for variations in floor height and toilet bowl-to-wall distance. When the floor flange is screwed onto a rigid pipe, floor height, floor flange and stack Tee must be nearly perfect.

PLASTIC PIPE

Plastic pipe is the coming thing. Weighing roughly 1/20 as much as galvanized of equal internal diameter, quickly joined with a daub of cement, more durable than other piping material in many ways, plastic pipe is for most applications the easiest pipe to work with.

A number of chemical combinations are used for making plastic pipe, and the pipe itself is manufactured in various lengths, diameters, and wall thicknesses. Each formulation has its own special properties, and while the statements following apply to plastic pipe and their fittings as a class, all plastic pipe is not alike and should not be treated alike.

In general, then, plastic pipe is far more resistant to corrosion than galvanized and even copper tubing. Plastic pipe can be buried in moist, acidic soil next to metallic pipe of any kind without fear of galvanic attack (corrosion). Galvanic attack is the process whereby two dissimilar metals in a conductive medium form a voltaic cell. Electric current flows between them and the more negative metal becomes the anode and corrodes most strongly.

When any two metals or alloys in the table below are immersed in an electrolyte (moist soil), the metal or alloy highest on the list becomes the anode and corrodes with little damage to the metal or alloy lower on the list, which becomes the cathode. The farther apart the two metals are on the list, the greater the voltage difference and the more rapidly the anode corrodes.

Magnesium (Mg)
Magnesium alloys
Aluminum (Al)
Zinc (Zn)
Iron (Fe)
Cadmium (Cd)
Steel and iron
Cast iron
Nickel (Ni)
Lead-tin
Lead (Pb)
Tin (Sn)
Hydrogen (H)
Brass
Copper (Cu)
Bronze
Copper-nickel
Silver (Ag)
Mercury (Hg)
Platinum (Pt)
Gold (Au)

Flexible types of plastic pipe (some plastic pipe is rigid) are particularly well suited to outdoor plumbing, that is, pipelines left exposed or only partially covered by soil and thus subject to frost. Flexible plastic pipe easily expands the 11 percent normal to water when it changes to ice. Plastic pipe is therefore also suited to cottages that are not heated in the winter. However, plastic fittings do not expand as readily as does the pipe, and it is therefore a mistake to leave plastic lines filled with water in the belief nothing can harm them. It is not so. To be safe, plastic lines should be drained, or at least opened. The value of plastic lines in frost-exposed installations is that whereas galvanized pipe will always burst, and copper pipe usually will, plastic pipe won't, no matter how low the temperature.

Plastic pipe has additional advantages over galvanized and even copper. Plastic is an insulator. Hot water flowing through plastic lines loses less heat to the surrounding air than it does flowing through metal, especially copper. By the same token, plastic pipes carrying cold water do not sweat like metal pipes do. As plastic is flexible it dampens sound waves rather than amplifies them. Whereas the sound of someone opening and closing a faucet in the far end of the house will carry through the house when the pipes are metallic, the same sound will be muted when the pipes are of plastic.

Plastic pipe interiors are very smooth, like that of copper tubing. It is therefore possible, in many instances, to use plastic

Table 7

MAIN TYPES OF PLASTIC PIPE

Material	Type	Applications	Operating Temperature	Joining Methods
ABS Acrylonitrile-Butadiene Styrene	Rigid	Drain, waste and vent (DWV) Building sewers and sewer mains Electrical and communications conduit Water piping	100°F (Pressure) 180°F (Non-Pressure)	Solvent cement Threading Transition fittings
PE Polyethylene	Flexible	Water service and gas mains Gas service and gas mains Chemical waste Irrigation systems	100°F (Pressure) 180°F (Non-Pressure)	Insert fittings Socket fusion (heat) Butt fusion (heat) Transition fittings
PB Polybutylene	Flexible	Water service and water mains Gas service and gas mains Irrigation systems	180°F (Pressure) 200°F (Non-Pressure)	Insert fittings Socket fusion (heat) Butt fusion (heat) Transition fittings
PVC Polyvinyl Chloride	Rigid	Water service and water mains Gas service and gas mains Building sewers and sewer mains Drain, waste and vent (DWV) Electrical and communications conduit Industrial process piping Irrigation systems	100°F (Pressure) 180°F (Non-Pressure)	Solvent cement Elastomeric seal (compression) Threading Mechanical couplings Transition fittings
CPVC Chlorinated Polyvinyl Chloride	Rigid	Hot and cold water distribution Chemical process piping	180°F @ 100 PSIG (type SDR-11)	Solvent cement Threading Mechanical couplings Transition fittings
PP Polypropylene	Rigid	Chemical waste Chemical process piping	100°F (Pressure) 180°F (Non-Pressure)	Mechanical coupling Socket fusion (heat) Butt fusion (heat) Threading
SR Styrene Rubber Plastic	Rigid	Septic tank absorption fields Sub-soil dewatering systems Storm drains	150°F (Non-Pressure)	Solvent cement Transition fittings Elastomeric seal (compression)

Courtesy Plastic Pipe Institute

pipe one diameter smaller than galvanized for the same job.

Like copper tubing, flexible plastic piping comes in rolls. Some types of plastic pipe can be had in 1,000-foot rolls.

Plastic's major advantage over metal is its absolute ease of installation. It is light, even in the larger diameters, and flexible pipe can be pushed or led through a wall more easily than soft copper. Flexible pipe can be cut with a knife or hacksaw, rigid pipe with a hacksaw. Pipes and fittings are literally glued together. A little cement is applied with a brush, the pipe inserted into the fitting. A minute later, the joint is ready to hold pressure. The joint so formed is so strong that it is sometimes referred to as a "liquid weld."

Plastic can also be joined with clamps or by threading. However, threading reduces the pipe's pressure rating by about 50 percent and is generally not done.

Type I PVC plastic pipe and fittings ready to be used for drainage piping in a new house. Fittings are cemented in place.

Table 8

PLASTIC VS. METAL PIPE

Weights (lb.) Per 100 Ft. of Pipe

			Cast Iron (5 Ft. Lengths)			
			Single Hub		No-hub	Plain End
Nom. Size, in.	PVC	Copper DWV	Service Weight	Extra Heavy	Service Weight	Sched. 40 Galvanized
1¼	42	65	—	—	—	227
1½	51	81	—	—	270	272
2	68	107	400	500	330	365
3	141	169	600	900	480	758
4	201	287	800	1200	740	1079

Courtesy Plastic Pipe Institute

CAST-IRON PIPE

This pipe has long been used for drain, sewer and vent lines, and may continue to be the material most often selected for this purpose. While cast iron rusts quickly, it is highly resistant to acids, alkalies, wet soil and the like. It is heavy, awkward to handle, but very strong.

Cast-iron pipe used for home service comes in two weights (wall thicknesses), service and heavy. The lighter weight pipe is more than satisfactory and should be used where local codes permit.

Cast-iron pipe usually comes in 5-foot lengths, but some supply houses carry lengths to 10 feet. One end of the pipe is always formed into a hub large enough to accept the other end of a following length of pipe. This end is always called the "spigot."

Cast-iron pipe may be cut with a hacksaw, chisel, or soil-pipe cutter. This is a special tool that enables you to cut the pipe with one swing of the large handle.

Cast-iron soil fittings. There are probably a hundred or more different cast-iron fittings currently manufactured not including size variations. Some fittings can be joined to the pipe only by lead caulking or screw threads, but others can be joined by both types of joints.

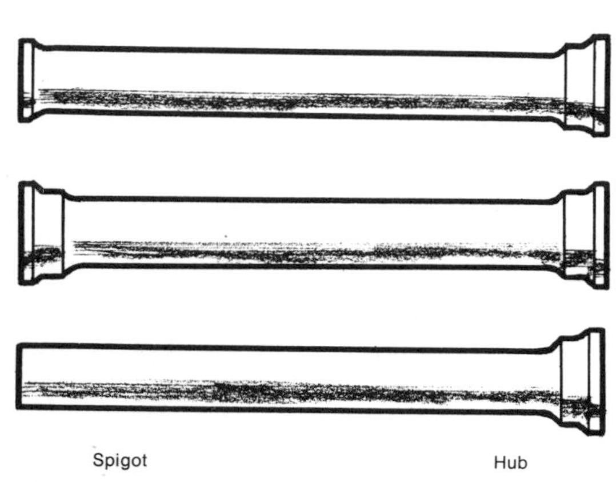

Standard cast-iron soil pipe (top) has a hub at one end and a ridge at the other. Modern soil pipe (bottom) has hub at one end, no ridge at spigot end. Pipe at center is variation of standard with double hubs.

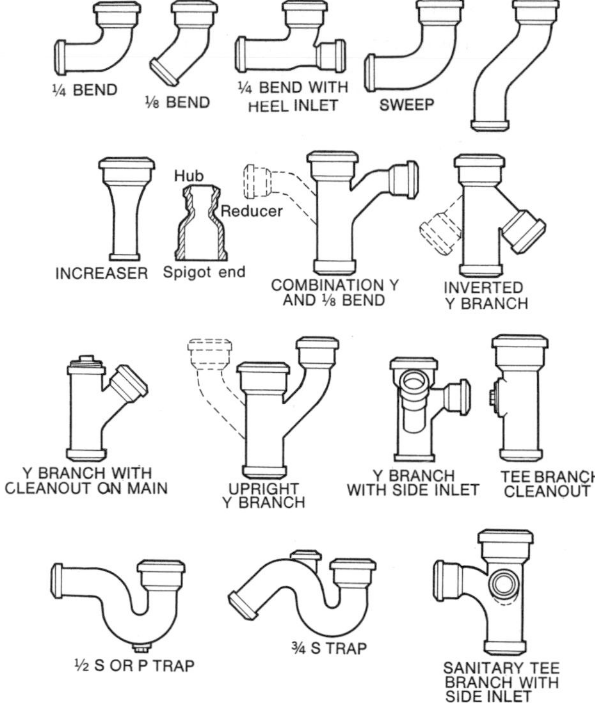

Examples of fittings used with cast-iron soil pipe.

Modern cast-iron soil pipe with hubs and straight spigots simplifies joint-making.

Table 9

CAST-IRON SOIL PIPE

Size	A	M	J	T	Weight (lbs.) per 5′ length Single Hub	Double Hub
2	3.06	2.75	2.38	0.12	25	26
3	4.19	3.88	3.50	0.18	45	47
4	5.19	4.88	4.50	0.18	60	63
5	6.19	5.88	5.50	0.18	75	78

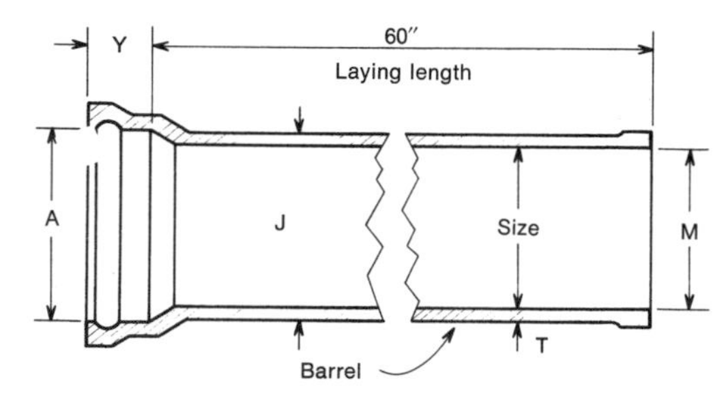

Measurements of cast-iron soil pipe. Use table above to determine various dimensions.

Table 10

CAST-IRON TEE DIMENSIONS AND WEIGHTS

Size (inches)	A	B	E	C	F	G	Weight (pounds)
2	3	4	4½	6	11½	7	11
3	3¼	4	5¼	6¾	12¾	7½	16¼
4	3½	4	6	7½	14	8	22½
5	3½	4	6½	8	15	8½	28
6	3½	4	7	8½	16	9	34½
3 x 2	3	4	4¾	6½	11¾	7	14
4 x 2	3	4	5	7	12	7	17¼
4 x 3	3¼	4	5½	7¼	13	7½	19¾

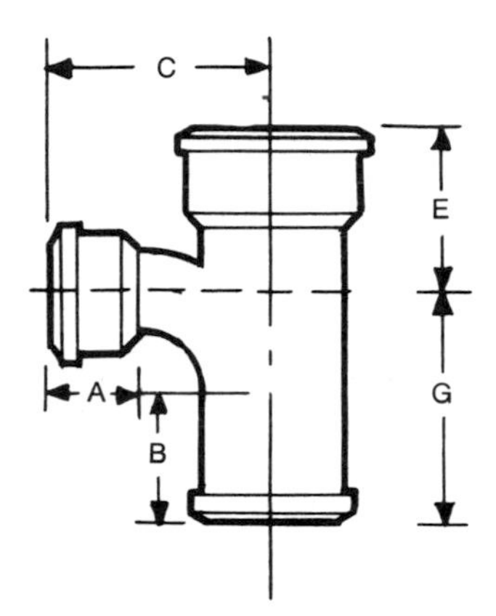

A cast-iron Tee fitting.

No one needs to know them all, but it is wise to glance at the more common cast-iron fittings illustrated here so you can see what is available and how they can be used in a drain system.

STAINLESS-STEEL PIPE

A comparatively thin-wall pipe of stainless steel has been recently placed on the market. Its cost is roughly about one third that of equal diameter copper tubing. No fittings are offered, and it is joined by soldering to standard copper tubing "sweat" fittings. Stainless requires an acid flux. The same propane torch and technique used for soldering copper is used, but the residue of the acid flux should be washed off. Flux remaining inside the pipe is carried away by the flow of water.

Stainless is stiff. It may be cut with a hacksaw or standard pipe cutter, such as used with copper, but a "hickey" (pipe bender) is necessary to bend it. Most building codes make no mention of it, but you can probably get it accepted if you bring a sample to your local plumbing inspector.

VITRIFIED CLAY PIPE

Red-brown in color and shiny with melted glass (glaze), vitrified clay pipe is the most nearly permanent piping material used in home plumbing systems. It is impervious to acids and alkalines, doesn't deteriorate with age and would be perfect if it could withstand shock and could flex a little. It can't. Vitrified clay pipe is therefore never used inside the house, nor exposed outside the

A length of vitrified clay pipe and an elbow of the same material.

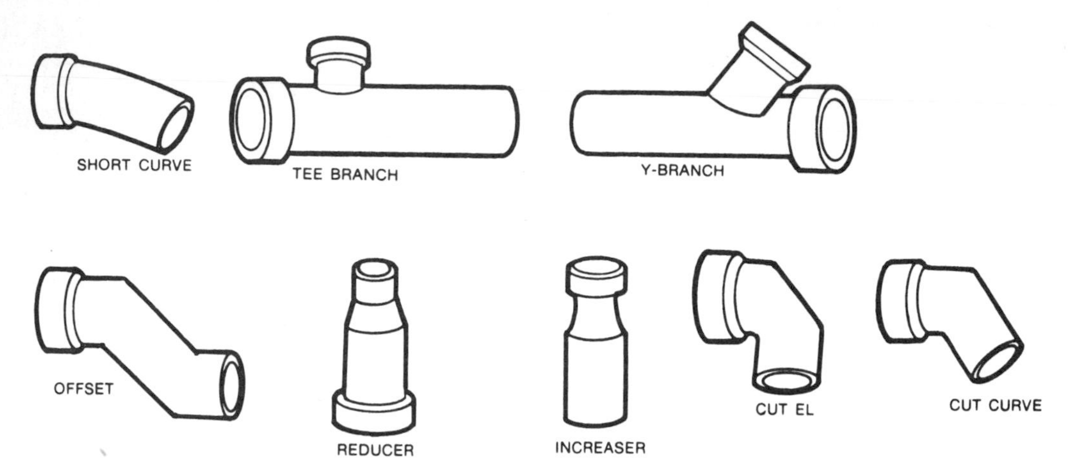

Fittings used with vitrified clay pipe.

house. Vitrified clay pipe is limited to sewer runs starting some 5 feet beyond the outside of the foundation wall, at a depth well below the frost line. When the sewer line must run beneath a road, that section of line is never vitrified clay pipe, but always cast iron.

Vitrified clay fittings. Only about a dozen or so clay-pipe fittings are normally available at a plumbing supply shop. However, as clay is never run indoors, the few fittings available can easily be adapted to the job. If you need an unavailable fitting you can make do with a cast-iron soil pipe fitting of the proper diameter. Handle it as you do the clay.

ASBESTOS CEMENT PIPE

Pipe made of asbestos cement, which is asbestos fibers mixed with portland cement, is used almost exclusively for sewer lines, but can also be used for vent lines. The pipe comes in 5- and 10-foot lengths and can be cut with a hacksaw.

Asbestos cement pipe may be buried directly in the earth. This pipe is used only for drains.

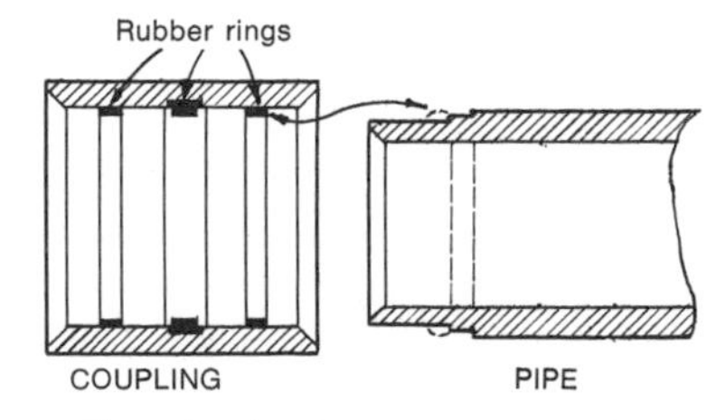

Cross section of asbestos cement coupling and pipe end. Coupling is simply forced over pipe end. Rubber rings (gaskets) in coupling seal the joint.

Fittings. These are currently available in limited number. If you are planning to use anything less common than a quarter turn or Tee you are advised to order the fittings before you start. When ordering, use the same system as for wrought-iron or cast-iron fittings—that is, a 4-inch quarter turn or elbow is designed to accommodate pipe with a 4-inch internal diameter.

BITUMINOUS FIBER

This is a lightweight pipe used for sewer and drain lines outside the house. It is made by combining fibers and various asphalt formulas under high pressure. It is not as strong as vitrified clay or asbestos cement pipe and not nearly as corrosion resistant. Its ability to withstand ordinary corrosion and attack by various detergents depends on the grade of fiber pipe you purchase. This is an important point to check. Another is its acceptance by your local plumbing inspector. Bituminous fiber piping is not welcome as a sewer pipe in all communities.

The pipe comes in two forms, most frequently in lengths of 8 and 10 feet: standard piping for sewer lines and piping with a row of holes on each side. The latter can be used to distribute rain water from a house leader across the surface of a lawn, or into a long, dry well dug beneath the surface of a lawn. And it can be used to carry waste liquids from a septic tank into a drain field.

Bituminous fiber fittings. Few types of fittings are available for bituminous pipe because the pipe is rarely connected in complex figurations. Most often it is run on a straight line or around a simple bend. As with the asbestos cement fittings, you are advised to make certain you can get what you need before you start the job. If you need to connect a bituminous pipe or fitting to clay or asbestos cement use a bituminous compound to make the joint.

Bituminous fiber pipe is light and easy to work with. Aluminum paint keeps the pipe from soiling your hands. Check if code permits this pipe to be used.

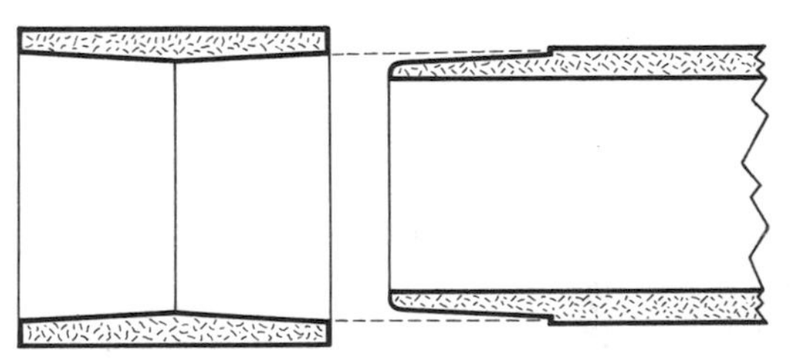

Cross section of bituminous fiber coupling and pipe end. Coupling is forced over pipe end. No gaskets are used. This pipe is usually run straight.

CONCRETE PIPE

Concrete pipe is inexpensive and well suited to the task for which it is used in home plumbing—carrying waste into septic tank drain fields. Concrete pipe is porous. Some of the liquid effluent entering the pipe from the septic tank can seep through the pipe walls directly to the earth. As the purpose of the drain field is to spread the liquid effluent as widely as possible for maximum absorption, pipes with porous walls help.

Concrete pipe is durable and inert and particularly well suited to immersion in water. Concrete cures slowly, and best in water. Final cure takes about twenty-seven years, at which time the concrete is about three times stronger than it was shortly after setting.

CLAY DRAIN TILE

Vitrified clay pipe is pipe made of clay over which a layer of glaze has been placed and the whole fired. That is what makes for the glazed—glasslike—coating. Clay drain tile is also a fired clay product, but it is not glazed. Neither does it have a bell and spigot. Instead, drain tile is a short—usually 1 foot—square-ended length of pipe, octagonal on the outside and smoothly round like standard pipe on the inside. Generally yellow or red in color, drain tile is used for septic-tank and house-foundation drainage.

Table 11

HOW PIPES ARE USED

Piping Material	How Used
Galvanized steel Wrought iron Copper	Water service main
Galvanized steel Copper Plastic	Domestic hot- and cold-water lines
Plated brass Stainless steel	Exposed domestic hot- and cold-water lines
Cast iron Galvanized steel Copper Brass Asbestos cement Plastic	Waste and soil mains inside the house, vent lines
Plated brass	Exposed waste lines
Lead Galvanized steel Cast iron Copper Plastic	Toilet to waste pipe connection
Black steel Copper	Heating lines (steam and hot water)
Vitrified clay Asbestos cement Bituminous fiber Plastic concrete	External sewer lines
Clay drain tile	Septic tank drain field

MORE ABOUT FITTINGS

The need for fittings becomes obvious as soon as one starts to run (install) pipe. Every time a rigid pipe has to make a sharp bend, a fitting must be used. Every time a pipe branches off into two and three pipes, a fitting is needed. Every time a pipe changes size (diameter), a fitting is required. And since plumbers have been at their trade since Roman times, fittings have been developed that meet almost every conceivable piping situation. When one fitting won't do the job, sometimes a combination of two fittings will.

As each fitting does a somewhat specialized task, it is important for you to have an over-all picture of the many types of fittings currently manufactured. Also, you should be able to describe the fitting you want, to help the clerk at the plumbing supply house find it for you.

Fittings fall into four classes according to strength, or weight.

lightweight	heavy
standard	extra heavy

When the finish is given, the fitting may be described as:

galvanized	rough
black	polished
plated	

Type of service would include one of the following:

gas	drainage
steam	vent
domestic hot water	storm water (rain water)
domestic cold water	septic tank effluent
hot water for space heating	

Shape and purpose of a fitting is described to some degree by the fitting's common name. Some of the better known and more frequently used fittings include:

coupling	return bend
nipple	cross
cap	extension piece
plug	cross over
angle	offset
bend	vent increaser
reducer	hose adapter
street elbow	adapter
union	strainer
union ell	transition
Tee	floor drain
Wye	toilet floor flange
bushing	flange

Fittings are also identified by the nature of their service, meaning the substance carried (liquid or gas), temperature and pressure, as for example:

gas	hot water for space heating
steam	drain
domestic hot water	vent
domestic cold water	storm water (rain water)

Additional identification includes the method used for joining the fitting to the pipe. If more than one method is used, to join two or more different pipes, that too will be included in the description of the fitting, though not always in direct terms. Connecting methods include:

screw thread	solder
flexible compression	clamp
metallic compression	flare
slip joint	flange
bell and spigot (caulk)	chemical bond

A fitting is also described by its size, which is almost always given as that of the two or more pipes it joins. For example, a coupling that connects two lengths of 1-inch pipe would be termed a 1-inch coupling.

In the case of a Tee, which always joins a branch pipe to the main pipe at a right angle (90 degrees), the pipe size or sizes given depend on the size of the pipe to be used and their relationship. If all the three pipe diameters are, say, 1½ inches, the Tee would be called a 1½-inch Tee. When the branch pipe is different, the size of the run —the pipe onto which the branch joins—

To connect the 2-inch galvanized pipe at the top to the 4-inch cast-iron soil pipe below required a cast-iron Tee with a threaded outlet into which a reducer was screwed, which "took" a close nipple followed by a 2-inch elbow. Four fittings to make one turn and joint. The elbow is a drainage fitting, as the galvanized pipe comes into it at a pitch.

would be given first. Assuming the run is 1-inch pipe and the branch is ¾-inch pipe, the Tee used would be called a 1-by-¾-inch Tee. If all three internal diameters are different, the size of the main pipe, the pipe from which the liquid or gas flows would be given first. The diameter or size of the pipe in line with the main pipe would be given second and the tap would be given last. For example, a Tee listed as having the size 2-by-1-by-¾-inch would connect a 2-inch pipe directly to a 1-inch pipe. The ¾-inch pipe would be the pipe at the right angle.

Thus a fitting might be described as a 3-inch vitrified, 30-degree Wye, or as a 1-inch galvanized Tee, or 2-to-1-inch brass reducer, a 1½-inch plated brass Tee with three flexible compression couplings, and so on.

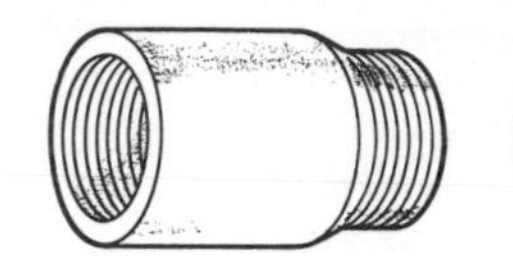
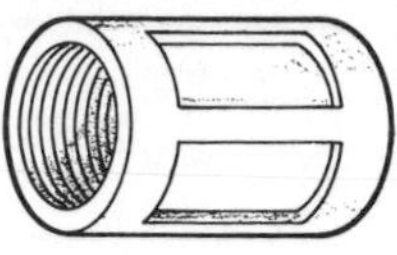

Standard coupling and an extension piece to connect a male pipe end to a female.

Couplings. These are short lengths of pipe internally threaded at both ends. Generally, when you purchase a full length of pipe, one coupling is thrown in free—usually it is on the pipe.

Couplings are almost always threaded with right-hand threads on both ends. When a right- and a left-hand thread is wanted, they may be obtained on special order. Such couplings are usually marked by projections or rings to distinguish them.

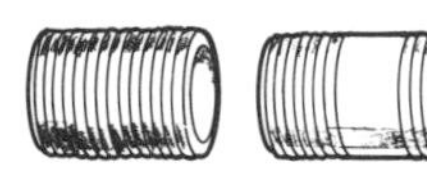
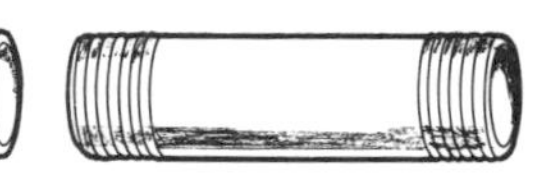
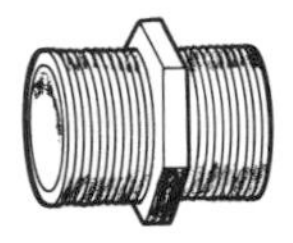

Four types of nipples used in plumbing. Nipple with hexagon nut has right- and left-hand threads.

Nipples. A nipple is a length of pipe not more than 12 inches long, externally threaded in the same direction (right hand) at both ends. A close nipple is so short there is no unthreaded area between its ends. A short nipple, or a shoulder nipple as it is sometimes called, has a small amount of unthreaded pipe between its ends. A long or extra-long nipple has a longer space bare of threading.

The close nipple might be used where two valves are to be placed very close to one another. The shoulder nipple would bring them close but would provide a separation.

The length of a coupling varies with its diameter, but is always designed to permit both pipe ends to meet or almost meet in the center. The size of a coupling is the size of the pipes it accepts.

Pipe extension piece. This is another form of coupling. Like the coupling, the two pipes it joins are in a straight line. But unlike the standard coupling, one end of the extension piece has a female thread, the other has a male thread.

Reducers. Like the standard coupling, the reducer has female threads at both ends and

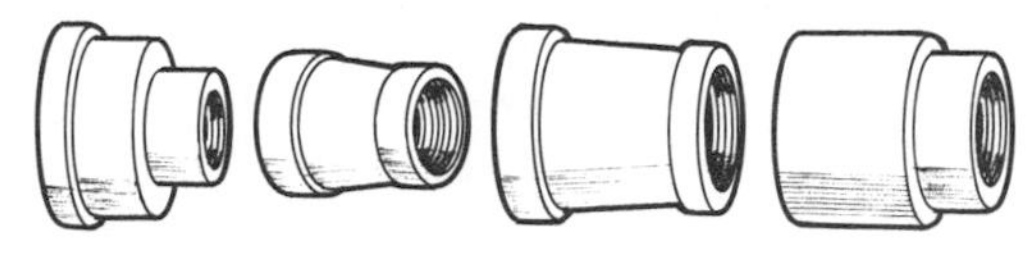

Threaded reducers for moving from one pipe size to another find frequent use on the job.

connects two lengths of pipe in a line. However, the reducer accepts pipes of two different sizes, hence its name. Generally, when the size of a reducer is given, the larger size is mentioned first.

Bushings. A bushing is also used to connect two pipes of different sizes, but requires less space. Most or all of the bushing screws into the larger pipe. The smaller pipe screws into the bushing. To use a bushing, the inside of the larger pipe must be threaded. As pipe is threaded on the outside at the factory, the easy solution is the use of coupling followed by the bushing. The alternative, which might be needed when space is tight, would be to cut an inside thread on the pipe.

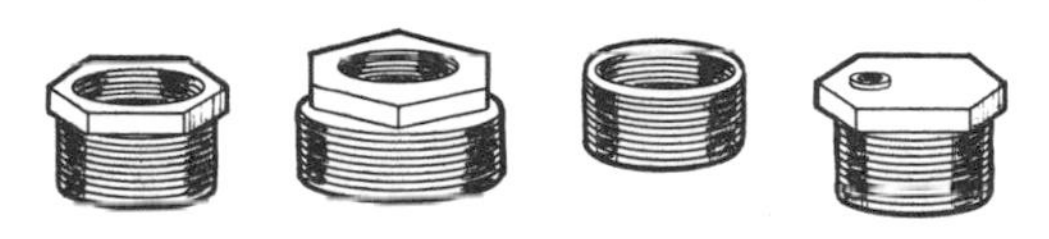

Various types of bushings.

The interior and exterior of a bushing are generally made in axial alignment. But when the pipe size difference is large, there is space for axial offset. Some are made this way. They are very useful when you wish to bring the smaller pipe closer to one side than would be possible when the smaller pipe joined the larger at its exact center.

Bushings are sized according to the pipe size of the male thread on the bushing itself. Thus if the hole into which the bushing is to be screwed accepts a 1-inch pipe, the bushing would be called a 1-inch bushing. If the smaller pipe's size is ¾-inch, the bushing would be a 1-inch, one-pipe-size, reducing bushing. However, to avoid confusion it is best to give both pipe or fitting sizes when ordering bushings. In this case, one would ask for a 1- to ¾-inch reducing bushing.

Unions. In the parlance of the trade, pipe runs are "made up" by screwing one length of pipe into a fitting, a second fitting onto the first length of pipe, followed by another piece of pipe, and so on. In the course of crossing a room, turning at a wall and then going up or in another direction, a number of pieces of pipe and a number of fittings are used. Every screw thread on the pipe and fittings is normally right-handed. Should you, for any reason, want to open the first joint, it would be necessary to unscrew every fitting and piece of pipe following that joint. Obviously, if the last screw-thread joint is under the kitchen sink, you have quite a job on your hands. A union eliminates this necessity, permitting you to "break" the pipeline without disturbing any of the preceding or following joints.

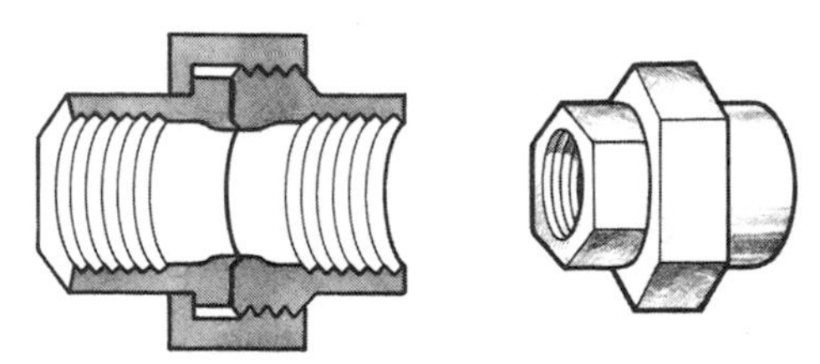

Section through an assembled standard union (left) and an exterior view (right).

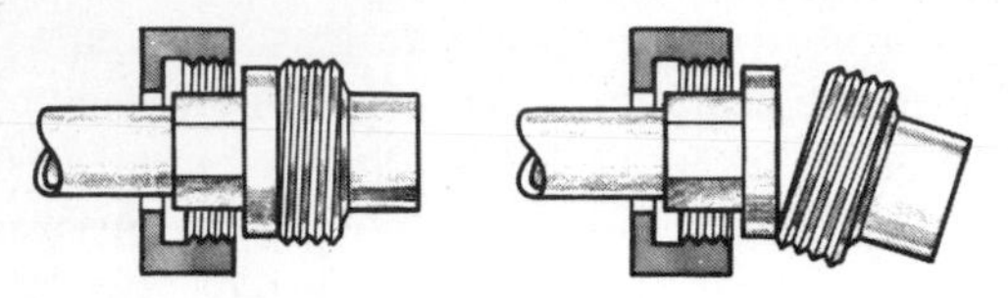

Parts of a union in correct alignment (left) prior to assembly, and parts in bad alignment (right)

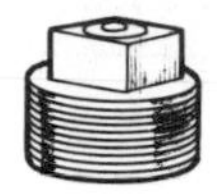
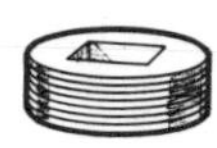

Three types of plugs.

Older types of unions come with a gasket between the parts of the fitting that mate. The newer or modern unions have no gasket but rely on two perfectly smooth metal faces making a tight joint.

For close work, unions are manufactured integral with Tees, street elbows and regular elbows. Unions are available for standard screw and sweated-joint piping. They are made from galvanized and nongalvanized malleable iron, brass and copper. No unions are made for use with cast-iron, clay, asbestos, fiber or concrete pipe. When a union is to be introduced into a plastic pipeline, two transition fittings may be used, or the union's parts cemented in place.

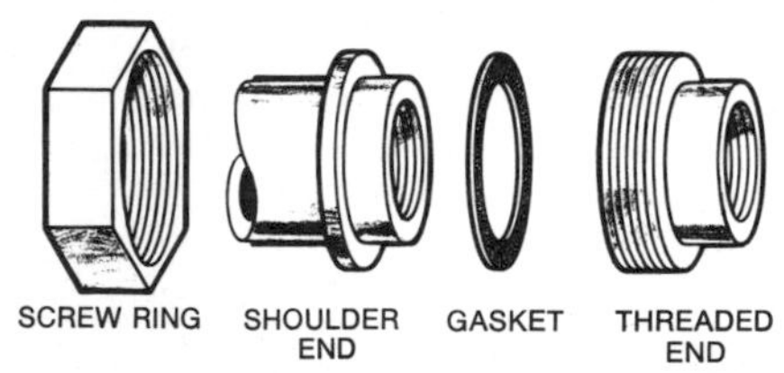

Parts of an old type of union requiring a gasket.

Unions combined with Tees and elbows: two with male threads, two with female threads.

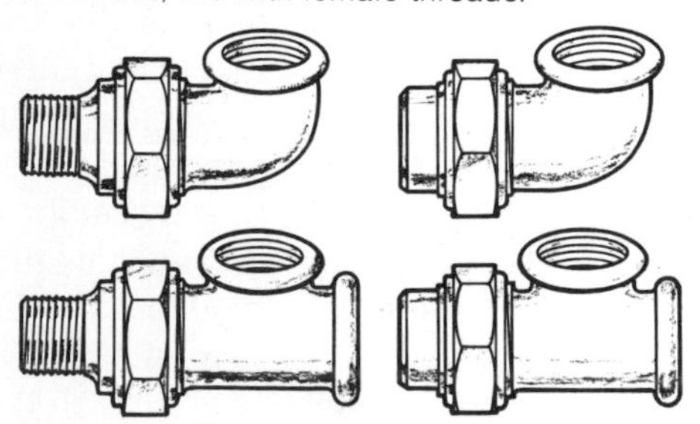

Plugs. A plug is a short, solid length of metal having a male thread and some means of being turned by a wrench. It is used to seal the end of a pipe or fitting having a female thread. It is made of cast-iron, galvanized and nongalvanized malleable iron or brass. When a cast-iron pipe is to be sealed, it is best to use a brass plug. Plug sizes run from ⅛ inch to 12 inches.

Caps. When the pipe to be closed has a male thread on its end a cap is used. Malleable iron and brass caps come in sizes to fit all pipes from ⅛ to 6 inches. Cast-iron caps can be purchased in sizes running from ⅜ inch to 15 inches. Like the plug, the cap is simply screwed in place with the aid of a suitable wrench.

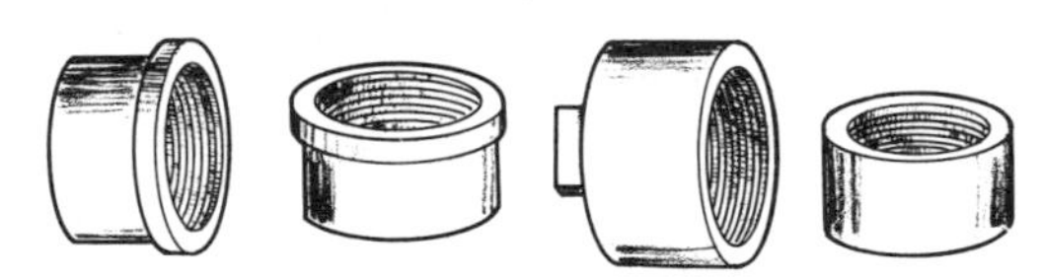

Four types of screw caps.

Elbows. When a rigid pipe has to change its direction, make a turn, an elbow is used. Prefabricated elbows are available for all types of piping material and for all the connection methods used. There are galvanized, black, copper, brass, plastic, lead, clay, cast-iron, fiber, asbestos and even concrete elbows for sale at the supply houses. They come with screw threads, sweat joints, compression, bell and spigot and every type of joint connection used. Some elbows are long and extend well beyond the turn. Others just make the turn, and still others are so short they barely accept the full thread length on the pipe ends. Some elbows terminate in flanges, others have lips or ears for fastening and some are reducing elbows, joining two pipes of different diameters at an angle.

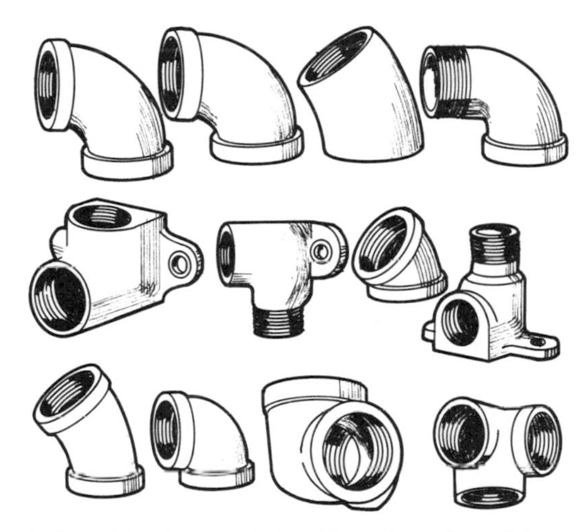

A few of the large variety of threaded elbow fittings presently available. Two fittings on the bottom right are side elbows with side outlets for three pipes.

The change in pipeline direction introduced by an elbow is measured in degrees. This may be measured by erecting a line orthagonal to the plane of each end of the pipe and measuring the angle between these two projections. Measuring the angle between the elbow's arm ends leads to error. Elbows with angles of 45 and 90 degrees, used for water, gas and steam, are called standard angles. Elbows used for the same service but having directional changes of 22½ and 60 degrees are classed as special angles. Cast-iron elbows for drainage work come in a larger range of angles: 5⅝, 11¼, 22½, 45, 60 and 90 degrees.

Elbows for screw-thread pipe have female threads at their ends. Elbows that are to be sweated have slip joints at their ends. Cast iron elbows may have hub and spigot ends and may have threaded ends, as the need may be.

Return bends. A return bend is a length of pipe with a female thread at both ends,

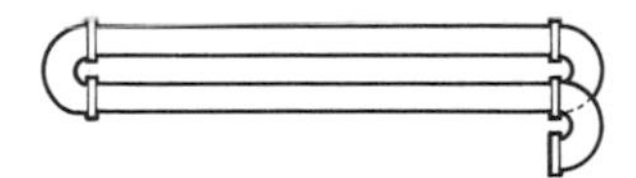

How return bends may be used to make up a small radiator in a hot-water or steam-heat system.

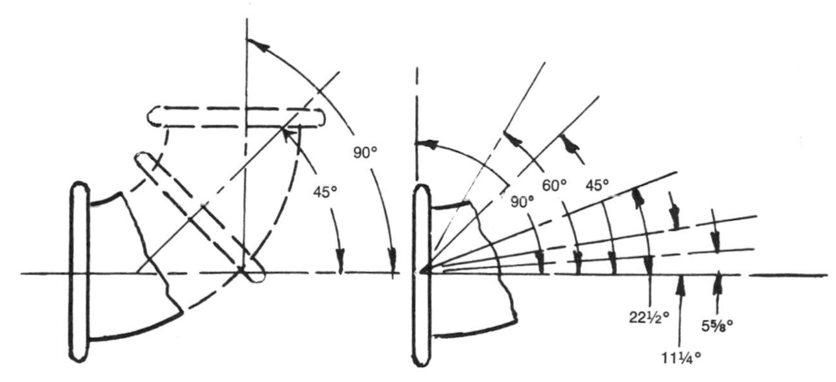

Angles at which elbows are currently manufactured and how they are measured.

both of which face in the same direction. That is to say the pipe makes a complete turnabout, a 180-degree directional change. Generally return bends are used to assemble coils for steam and hot-water heating, to substitute for a cast-iron radiator.

Back and side-outlet return bends. These are return bends that have an opening for a third pipe. When the third opening is at an angle to the other two, the fixture is called a side-outlet return bend. When the third outlet is facing in a direction opposite to the other two openings, it is a back-outlet return bend.

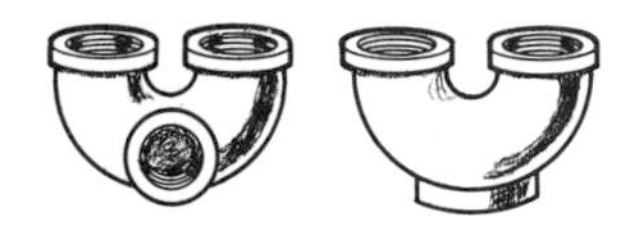

Return bends of cast iron with side and back outlets.

Side-outlet elbows. Elbows having a third opening for a third pipe are called side-outlet elbows. When a side-outlet elbow is not available it can be duplicated with an ordinary elbow and a Tee.

Tees. Manufactured in all the metals (except lead), and in plastic, vitrified clay, bituminous fiber, and asbestos cement. They come in a wide range of sizes in metal. Sizes and variations are limited in clay and fiber.

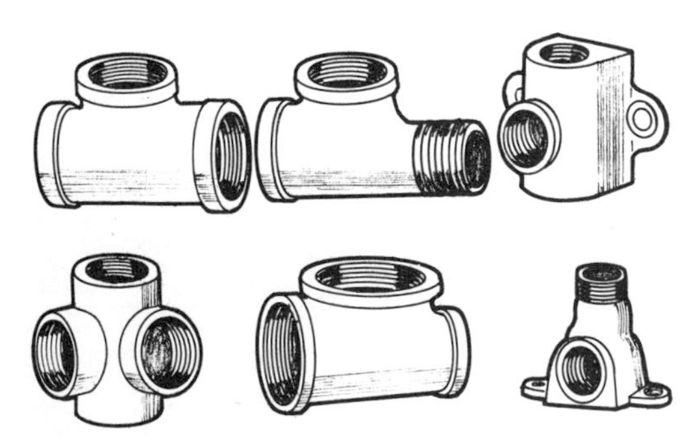

Examples of threaded Tees in cast iron or brass. Some are reducing; two have male *and* female threads for added versatility.

To specify a Tee size, the two in-line pipe sizes are listed first, followed by the size of the branch pipe. It helps to draw a little sketch.

The branch of a Tee is always at a right angle, 90 degrees to the main pipe run. When the branch or side outlet is set at an angle other than 90 degrees, the fitting is no longer classified as a Tee, but is considered a Wye.

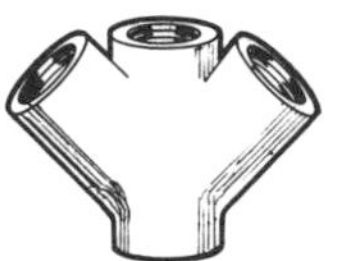

From left: single Wye, double Wye and a cross-connecting fitting.

Wyes. Some of these fittings look like the letter "Y"; others have two branches at an angle to the main run. They are called double-Wyes. Single Wyes with branches set at 45 degrees, with branches the same size as the main run and smaller, are regularly stocked in sizes from ½ to 4 inches. Double Wyes with both branches set at 45 degrees are commonly available in sizes ranging from ½ to 2 inches. These sizes by no means include all Wyes, but are most common. Again, Wyes, single and double, straight and reducing can be had in all the metals (excepting lead), plastic, vitrified clay, bituminous fiber and asbestos cement. Cast-iron Wyes are available with all hub and spigot ends and with a combination of hub and spigot ends and threaded ends.

A 45-degree, 4″×3″ hubbed cast-iron Wye (left) and a simple, quick sketch that you can make to positively identify the same Wye for the plumbing supply shop.

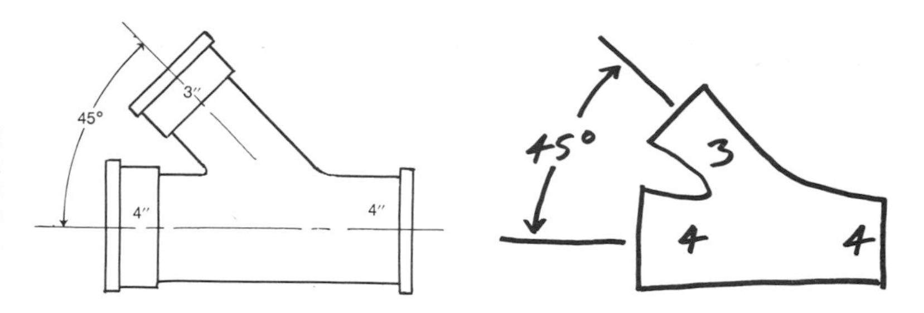

Crosses. A cross fitting connects four pipe ends together at right angles to one another. They are made in a number of sizes and material. The connected pipes are usually on a single plane. That is to say, the four pipes can lie flat against a floor or wall.

Drainage fittings. These differ from other screw fittings in two ways. Their female threads are always recessed and the opening that is to accept a horizontal drainpipe is always pitched to provide a drop of ¼ inch to the foot. The purpose of the recessed threads is to bring the inner surface of the pipe flush with the inner surface of the fitting. In an ordinary water pipe fitting, these two surfaces are not flush. When water alone flows, there is a little friction at that point, but no other problem. When waste water flows, the hair and lint in the waste tend to collect at that little protrusion and will build up with time and eventually seal the pipe.

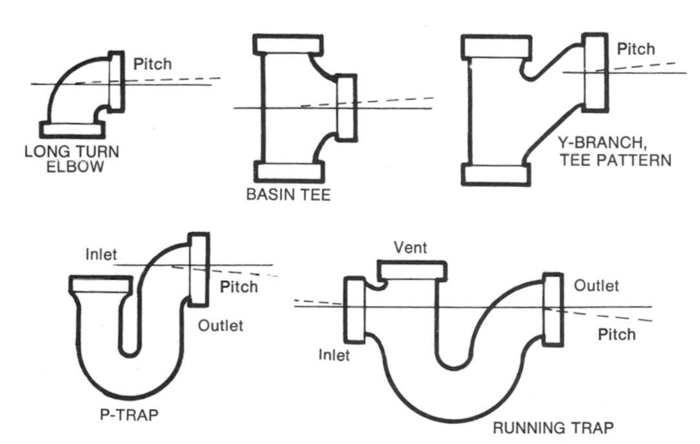

Drainage fittings are always pitched to discharge waste. Solid lines indicate horizontal; dotted lines indicate the pitch or angle at which the pipes connect to the fittings. Ordinary fittings do not provide the pitch necessary for drainage.

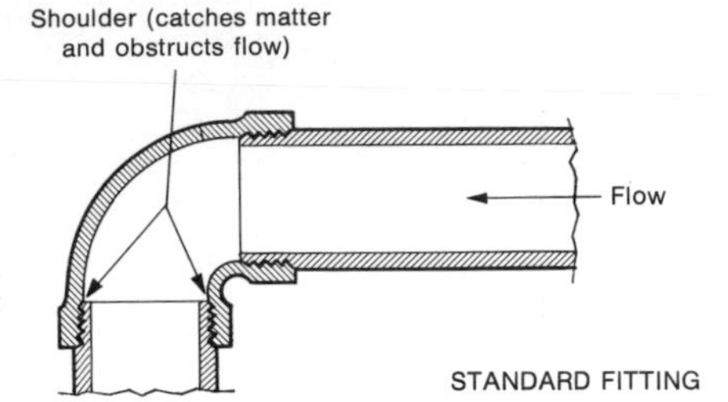

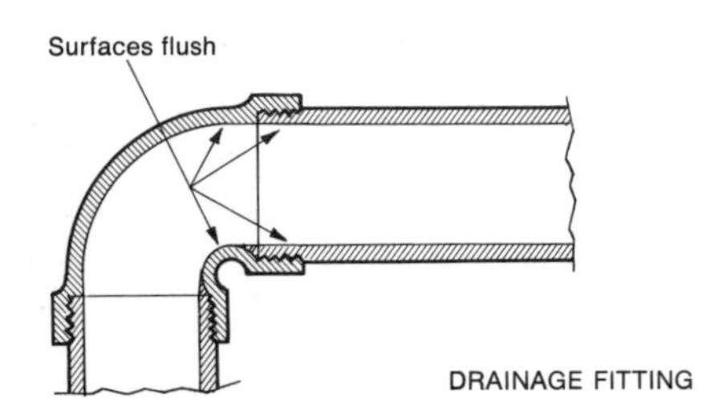

Difference between fitting and drainage fitting. Note ridges on interior of standard elbow which impede flow. Surfaces of drainage fitting are smooth.

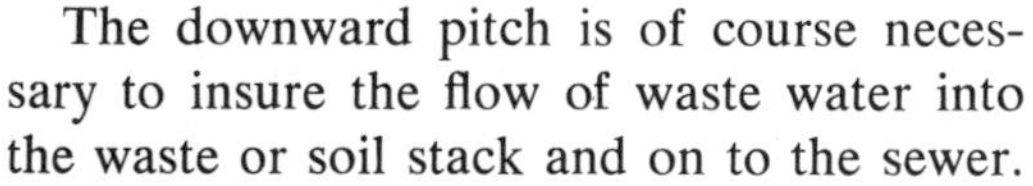

The downward pitch is of course necessary to insure the flow of waste water into the waste or soil stack and on to the sewer.

Drain fittings that are hidden from view are generally made of cast iron and malleable iron. Drain fittings terminating at a sink or other fixture are often made of plated brass.

Special drainage fittings. There are any number of drainage fittings manufactured for special purposes. Some will reduce the number of joints needed for drain and vent lines. Others simplify the hanging of wall-mounted toilets and other fixtures. When you have a piping setup that requires a large number of fittings, or the available standard fittings do not do the job easily, it is wise to bring a sketch of your particular layout to the plumbing supply house. Very often they can suggest a special fitting that will reduce cost and labor.

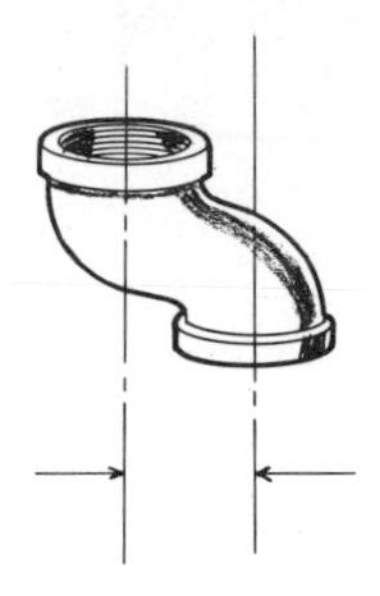

Typical cast-iron offset fitting. Note how the pipe-run offset is measured (arrows).

Offsets. When a pipe has to be moved to one side without changing direction, an offset can be bent in the pipe, or an offset fitting can be used.

Vent increaser. To reduce the possibility of the exposed, top end of a small vent pipe becoming blocked with ice and snow, its final diameter is often increased with a vent increaser fitting.

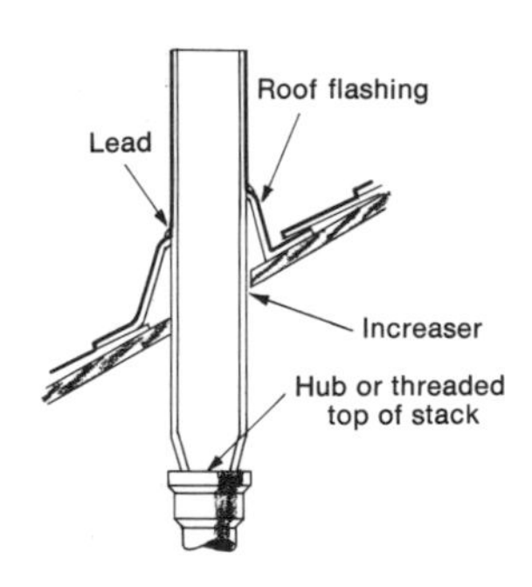

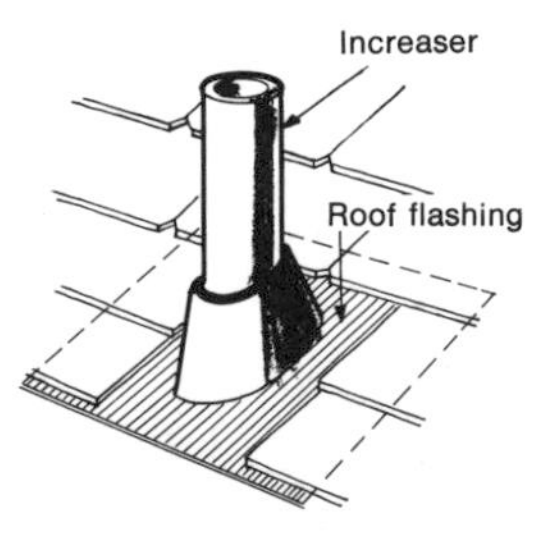

A vent increaser connected to the top of the main soil stack permits easier egress of sewer gases and also reduces chance that frost and a heavy snowfall will seal the stack's opening. Note how upper side of flashing goes beneath a shingle while the lower side goes on top of a shingle.

Hose adapter. This is a threaded length of pipe that permits a garden hose fitting, having very coarse threads, to be connected to the end of a pipe having standard pipe threads.

Adapter. A pipelike fitting used to connect plastic pipe to metallic pipe, or to a fitting having standard pipe thread.

Transition. An alternate name for an adapter. It provides a transition between plastic and nonplastic pipe and fittings.

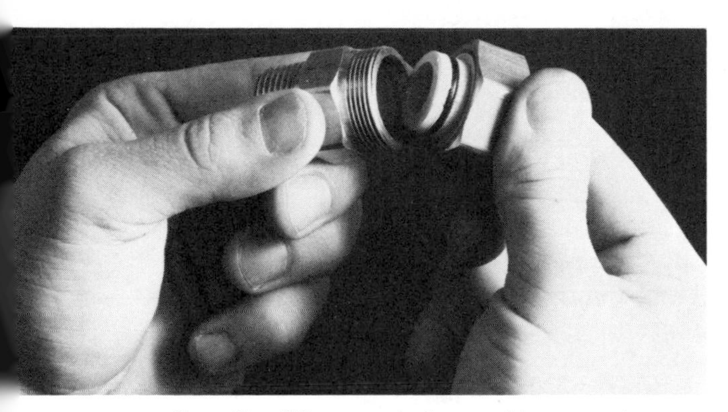

Transition fitting or adapter used to connect screw-thread to plastic pipe. Note O-ring that seals joint.

Floor drain. Generally a cast-iron bowl with a pierced cover and a drain hole at its bottom. It is installed in a concrete slab, its cover flush with the top of the slab. It is used to drain surface water from the floor.

Toilet floor flange. A circular, flat casting generally of brass about 8 inches across. It is fastened by screws to the bathroom floor. The center of the flange may be threaded to accept a threaded soil-pipe elbow, or the edge of the central hole may be beveled to be soldered to the end of a lead soil-pipe bend. The toilet sits atop the flange, the toilet's horn sealed to the flange and soil pipe by means of a gasket.

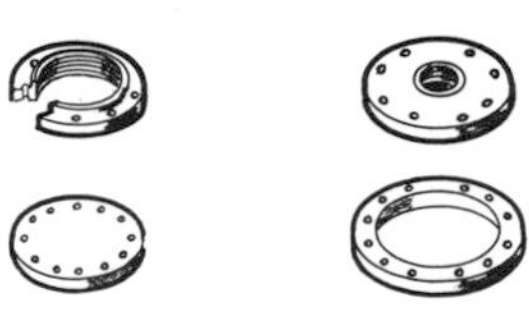

Four types of flanges.

Flanges. Flanges are metal discs used to connect the ends of two pipes. One is screwed on each pipe end, a gasket is placed between the two, and bolts are passed through matching holes. When the facing flanges are pulled tightly together, a leak-proof joint is formed. Flanges that have no central holes are called blind flanges. They are used to close and seal a pipe or fitting end.

Speedee fittings. This is a trade name for a group of special compression-type fittings preformed on the end of a length of pipe. They are used to connect a water line to a toilet tank or faucet. Tanks and faucets are made to accept their own particular type of pipe termination and no other. Older faucets and utility faucets, such as are found in the basement, are usually screwed to the end of a threaded pipe.

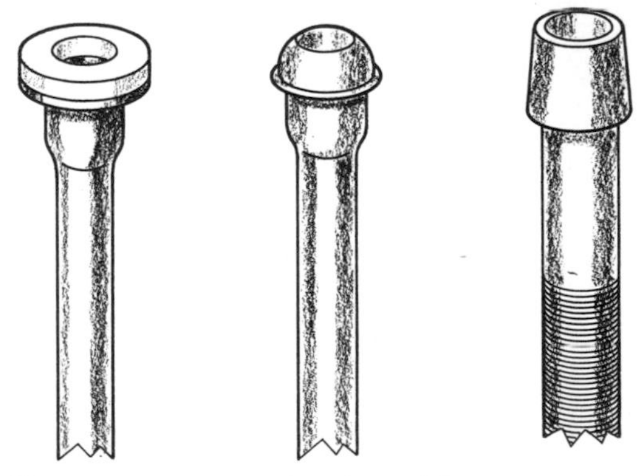

Three common types of preformed fittings on water-supply pipes. Nuts that slide up and hold ends in place are not visible. Pipe on left, which comes with in-place gasket, is used with a ballcock valve. Center pipe is used with modern faucets. Pipe at right terminates in a ground joint, is used with older faucets.

4

Cutting and Joining Pipes

The way pipe is cut and prepared for joining depends on the material which the pipe is made of and the method to be used for joining. As a rule, certain types of pipe are always joined the same way, but there are exceptions. For example, galvanized steel pipe is usually joined by screw threads, but on occasion it may be joined by caulking. Cast-iron pipe is usually joined by caulking, but sometimes it can be threaded and joined to a fitting or following length of pipe. The cutting and joining methods described in this chapter are those most commonly used for the various pipe materials.

MEASURING PIPE FOR CUTTING

There are several ways you can accurately determine necessary pipe length. One is to measure the distance between the two fittings and *add on* the two lengths that will be "lost."

Another method consists of visually estimating the dimension. With one fitting attached to its pipe and the other fitting positioned where it needs to be, place the end of your rule in the first fitting just as far as the pipe will go. Move the free end of your rule against the side of the second fitting. Read the dimension off by estimating how far the pipe will enter the second fitting. This is not very difficult to do as all you need to keep in mind is that ¾-inch galvanized pipe enters a fitting about ½ inch and 1-inch galvanized loses or enters the fitting about $^{9}/_{16}$ inch.

Still another method, which is even easier in some instances, is to actually screw or slip the end of the pipe into the first fitting and then hold the second fitting up against the side of the pipe and mark your cut line directly on the pipe.

GALVANIZED, BLACK AND BRASS PIPE

Pipes made of galvanized steel, nongalvanized steel (black pipe) and brass are almost identical in dimensions. They are therefore cut and threaded the same way with the same tools.

Three Ways to Measure Pipe

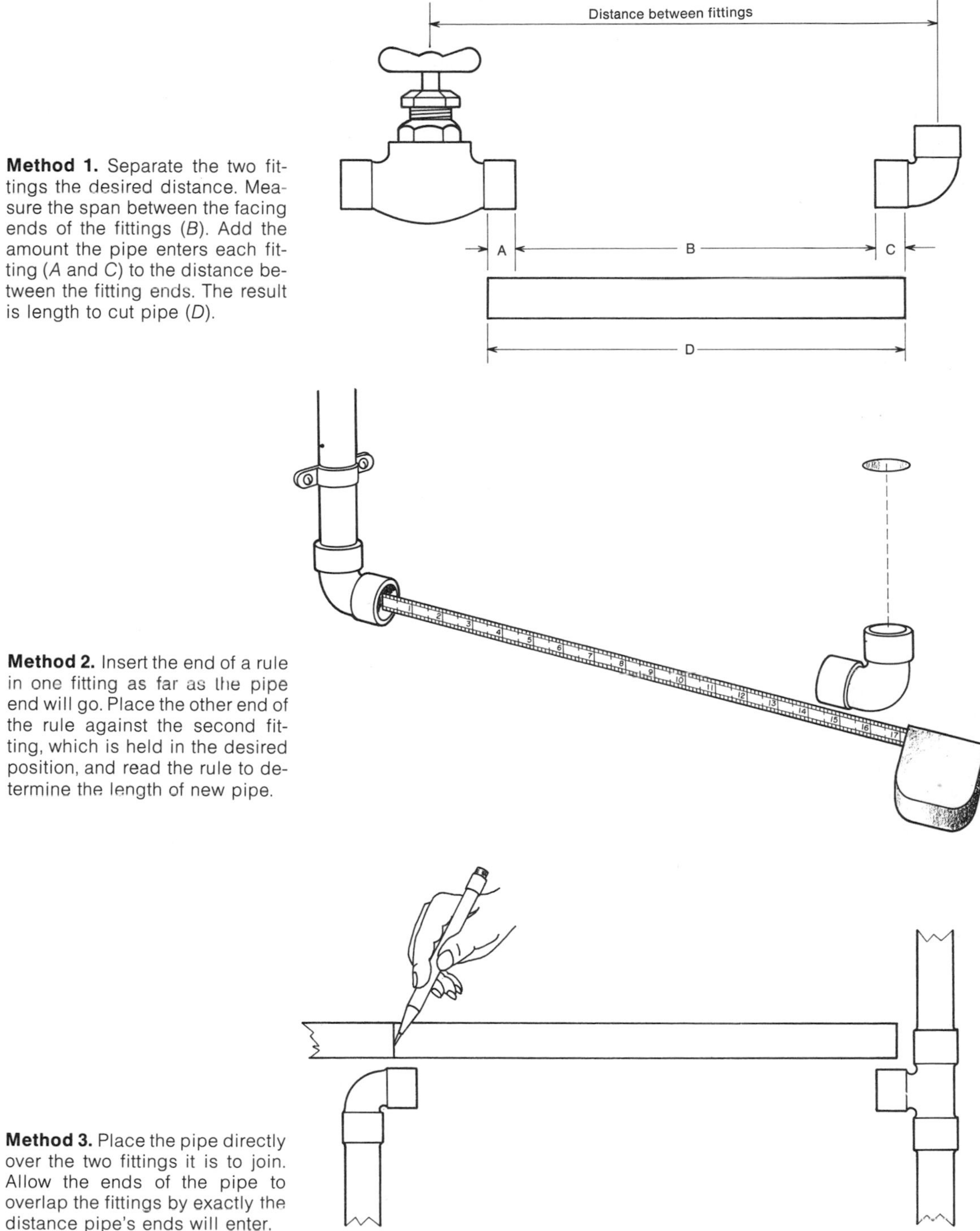

Method 1. Separate the two fittings the desired distance. Measure the span between the facing ends of the fittings (*B*). Add the amount the pipe enters each fitting (*A* and *C*) to the distance between the fitting ends. The result is length to cut pipe (*D*).

Method 2. Insert the end of a rule in one fitting as far as the pipe end will go. Place the other end of the rule against the second fitting, which is held in the desired position, and read the rule to determine the length of new pipe.

Method 3. Place the pipe directly over the two fittings it is to join. Allow the ends of the pipe to overlap the fittings by exactly the distance pipe's ends will enter.

With a hacksaw. Pipe made of these metals can be cut with a hacksaw. It is easier and more accurate to use a wheeled cutter, but if you only have to cut one or two pipes, there is no point in purchasing or renting one.

Before cutting, the line of cut should be clearly marked around the circumference of the pipe. The easy way to do this is to wrap some tape around the pipe, taking care not to stretch the tape while doing so. The pipe may be held in a pipe vise or a standard vise with the aid of some blocks of wood or a simple jig. Simply clamping the pipe between the flat, parallel jaws of the vise will not do. When sufficient pressure is applied to hold the pipe firm, it will be distorted. In the case of brass, the pipe may not revert to a perfect circle when released.

Use a coarse-toothed hacksaw blade. Press lightly downward on the forward stroke. The teeth should be pointing forward. Let the blade ride on the return stroke. Do not press down very hard or the blade will turn as it cuts.

Cutting done, use a file to remove any external burrs. Usually the cut section falls down before it has been completely severed from the main section, leaving a bent lip of metal. This must be removed. Use a round file to remove any burrs present on the inside edge of the pipe. Or use a pipe reamer mounted on a brace.

With a wheeled cutter. There are two types of cutters: one has a single cutting wheel, the other has two. The advantage of the two-wheel cutter is that it can do its work with no more than a 90-degree swing. This permits you to cut in-place pipe next to a wall or between joints. The one-wheel cutter requires a longer swing.

To use either cutter, place the pipe in a vise, if it is not in its place in the run. Position the tool over the pipe and turn the handle until the cutting wheel presses lightly against the pipe on the desired line of cut. Swing the handle completely over the pipe (or as far as it will go). Next, give the handle a fraction of a turn, pressing the cutting wheel more firmly into the metal, and swing the handle around the pipe. The cutting wheel digs in and once again loosens. Tighten it again and repeat until the wheel (or wheels) severs the pipe. If the wheel is not tightened sufficiently each time it comes around, there will be a lot of wasted motion swinging the tool around the pipe. If the cutter wheel is too tight, it will cut faster, but the tool will be more difficult to turn, and the ridge normally raised on the outside of the pipe, along both sides of the cut, will be higher than usual.

Power cutting is done the same way, but the muscle is furnished by a machine, which can be rented at a local Rent-All Shop.

With a motor-driven pipe cutter. With the motor switched off, a square-ended wrench, which usually comes with the machine, is used to open the three vise jaws in the center of the device. The pipe to be cut is slipped through the space between the jaws, which when tightened move towards a common center. There are two guide bars, one to either side of the center of the machine, about a foot down. One is pulled out as far as necessary. The cutter is positioned on the pipe, tightened lightly, and its handle rested against one guide.

Rotational speed is very slow but powerful. Don't get your hand caught between the tool and the guide bar. Don't get your sleeve caught on the rotating pipe; your arm can be severely injured.

As the pipe rotates, the cutter wheel digs in. As it digs, slowly and gently turn the cutter handle to the right, moving the cutter wheel further towards the center of the pipe. The pipe is cut clean and square. Shut off the machine and wait until it stops completely before attempting to insert the wrench and open the vise jaws.

After pipe has been cut, no matter what tool is used, remove the internal burrs with a reamer. Use a flat file to remove the external lip.

Cutting Thread

Male thread (external) is cut with a die held in a handle called a stock. The die looks like a large nut with slots and holes cut into it. It is made of hardened steel and differs from dies made to cut thread on bolts. Standard bolt-cutting dies cannot be used to cut pipe thread.

To cut thread on the end of a length of pipe, lock the pipe firmly in a vise, and slip the stock and die over the end. The guide, or bushing, on the stock goes first, the die itself last. There can be no mistake about this; the stock and die will not rest on the pipe if an attempt is made to start the die from its wrong side. The guide or bushing should be just a little large than the pipe size. In some stocks it is possible to change die sizes without troubling to change the guide. If the guide is for a larger size pipe, it is possible to cut the threads incorrectly.

Give the pipe end and the die a "shot" of cutting oil before starting. The cutting oil lubricates the die, holds its temperature down, prolongs its life and makes for a cleaner, smoother cut. Without oil, the thread may be ragged, uneven and even useless.

Now, press the die against the pipe end and give the stock handle half a turn or so to the right; that is, your right as you look at the facing end of the pipe. Back the die off about a quarter turn (turn it to your left) to break off the chips, then turn the die to the right, forward, another half turn. Every two complete turns or so, poke the spout of the oil can between the teeth of the die and lubricate the pipe with more cutting oil. Most of the oil will run right down, so if you are working indoors it is wise to place a catch pan underneath. Following the forward, backward, lubricate procedure described, the pipe is threaded until two full threads emerge from the die. At this juncture the stock is backed off and the die removed.

You need five full-depth threads. A few more do no harm; too many more and you will not be able to tighten the pipe. Too few threads and there will not be enough pipe in the fitting to hold firmly.

Table 12 shows how much longer you must make the pipe when cutting to allow for the threaded end that enters the fitting. For example, you are cutting 1-inch galvanized pipe to go between two screw fittings 3 feet apart. According to column B, .682 inch of 1-inch pipe enters the fitting. Therefore you need .682 inch or ¾ inch extra on each end of the pipe or a total of 3 feet 1½ inches of pipe to join two fittings the ends of which are 3 feet apart.

The table can also be used to help you cut thread. For example, correctly cut thread on 1-inch pipe can be hand threaded about ½ inch (column B). If your fitting won't go that distance you haven't cut enough thread. If it goes much farther, you have cut too much thread and may not be able to tighten the joint properly.

You cannot cut clean thread with a worn or damaged die. If you make certain the die is well lubricated, and make certain you don't have an old piece of pipe with carbon (hard) spots, then the die is at fault and must be replaced. It is easy enough to determine whether or not the die has cut clean

Cutting and Threading Pipe

1. Put the pipe in a vise, measure the length you want to cut, and mark the pipe.

2. Place the cutter on the pipe and turn handle, bringing the cutting wheels against the line of cut.

3. After pipe is cut, use reamer to remove internal burrs left by the cutter.

4. If the cutting wheel has raised a lip on the outside of the pipe, remove it with a flat file.

5. Slip the threading die over the end of the pipe, making sure the guide, or bushing, goes first, and rotate the handle.

6. Stop every once in a while to give the die and partially cut thread a shot of cutting oil.

7. If you are using a ratchet stock, as shown, continue slowly in the same direction, stopping frequently to lubricate the die. If you are using a two-handed stock, you can back up a fraction every once in a while to clear the chips.

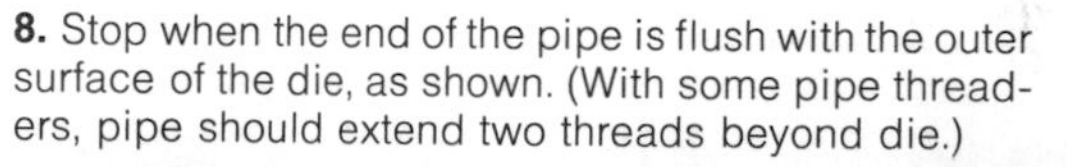

8. Stop when the end of the pipe is flush with the outer surface of the die, as shown. (With some pipe threaders, pipe should extend two threads beyond die.)

threads. Try the pipe end in a known-to-be-good fitting. Look at factory cut pipe thread and compare the depth of cut with what your die has produced.

Pipe with shallow thread does not enter the fitting to a suitable depth, tightens up very quickly, and usually leaks despite all quantities of pipe dope.

Pipe thread is unlike bolt thread in one important way. Bolt thread is exactly the same size for its entire length, even if the thread runs the full length of the bolt. Pipe thread is not. The die is shaped so as to produce a taper. Not only are the threads nearest the end of the pipe deeper, the diameter of the pipe is a fraction smaller after it has been threaded. This is why tightening a pipe into its fitting seals the joint. This is why a joint should not be made up too tightly. Doing so permanently expands the female fitting. The joint will hold. However, should the joint need to be backed off a little, for one reason or another, it will surely leak. A joint brought to normal tightness can be backed off a fraction of a turn without leaking.

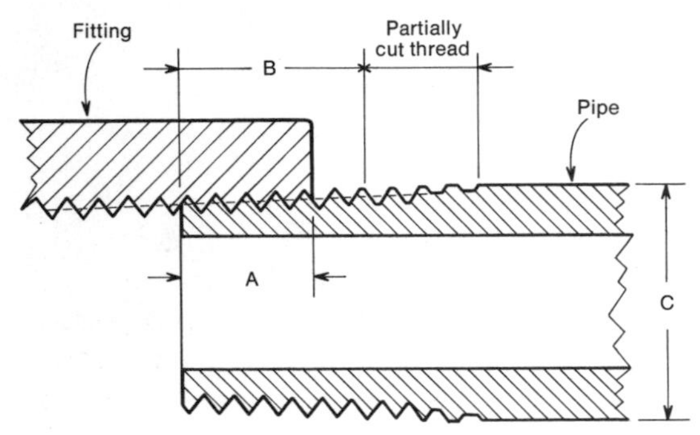

Cross section of a standard pipe thread, with major dimensions as shown in table below.

Two types of pipe stocks are presently in use. One has two handles in a straight line. The other has one handle attached to the die by a ratchet. It is sometimes possible to use it to thread a pipe that has been cut and left in position near a wall. Both types may be used with a power-driven pipe cutter.

To use this cutter, open the three-point vise, slip the pipe inside and tighten the vise. Squirt cutting oil on the end of the pipe. The die is *hand started* for the first partial turn. The motor is geared down so far that the vise won't turn while you start the die. Once the die has caught hold—you can feel it bite in—pull out the guide bars and rest the stock handle against the proper side. Start the motor. As it turns, squirt lubricating oil into the die. Although the power-driven die does not back up, the speed at which it turns is so slow that, with sufficient cutting oil, a clean thread is cut. Some plumbers stop the machine and blow the chips out every few turns.

Should you have difficulty in starting the die, the cause may be a pipe with its end cut at an angle (you will have to recut the pipe end), a ridge or lip at the end of the pipe you have failed to file down, or a worn

Table 12

PIPE THREADS—MAJOR DIMENSIONS

Pipe size, inch	Outside diam. of pipe, inch C	Threads per inch	Hand Tightened Length A		Wrench Tightened Length B	
			Inch	Threads	Inch	Threads
1/16	0.3125	27	0.160	4.32	0.2611	7.06
1/8	0.405	27	0.180	4.86	0.2633	7.12
1/4	0.540	18	0.200	3.60	0.4018	7.23
3/8	0.675	18	0.240	4.32	0.4078	7.34
1/2	0.840	14	0.320	4.48	0.5337	7.47
3/4	1.050	14	0.339	4.75	0.5457	7.64
1	1.315	11 1/2	0.400	4.60	0.6828	7.85
1 1/4	1.660	11 1/2	0.420	4.83	0.7068	8.13
1 1/2	1.900	11 1/2	0.420	4.83	0.7235	8.32
2	2.375	11 1/2	0.436	5.01	0.7565	8.70
2 1/2	2.875	8	0.682	5.46	1.1375	9.10
3	3.500	8	0.766	6.13	1.2000	9.60
3 1/2	4.000	8	0.821	6.57	1.2500	10.00
4	4.500	8	0.844	6.75	1.3000	10.40
5	5.563	8	0.937	7.50	1.4063	11.25

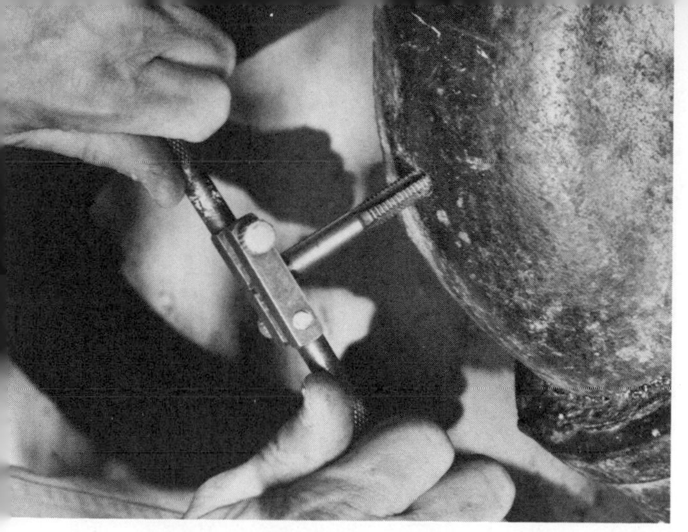

Using a machine-thread tap to thread a hole in a cast-iron elbow. The threaded hole will then accept a machine bolt. Do not use too much pressure on the tap, which is easily broken, and be sure to use plenty of lubrication.

Table 13

PIPE AND DRILL SIZES

Pipe Size	Tap Sizes (Threads per Inch)	Drill Size
1/8	27	11/32
1/4	18	7/16
3/8	18	19/32
1/2	14	23/32
3/4	14	15/16
1	11 1/2	15/32
1 1/4	11 1/2	1 1/2
1 1/2	11 1/2	1 23/32
2	11 1/2	2 3/16
2 1/2	8	2 5/8
3	8	3 1/4
3 1/2	8	3 3/4
4	8	4 1/4

die. Generally, a worn die is the last obstacle to difficult starting; it starts easily enough, but does a poor job.

Tapping a Hole

When thread is cut in an exterior surface such as a pipe or bolt, a die is used. When interior thread is cut, as for example in the length of a hole, the hole is said to be tapped, and the tool used is called a tap. It looks like a tapered bolt with three or four slots along its length.

Like the die, the tap cuts a spiral groove in the metal. Like the die, the tap must be lubricated and backed off a little as it is turned forward, into the hole.

Taps, however, have several special problems. A tap is not as strong as a die. Whereas it is difficult to impossible to break even a small die by forcing it onto the pipe, it is not too difficult to break a small tap. It has a certain degree of spring, but when this is exceeded, the tap will break. So when using a tap, don't keep trying to turn the handle when the tap itself stops turning. Back it off, clear the chips and start again.

Another problem is that of hole size. Table 13 lists common drill sizes and associate pipe tap sizes. When the correct pair are used in cast-iron or steel pipe there is no problem. However, there is a tendency for beginners to believe a larger tap can be forced, because it can be started and will turn a little. It can't; it will break.

Joining the Pipe

Galvanized, black and brass pipe are usually joined by a screw joint. The male end of the pipe or fitting is simply screwed into the female threaded fitting and tightened. *However, the joint should never be made up dry.* The male threads should always be given a generous coating of pipe-joint compound, or pipe dope as it is usually called by plumbers. Pipe dope acts as a lubricant, sealant and rust preventive. Made dry, a joint may leak. Made with pipe dope on the threads, the same joint requires less torque to tighten and will not leak. Without the pipe compound the joint will be almost impossible to take apart. With the dope, the joint will easily come apart twenty years later.

Joining Threaded Pipe

1. Apply pipe-joint compound to male thread. Compound acts as a lubricant and sealant.

2. Start the coupling carefully to make certain it is not cross-threaded.

3. Dope the male end of the following length of pipe and screw it into the other end of the coupling.

4. Now "make up" the joint by tightening the second length of pipe with a stillson wrench. On the 30-inch wrench shown, about 20 pounds of pressure is all that is needed when the joint has been lubricated with pipe dope.

COPPER TUBING

Copper tubing differs from brass and copper pipe in that the wall of the tubing is considerably thinner than that of the pipe; therefore it is not usually threaded for joining. Instead copper tubing is joined by sweating (soldering), flexible and metal compression joints and flare joints.

Using a tubing bender. Unlike copper or brass pipe, copper tubing is easily bent with hand tools. At the same time care must be exercised to prevent the thin walls from kinking or flattening. This is easily done with a tubing bender which consists of a spiral tube of wire. The copper tube is either inserted inside the tool or the tool is placed inside the tube, depending on relative diameters. Then the tube is slowly bent by hand. The only precaution here is that you get a tubing bender of the correct diameter. If there is too much space between the tool and the tube, the tube will kink.

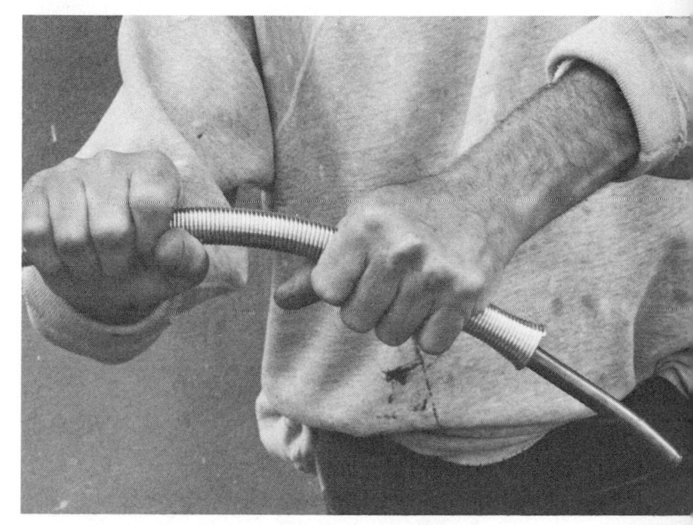

To bend copper tubing, slip the tube inside a pipe bender. For large-diameter tubing, the spiral-wound wire bender can be slipped *inside* the tube.

Cutting with a hacksaw. As even rigid copper tubing is very soft, it cannot be held in a vise while it is cut with a hacksaw. You need a jig that resembles a carpenter's miter box to hold the tube safely and firmly. When cutting tubing to be sweated or joined by means of a compression ring and joint, a little inaccuracy in cutting is permissible. When the tubing is to terminate in a flare fitting, the cut must be smooth and perfectly square.

To cut tubing quickly and easily, use a tubing cutter as shown. Don't tighten the cutting wheel too much as it will kink the tube.

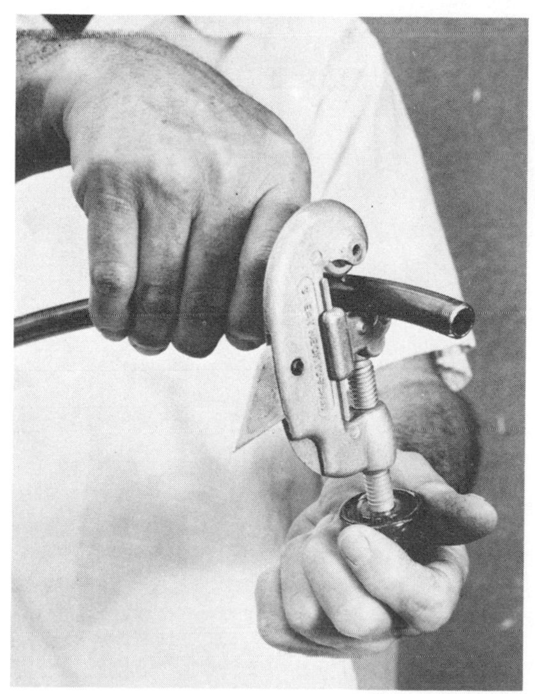

Using a tubing cutter. To use the cutter, the pipe is hand held. The device is slipped over the end of the tube, its handle is turned to bring the cutting wheel or wheels firmly against the copper. And as the device is rotated around the tube the pressure of the cutting wheel is gradually increased until the tube is severed. Excessive pressure of the

Using a large cutter to cut large-diameter copper tubing on a roughing-in job.

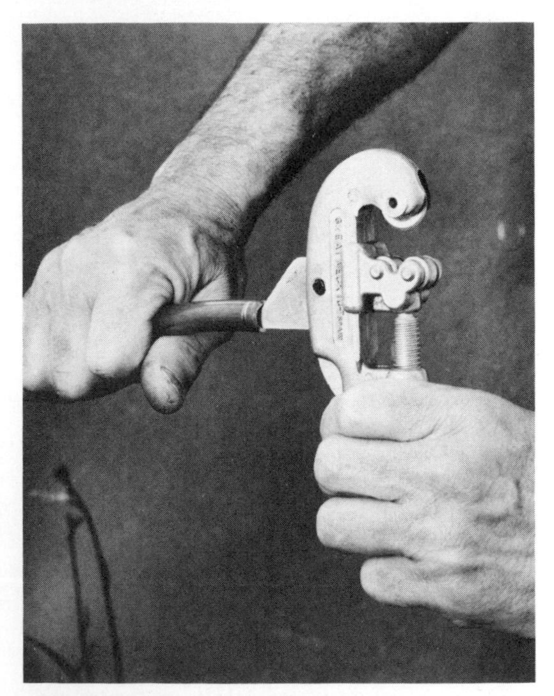

After the tubing has been cut the inside edge must be reamed. The burr that is always formed stops an inordinate amount of water. Most tubing cutters, like this one, are equipped with a small reamer.

cutting wheel against the tube may distort or flatten it. Should this occur, don't waste time trying to straighten the walls of the tube; discard the flattened section.

A metal compression ring will not slip onto a slightly flattened pipe; the fit is too close. If you force the ring on, chances are that the final joint will leak; they are that finicky, though they will hold water at high pressure with only finger tightening when they are perfect. The same is true of the flare fitting, though it is not quite as critical. If the portion of the tube that is flared is slightly distorted, the flare will not be perfect and the joint will leak.

After the tube has been cut, the inside edge must be reamed free of burrs. A wheeled cutter usually has a small triangle of metal attached; this is the reamer. The accompanying photo shows how to use it. If you have cut the tube with a hacksaw, use a small rat-tail file to get the burrs out.

If the cut end of the tube is to be used with a metal compression ring, check to see that a ridge has not been raised around the end of the copper by the cutter. Just try slipping the metal ring on. If it goes, fine. If not, carefully file the lip down. It only takes a few thousandths of an inch to stop the ring.

When using the hacksaw or wheeled cutter on soft tubing take care not to bend the tubing while holding it, especially near what will be the cut end. A slight bend can interfere with a compression ring's seal.

Soldering

Soldering is a very old method of joining metals. The name of the first man to solder has been forgotten, but the technique is over three thousand years old.

To solder, the surfaces to be joined are cleaned of all dirt, grease, oil and oxide, covered with a layer of suitable flux and spaced a few thousandths of an inch apart. The joint to be is heated to a temperature well above the melting point of the solder that is to be added. The flux responds to the heat and reacts with the contiguous metal. The flux removes and displaces whatever surface oxide remains on the facing surfaces and protects the now perfectly bare metal from contact with the air and oxidation. The solder is touched to the joint. It melts and is drawn between the surfaces by capillary attraction (molecular attraction between the molten molecules of the solder and the hot metal walls of the joint). The molten solder displaces the flux and dissolves the submicroscopic high spots it encounters as it flows along and also fills in the equally small holes it finds in its path. This process is called wetting. It only takes a few seconds, and if one examines the flow of molten lead across a fluxed copper surface, the appearance of wetting is quite obvious.

Wetting is the basis of soldering. The molten solder must make intimate contact with the perfectly bare, hot base metal. If the solder is hot but the base metal is cold, the solder will not dissolve the peaks and fill the valleys. If there is any grease or dirt on the base metal, the flux will not touch the base metal. If the flux is not present, or is isolated from the base metal, the flux cannot do its job, which is to remove the layer of oxide and then protect it from reoxidation.

Molten solder is drawn into the joint by capillary attraction. That means the entire joint must be hotter than the molten solder; if it is not, the molten solder will cool, solidify and block the flow of solder into the balance of the joint.

If no flux at all is used, the solder will flow and may even adhere in spots; but when cold, the solder can be pulled off. Joints made without flux are sometimes strong, but never leakproof. When insufficient flux is applied, the quality of the joint will suffer. If too much flux is applied, no harm is done; just a little flux is wasted. If the fluxed joint is kept hot too long before the solder is applied, the flux may run off or char, in which case it becomes useless.

A plumber solders copper tubing on the job, using scraps of wood to support the hot tubing.

Solder in wire form, composed of equal parts of lead and tin, is preferred. Bar solder, which is cheaper, is much too awkward for pipe joints. Solder with a 40/60 composition (40% tin, 60% lead) is acceptable. It has a slightly lower melting point and is therefore easier to use; it is also cheaper. However, it is also a little softer than 50-50 solder.

Rosin-core solder is not ordinarily used for plumbing joints because they are so large. The rosin leaves the solder wire as it melts and will not always flow inside the joint and flux it properly. For the same reason, acid-core solder is not ordinarily used for plumbing work.

If nothing else is available, use the rosin-core solder, but hold the heat on the joint a little longer than usual. If nothing but acid-core solder is on hand, that too may be used, but neither are preferred to solid-wire solder and hand-applied flux.

A rosin type of flux is used for soldering joints in copper tubing. There is also an acid type of flux that may be used, and it has its value on pipe that is heavily corroded and cannot be properly cleaned. The rosin type of flux is generally identified as "non-corrosive" rather than rosin base.

Use a small brush; a toothbrush will do, or a small finger to apply the flux paste evenly to both surfaces to be soldered. Incidentally, to solder stainless steel, use an acid-base flux especially formulated for this metal.

Neither the residue of the non-corrosive or acid type fluxes have to be removed after the joint is completed. The rosin stuff is dormant, and though the acid will leave a little green area, it won't eat the tube through by any means.

Using a propane torch. A propane torch with a replaceable tank is the most practical source of soldering heat for plumbing work. The largest soldering iron isn't nearly large enough for copper tubing. A gasoline torch is smelly and dangerous. Its flame contains carbon and other contaminants which tend to interfere with soldering. It generates poisonous fumes. The propane torch does not.

The propane torch with the flexible hose connected to a fuel tank is very useful if you are going to do a lot of soldering, but the hand-held torch has sufficient capacity and produces a sufficiently hot flame for almost all plumbing work. Use the largest nozzle to give you maximum heat. You will find it cheaper to buy small propane fuel tanks as you use them up than to rent a larger torch, even if you do save on the fuel.

To light a torch, screw the top half—the aluminum section with the valve—into the tank. Light a match, *then* open the valve a crack. Let the match burn a few seconds. Then open the valve a bit more. Opening the torch valve all the way will not increase the flame beyond a certain point. If too much gas emerges, the pressure will blow out the flame.

The torch will not work when it is tilted beyond the point where the liquid propane closes the orifice. The flame will go out.

When finished, close the valve just enough to stop the flame, then unscrew the valve body from the metal tank before storing.

If the flame appears ragged or flickers, try tapping the side of the nozzle lightly while the flame is on. This may dislodge the dirt. If that doesn't do it, shut off the flame, let the torch cool and look for dirt on or near the orifice. Don't poke a wire into the hole, because the hole is very fine and if it is enlarged the torch will not work properly.

CAUTION. Always shut off the torch when you are not using it. The flame is invisible in daylight and soundless. It is easy to reach through it and burn yourself. Always think about the results of the flame on material beyond the tube or pipe you are heating. If you are working near rough framing, you can let the lumber char a bit, but always wash down the charred wood afterward, to make certain no embers remain alive inside. They can linger for hours, coming to a blaze when a wind arises.

If you are working next to a finished surface, protect it with a large baking tin or a sheet of asbestos. Always keep checking behind the protection. The tin may protect the material from the flame, but when the tin becomes hot it will transfer heat to the material it is protecting. So when you are through soldering, cool the tin or what have you with water and remove it. Do not let it cool against the surface to be protected.

Sweating a joint. Begin by slipping the fitting to be sweated over the pipe end to make certain there are no bumps or kinks that will prevent a good solder joint. If a valve is to be sweated, the valve should be taken apart and its stem, along with its washer, removed. Placing a damp rag on a valve doesn't work; the water keeps the valve cool and prevents the solder from properly melting. Now disassemble the dry joint.

Clean the outside of the last inch of pipe with fine sandpaper or fine steel wool; also clean the inside end of the fitting. Apply flux to both the pipe and fitting, and assemble the two pieces. *Do not hold the joint,* even at a distance down the pipe. Copper is an excellent conductor of heat and you may burn yourself, or drop the joint and ruin it. Support the pipe on bricks or hold it in a clamp or a vise.

Apply the flame of the torch to the heaviest piece of metal. Torch flame is hottest a couple of inches in front of the nozzle. When you believe the heaviest piece of metal, which in most cases would be the fitting or the valve, is hot enough, shift the torch to the pipe. Don't play the flame at the edge of the joint, but about an inch away. If there are two or more pipes, concentrate on one at a time. Do not try to make all the joints at once. Test the temperature of the fitting by touching the solder wire's tip to it. If it melts, it is hot enough. Now play the flame over the pipe about an inch from the fitting end. Try to avoid heating the flux itself. At no time play the flame on the solder itself or touch the solder to the spot just heated by the flame; you will get a false reading.

When both the fitting and the pipe are hot enough, as evinced by readily melted solder, bring the flame back on the fitting, away from the end, and press the tip of the solder against the crack between the pipe and the fitting. As the solder melts, capillary action will draw it inside the crack. Keep feeding the solder until a ring of solder appears around the joint edge. Now, without pausing, go on to heat and solder the second and third pipe that may join the fitting.

When the bright circle of molten lead appears all around the pipe at the edge of the joint, the joint is filled with lead. It is com-

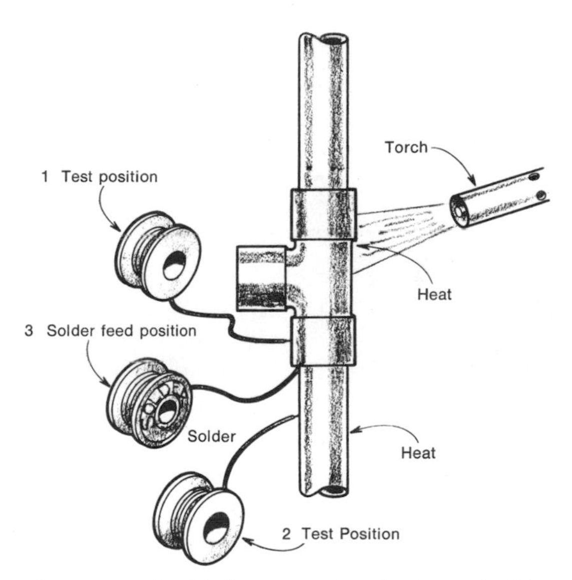

Sequence of soldering procedure is shown in this drawing. (1) The heavy part, the fitting, is always heated first and the temperature measured by touching it, on opposite side, with the tip of the solder. If solder melts, the fitting is hot enough. (2) The lighter part, the pipe, is heated next, and again temperature determined with tip of solder. (3) When both parts are hot enough to easily melt the solder, the tip of the solder is pressed against the joint until enough molten metal has been drawn inside to fill the space between pipe and fitting.

Soldering a Valve to Copper Tubing

1. Clean the ends of both tubes with steel wool.

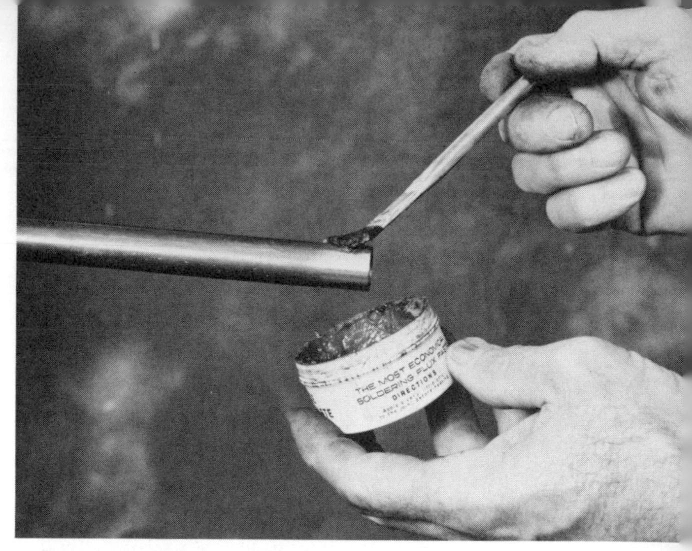

2. Coat the ends of the tubes with soldering flux.

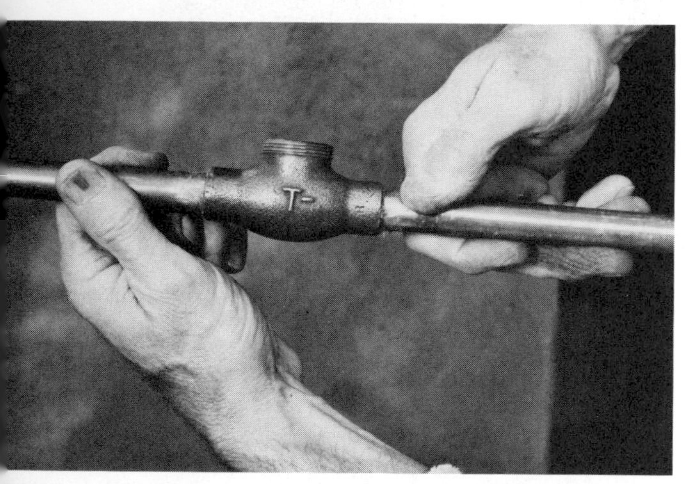

3. Disassemble the valve and insert the tubes in the body.

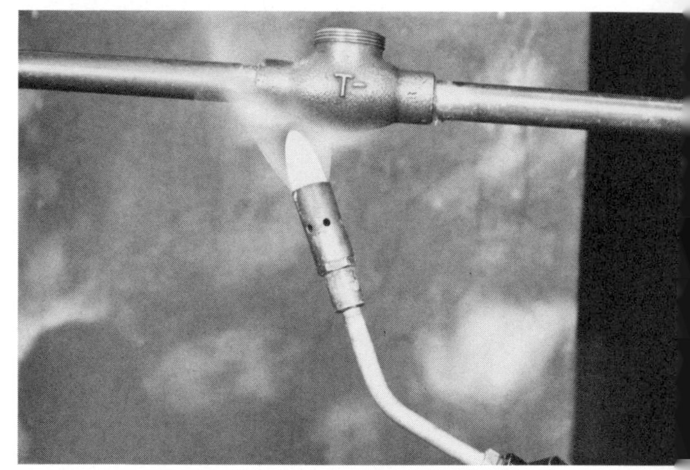

4. Heat the valve body, the heaviest part, first.

5. Test for soldering temperature on opposite side of valve.

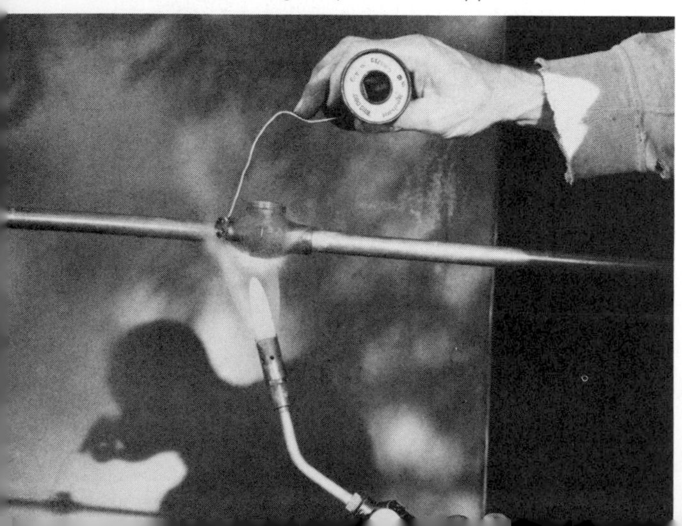

6. Heat tubes; then test them on the side opposite flame.

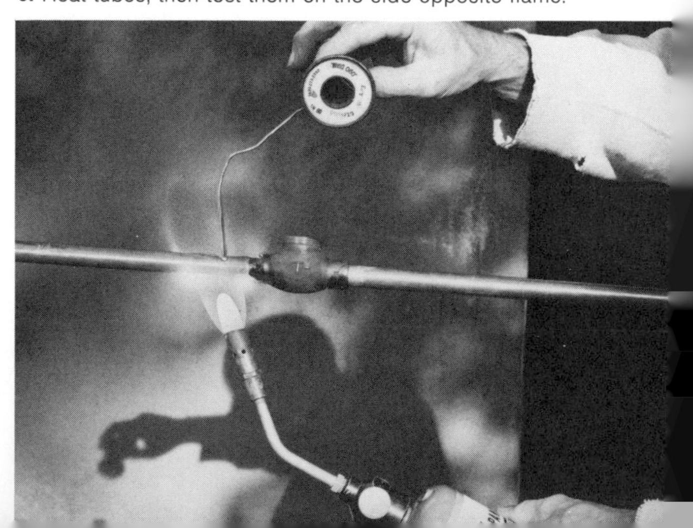

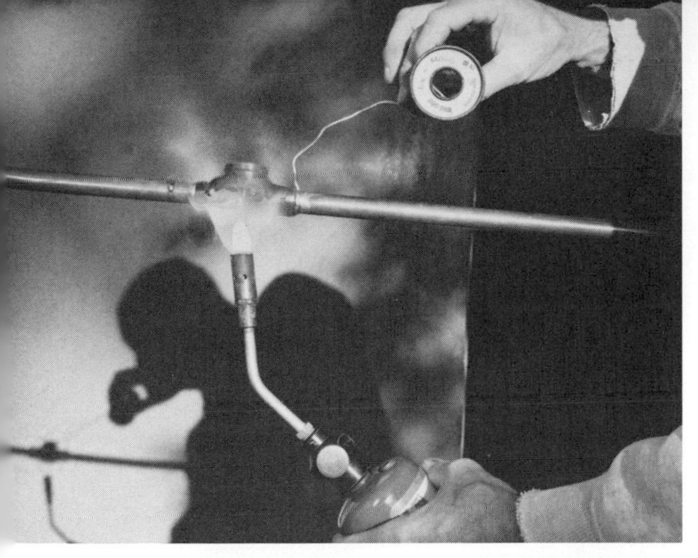

7. With flame directed onto valve body (heaviest part), touch tip of solder to joint. As solder melts, it will be drawn into joint by capillary action. When one side is filled, feed the other.

8. The finished joint. Note how shiny it is, but with hardly any solder outside the joint. Small glob of solder below the joint indicates that solder was fully liquid when joint was made.

plete. However, to be certain, some mechanics heat the fitting itself another sixteen to thirty seconds. Should you overheat the joint, solder may run out. If that happens, let it cool a bit and feed some more solder into it.

When the joint is finished, let it alone. Do not move it. Do not worry about dripped solder or too much solder. It has no effect on joint strength. Wait until the bright circle of metal dulls; you can see this happen. Wait another thirty seconds or more—if the fitting is heavy—and then if you wish to handle the joint, slosh some water on it. Once the solder has solidified, nothing will harm it. Move the joint before or while the solder is hardening and you have what is called a "cold solder" joint. It may look good, but it is very weak. The surface of a "cold joint" is crystalline in appearance.

Do not try to solder two tubes (or solderable steel pipes) of different sizes together without an intermediate fitting. Slipping the smaller tube within the other and filling the intervening space appears to be easy. It is, and it is not too difficult to lift the center tube to get solder all around it, but the resultant joint will be very weak. Unless the job is done perfectly and one pipe enters the other for a couple of inches, the joint may come apart under pressure.

If a sweated joint leaks, try heating it up and feeding it more solder. If that doesn't cure the leak, the joint has to be taken apart. The cause of the leak may be lack of flux, too tight a fit or poor soldering technique.

Do not attempt to solder pipe that is filled with water or has a little water and is closed at both ends. A pipe or tube filled with water will not permit the metal to come to solder temperature; all the water would have to boil off first. A pipe that is partially filled with water heats rapidly. The water quickly turns to steam and develops high pressure. If the line is closed and you heat it, it will burst and possibly explode. So always make certain all valves and faucets in the line are open and drained before you start applying heat.

Although capillary action will draw solder uphill into a joint, don't depend on it. Should you need to solder overhead, keep to one side. Wear gloves and don't be surprised if molten solder or hot flux drips down once in a while. If you must solder in a difficult position, first flux, heat and cover the parts with a generous layer of solder. Then position and heat them. When the solder on the pipe and fitting has melted, push them together. There is less chance of solder and flux drip this way.

Solderable Galvanized Steel Pipe

Solderable galvanized pipe can be cut with the same tools and methods as standard galvanized and black pipe. Joints between pipe sections are made with special solderable fittings. These are soldered with the same techniques used for soldering copper tubing and pipe. The pipe ends must be clean, fluxed with suitable flux and treated exactly as described for copper. However, although the steel does not conduct heat away as rapidly as does copper, the steel pipe is considerably more massive and requires the application of heat for a much longer time. Just make certain the base metal is thoroughly hot before you apply the solder and there will be no difficulties.

Metal-ring Compression Joints

Two lengths of copper tubing can be joined by means of a metal-ring compression joint. This joint consists of a flange nut, a metal ring, and a fitting, which may be a coupling, Tee, elbow or part of a valve. To make the joint, cut the tubing to the length you require, taking care to make the cut just as square as you can. Next, remove all external and internal burrs from the tube end. Slip the flange nut and the metal ring over the tube end. Now insert the tube end and metal ring into the fitting and screw the flange down over the ring and onto the fitting.

If you have made certain that the last inch or so of tubing is perfectly clean, round and free of burrs, that there were no metal chips or dirt in the fitting, and if the fitting has not been deformed by previous overtightening, you can hand-tighten it and it will hold water under pressure.

If a metal-ring compression joint has been properly made and doesn't leak, you will have no difficulty taking the joint apart and putting it together again. If the joint was overtightened it must be reassembled exactly as it was before. Overtightening the flange nut distorts the metal ring and tubing. In fact, you will find that the ring is permanently compressed around the tube end. If the joint isn't reassembled exactly as it was before, it will usually leak. Sometimes a seemingly perfect joint will leak upon reassembly even if you put it together correctly. When this occurs try the suggestions offered in Chapter 8, Repairing Leaks in Pipes, or start afresh with a new pipe and metal ring.

A valve joined to a tube with a metal-ring compression joint. The end of the tube is inserted into the fitting; the ring is slipped down and the flange nut tightened over the ring.

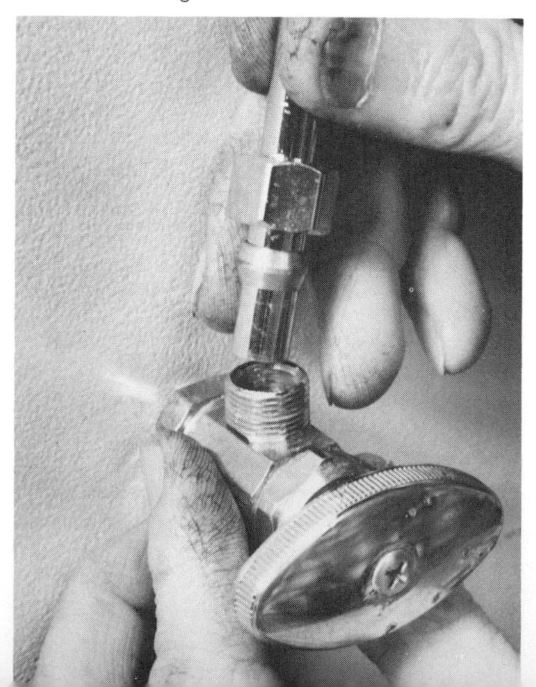

Flexible-ring Compression Joints

In home plumbing systems, flexible-ring compression joints are to be found mainly on exposed waste pipes beneath the kitchen sink and wash basin where thin-wall brass pipe is used. Generally the flexible-ring joints are used to connect the waste pipes to a cross or a Tee fitting. Sometimes the fitting will be fine-threaded to accept a threaded length of thin-wall pipe. This thread is "machine" thread and almost always cut at the factory. The purpose of the threaded joint is to give the assembly a degree of rigidity, which is not possible with a flexible-ring compression joint.

A flexible-ring joint consisting of a Tee, a flexible ring, a thin-wall brass pipe and a flange nut. The ring is compressed against the Tee by the flange nut.

The flexible-ring joint completed by screwing the flange nut tightly to the threaded side of the Tee. Drawings at right show interior of two types of joints.

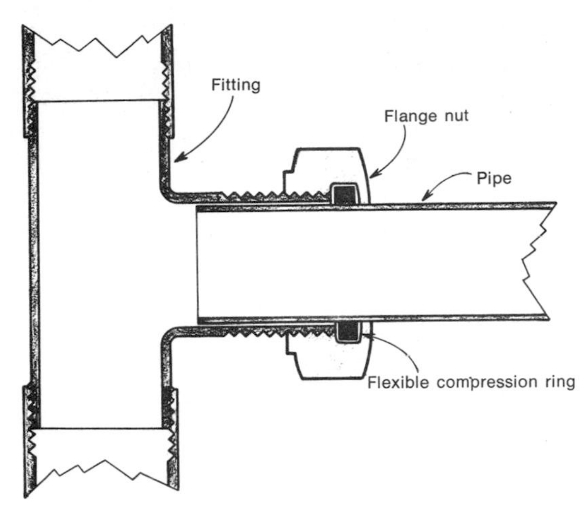

Cross section of a flexible-ring joint without an inner stop. This makes it possible to vary the depth the pipe enters the fitting.

Several types of joints are used. Examples are shown in the accompanying drawings. One is a slip joint. The other is on the same principle but doesn't permit the inner pipe to slip in and out of the fitting. The third utilizes a pipe with a preformed lip at its end. It is actually a kind of union.

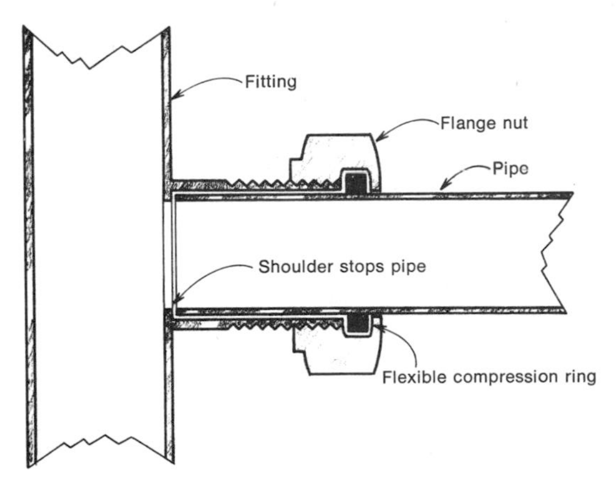

Cross section of a flexible-ring joint with an inner stop. This limits the distance the pipe can enter the fitting, but makes for a "stiffer" joint.

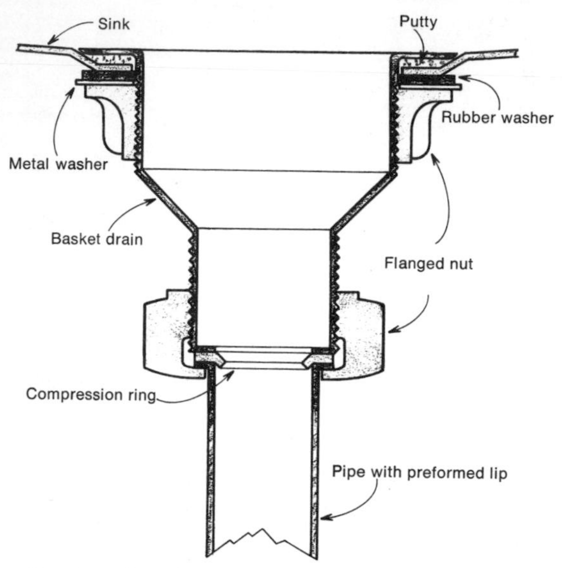

Cross section of a union type of a flexible-ring joint. Note that the dished ring goes between the top of the pipe's lip and the underside of the basket drain. A simple, flat rubber washer of the correct size will also work here.

A few of the many different flexible compression rings, gaskets and O-rings used for plumbing joints. O-rings have a circular cross section.

To assemble a flexible-ring compression joint, cut the end of the thin-wall pipe fairly square to the necessary length. Remove all internal and external burrs from the pipe's end. Slip the flange nut, followed by the flexible ring, over the pipe end. The compression ring has a square cross-section, and you will know you have the correct size if you have to stretch it just a fraction to get it on. (O-rings have circular cross sections.)

Now slip the end of the pipe and the flexible ring into the fitting. Bring the flange nut down against the fitting and give it a turn or two. Next, push the pipe into the fitting either as far as it will go or as far as you want it to go, making certain that you have at least ½ inch of pipe past the ring and into the fitting. This done, turn the fitting, if you have to, and tighten the nut. Generally, all you need do is make it finger tight. Tightening the flange nut forces the compression ring to seal the pipe to the fitting.

The "union" type of flexible-ring compression joint consists of a length of thin-wall brass pipe with a preformed lip about ¼ inch wide at one end, a flange nut and a flexible ring. This type of joint is used to connect a drainpipe to the strainer at the bottom of a sink or tub. The union type of joint differs from the other joints just described in that the flexible ring goes between the lip on the end of the pipe and the bottom of the strainer. If you position the ring behind the lip, the joint will leak because you have a bare metal-to-metal connection.

To assemble the union type of flexible-ring compression joint, slip the flange nut onto the thin-wall brass pipe so that the preformed lip rests inside the nut. Now position the flexible ring against the lip and inside the flange nut. The old-style flexible rings are simple rubber washers of the correct size. The new rings are shallow, dished

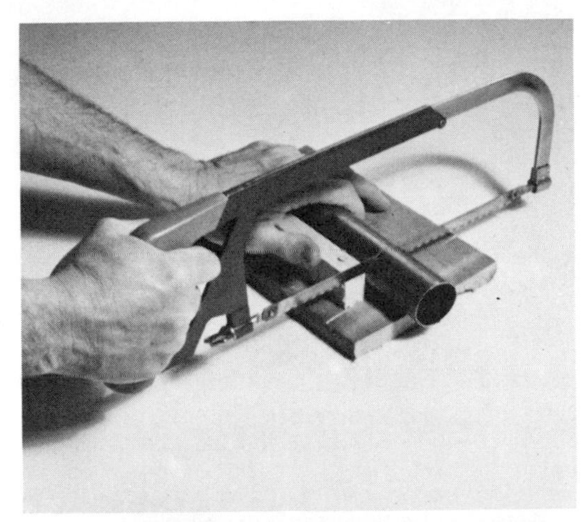

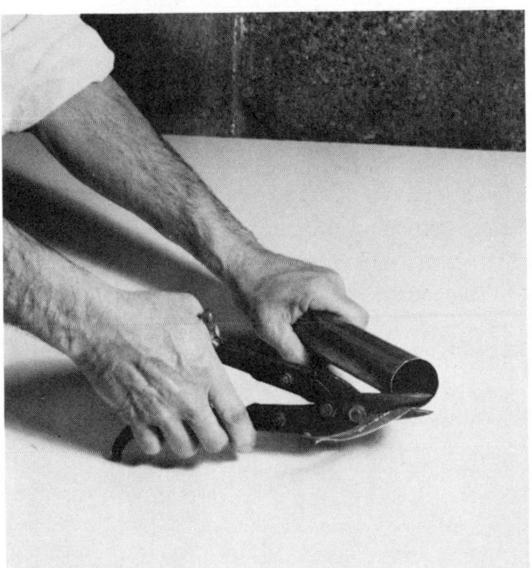

Using a hacksaw and a homemade jig to cut large-diameter, thin-wall brass pipe (top). Compound-gear tin snips can be used to remove a thin strip from the end. You don't need a perfectly square pipe end to make a slip joint.

rings of plastic. Position the plastic ring so that it nests in the pipe end. Now bring the pipe and the flange nut up against the bottom of the drain fitting and screw the flange nut in place. Use a wrench to lightly tighten. Do not make overtight or you will distort the fitting.

Flare Joints

Flare joints are made in copper tubing and stainless-steel pipe. A flange nut is slipped over the tube end. The end is placed in a flaring tool and flared. The flared end is placed against the body of the fitting, and the flange nut is tightened onto the fitting. With experience, you can make a flare joint in a couple of minutes.

Flare fittings for tubing are made in Tees, Wyes, angles, bends, combined with valves and joined to other terminations so that a flare joint fitting can be coupled to screw thread.

Flare joints have several advantages. The tools needed are simple, inexpensive and portable. No heat is needed, as for sweated joints. And they can be disassembled when necessary, generally with far less worry about reassembly and leakage than metal-ring compression joints.

However, flare joints have several disadvantages. They are not acceptable to all plumbing inspectors. They cost more than sweated fittings or screw fittings. They require more practice to make than the other two fittings. (Opening and closing a flare fitting doesn't require any practice, but making a perfect flare does.)

Normally, flare joints are made only in soft copper tubing. The flaring tool, however, can be used to flare rigid copper tubing if it is sturdy enough. The flaring tool used for stainless steel is identical in operation, but much heavier.

The trick to making a watertight flare the first time lies in preparing the tubing end. Not only must the end be cut perfectly square across, but all internal burrs and external ridges must be removed. However, great care must be used with the reamer not to remove anything but the burrs. Do not bevel the inside of the tubing, nor form a

How to Make a Flare Joint

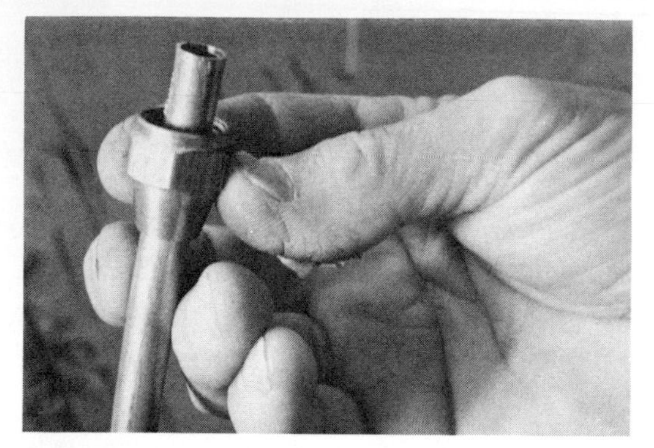

1. Slip the nut, threaded end outward, over the end of the tube.

2. Clamp the tube firmly in the flaring-tool vise, the end flush with the surface. Use a file, if necessary, to square the end of the tube. (Don't forget to deburr the inside end.)

3. Put a drop of oil or a touch of wax on the tip of the ram and screw it down slowly and carefully.

4. The pipe end is flared when you remove the tool. You can now assemble the joint. The fitting shown is a coupling for joining two tubes.

ridge on the outside of the tube. Also, make certain the ram end and the vise behind the tubing are perfectly clean. A grain of sand will cause the mating surfaces to be uneven and the joint will leak.

Flare joints are made in plastic pipe the same way, but the flare is made very slowly to avoid cracking the pipe. Warming the plastic just before flaring helps to avoid cracking both rigid and nonrigid tubing.

Thin-wall brass pipe should be cut with a fine-toothed hacksaw. The problem is how to hold the pipe firmly and how to keep the hacksaw cutting straight. A strap wrench can hold the pipe immobile on the workbench. Next best is to fasten a length of 2x4 to the bench top and press the pipe against the wood stop. To cut straight you need a sharp blade and a light touch. Don't press on the hacksaw; keep checking on its progress. Make certain you have a bright, clear guideline. If you need to remove only a fraction of an inch of metal from the end of a thin-wall pipe, use curved, compound-gear tin snips.

PLASTIC PIPE

Flexible plastic pipe can be cut with a knife or a saw. The rigid pipe can be cut with a conventional pipe cutter or a saw. Any type of saw can be used: hacksaw, crosscut saw or backsaw. For best results the cut end should be square, so use a miter box. After cutting, ream the inside of the pipe with a pipe reamer or a small penknife.

Types of joints. Flexible plastic pipe is most often joined by insert fittings and transitions fittings. An insert fitting is a short length of metal or hard plastic pipe that is inserted into the end of the cut pipe. The fitting is held in place by an external clamp that looks very much like the clamp used on an automobile water hose. A transition fitting is very similar except that the fitting end not inserted into the plastic pipe has threads to accept screw-threaded metal pipe or fittings.

To disassemble an insert or transition fitting, the clamp is loosened and the fitting is removed. Note that whereas the plastic pipe is flexible the fittings used are not.

Rigid plastic pipe is most often joined by solvent "welding," so called because the solvent or cement literally welds the pipe to the fitting, which is made of the same plastic material. Rigid plastic pipes are joined to threaded pipe by transition fittings, threading and mechanical couplings. The transition fitting is somewhat similar to a flexible-ring compression joint. The flange nut is slipped over the pipe and, followed by an O-ring (usually called an elastomeric seal by the makers of plastic pipe). Then the pipe end and ring are slipped into the end of the transition fitting. When the flange nut is tightened atop the ring it expands and seals the pipe to the fitting. The other end of the transition fitting is threaded to accept threaded pipe or a threaded fitting.

Some transition fittings are short lengths of preformed plastic pipe which are cemented onto the pipe end, then joined to the balance of the fitting that has the screw threads.

Mechanical couplings used with rigid plastic pipe are similar to those used with No-hub cast-iron soil pipes. A flexible rubber or plastic sleeve is slipped over the two pipe ends, which are butted against one another. Then a flexible metal cover is wrapped around the sleeve and held in place with two strap clamps.

Rigid plastic pipe is not very often threaded as threading reduces its strength about 50 percent. Should you need to join rigid plastic pipe to metal pipe or a fitting by threading, it is best to thread the outside of the plastic pipe, as it will then be reinforced by the pipe or fitting it enters. Dies designed for cutting threads in plastic pipe produce the best results, but standard metal-cutting pipe-thread dies can also be used.

Solvent cementing. Although joining plastic pipe to its fittings by solvent cement is very simple, a number of steps are involved.

Cut the pipe to the correct length, including the amount of pipe that will enter the fitting. Try the fitting on the end of the pipe. If it is too loose the cement will not properly seal the joint. If the fit is too tight the cement will be squeezed out.

Adjust the fitting to the correct position and mark the fitting and the pipe so you can quickly reposition them after applying cement. Once the cement sets—and it only takes minutes—you cannot rotate or remove the fitting.

Remove the fitting from the pipe end and check to see that the pipe has been reamed. Even a small burr can reduce water flow considerably.

To prepare the inch or so of pipe end that

Cutting and Joining Plastic Pipe

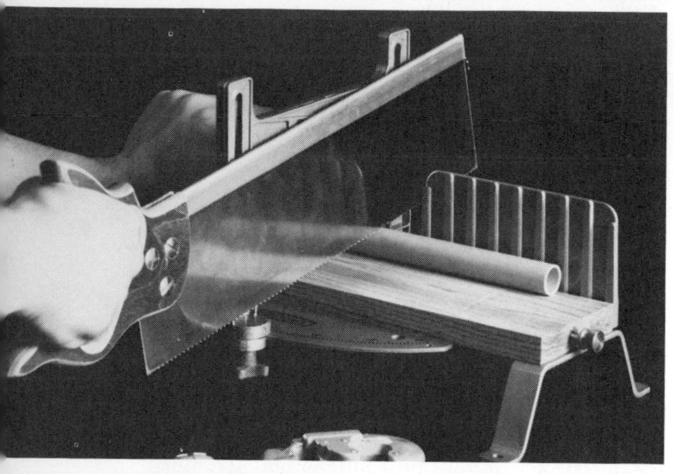

1. Cut the pipe end as squarely as possible, preferably with the aid of a miter box. Any saw can be used.

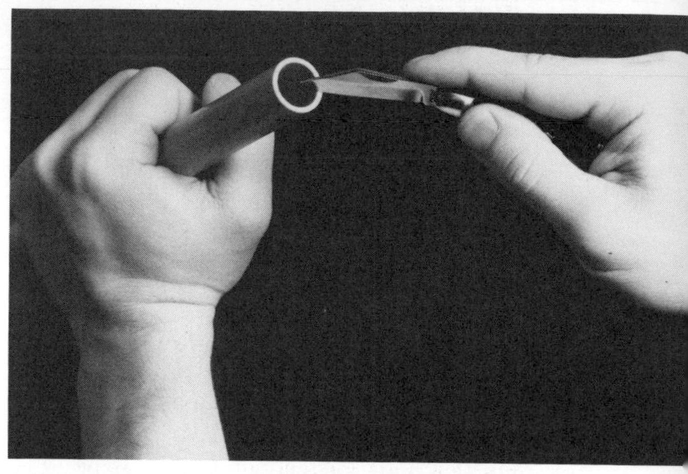

2. The inside end of the pipe is either reamed with a standard reamer or deburred with a penknife.

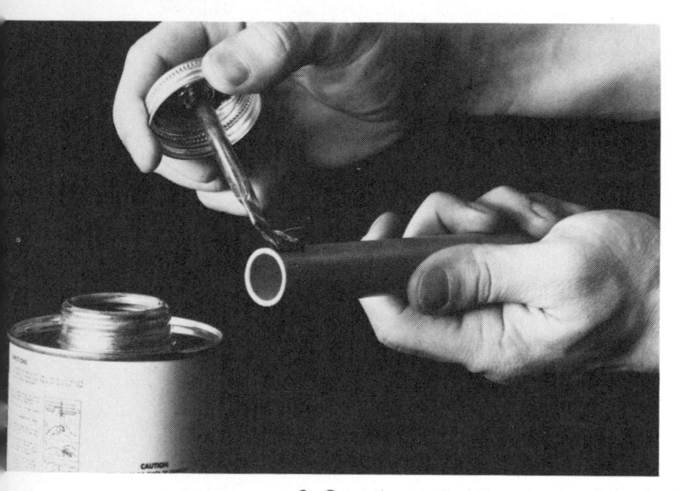

3. Coat the end of the pipe and the inside surface of the fitting with pipe cement.

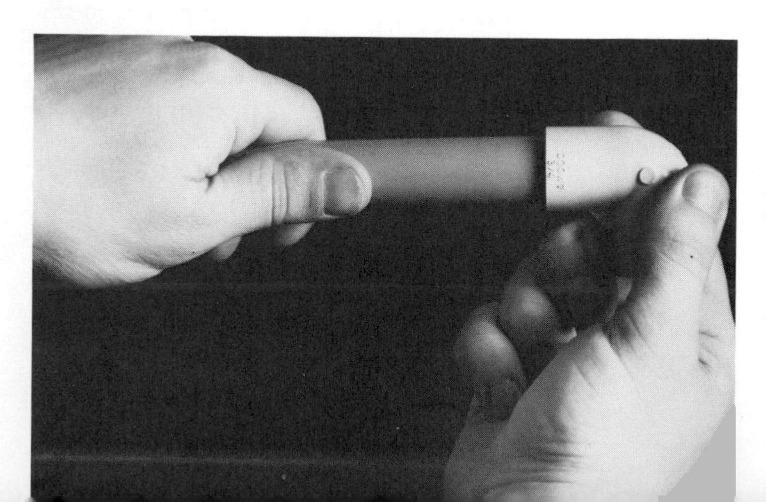

4. Assemble the joint and give the fitting a fraction of a turn to make certain the cement has been evenly distributed. Allow to harden in correct position. *Courtesy B. F. Goodrich Chemical Co.*

enters the fitting, make certain it is perfectly dry and free of grease and oil. Water will slow setting and grease or oil will prevent adhesion. Next, remove the shine from the end of the pipe to improve cement adhesion. You can use sandpaper, primer, or a thin coating of cement.

A word about primer, cement and their application. Always use the correct cement and primer for the type of plastic pipe you are running. Always check the cement you are going to use before using it. If it is thick or ropy it has either dried out or is too cold. Always use a brush as wide as the width of the joint. You must apply the cement quickly and evenly. If you have to "paint" the joint with a lot of little strokes, the cement will be too dry when you assemble the joint.

If you are using primer, apply a thin coat to the outside of the pipe and the inside of the fitting. Wait five to fifteen seconds or so until the plastic has softened sufficiently to permit you to scrape a hair's thickness off. Make one test at first and then you will know how long you have to wait each time. The colder and wetter the day, the longer you have to wait.

If you are using the cement itself as a primer, apply only a very thin coating.

When the pipe is ready, apply a thin coating of cement to the inside of the fitting and a thicker but even coating of cement to the outside of the pipe. Assemble the joint. Give it a quarter turn or so to line up the marks you have previously made.

If you have sandpapered the pipe end and do not want to use primer or a thin layer of cement, apply a thin layer of cement to the inside of the joint, a thicker, even layer to the end of the pipe and assemble. There is no rush to make the joint if you merely sandpaper. Just don't let it get dirty.

Most instructions now suggest wiping off the excess cement. But if the joint moves while it is drying it will be weakened, so hold it immobile for at least twenty seconds. Then allow it to dry another three minutes before starting on the other end of the pipe or fitting.

A properly made solvent cement joint will have an even bead of cement all around the joint edge just like a proper solder joint. If no cement is showing there is a chance you haven't used enough cement. Apply a little more on the next one.

Wait at least an hour in mild to warm weather before letting water into the pipe. In cold weather, wait longer. In general it it best to wait overnight before conducting any pressure tests.

CAST-IRON PIPE

It is not difficult to cut and join cast-iron pipe, but it is hard, time-consuming work. The pipe is heavy and thick, and lead-caulked joints take time to make.

The easy way to cut cast iron is to purchase or rent a soil-pipe cutter, sometimes called a pipe cracker. The chain of the cutter is wrapped around the pipe and tightened, instantly severing it.

The trick to cutting cast-iron pipe without a pipe cracker and without cracking the pipe lies in proper support. The pipe cannot be rested on its hub, but must be raised high enough to let the hub clear the ground. Pipe support must be resilient; use a block of 2-by-4 lumber or a pile of dirt and place it directly under the line of cut.

Mark the line of cut clearly. Measure on both sides of the pipe to establish the cutting line, making sure the line is square. Make certain you leave enough metal for the spigot end to enter the hub of the next section of pipe or the fitting.

Cutting cast-iron soil pipe with a pipe cutter, or "cracker." The cutting chain is wrapped around the pipe and tightened (left). Then the handles are brought together by stepping on them and, snap, the pipe is cut.

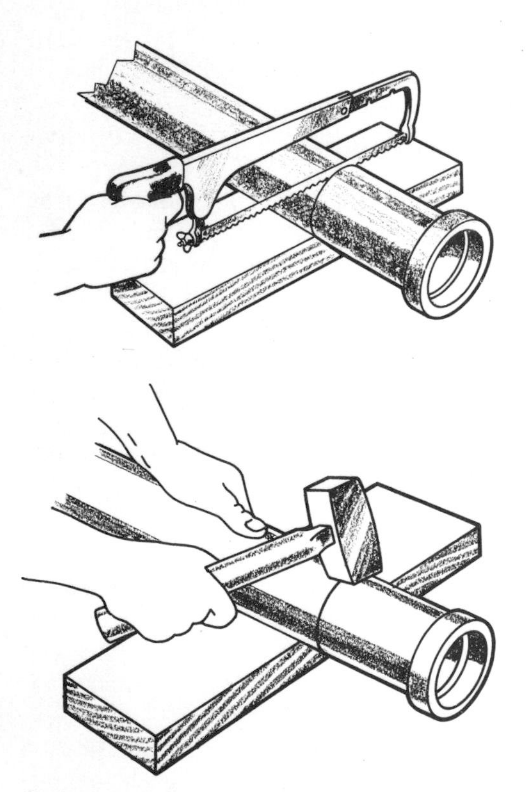

Cutting cast-iron soil pipe with a hacksaw and hammer. A groove is cut around circumference of pipe (top). The pipe is supported on a block of wood, with the saw cut a little beyond the support. Then the pipe is repeatedly tapped with a hammer alongside the saw cut until a crack forms which splits the pipe.

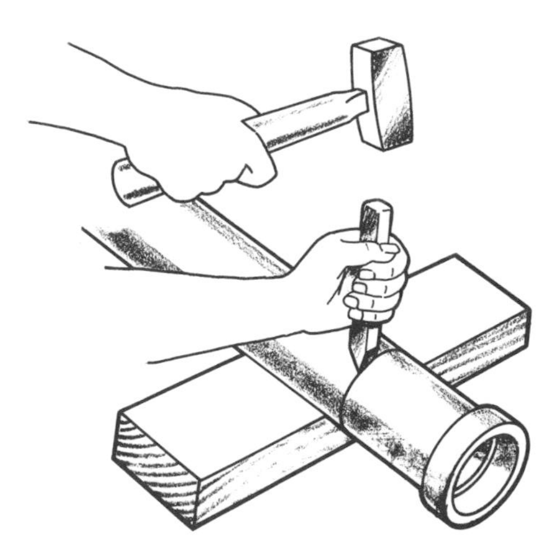

To cut cast-iron soil pipe with a hammer and chisel, the pipe is supported by a block of wood and a shallow dent is made around the circumference of the pipe along the desired line of cut. The dent is gradually deepened by repeated blows of the dull chisel until the compressed iron fibers cause the pipe to fracture along the line of pressure.

Use a coarse-toothed hacksaw and cut a slot about $1/16$ inch deep around the pipe along the line you have drawn. Then move the cut off the support and gently tap the pipe on the far side of the cut, turning the pipe with each stroke of the hammer, until it cracks on the line you have cut.

Alternately, you can crack the pipe by rapping it with a short, dull chisel. Work the chisel around the line until the pipe cracks. This is the method used by professional plumbers. If you wish, you can start the cut with a hacksaw, to mark it clearly, then continue with the hammer and chisel. When doing this, the underside of the pipe must be supported.

If the pipe cracks only part way, don't try to break it off, but continue to strike the uncracked section with your chisel. When the pipe separates, the edges will be rough, but this is unimportant. If the pipe cracks slightly off the line, it doesn't matter; however, if a fairly large section remains on one piece, don't attempt to chip it off. Use the hacksaw. Pipe cracks off the line because of internal flaws. If you rap along the cut without the support of the rest of the pipe, you will break off a large chunk. Hacksawing is harder, but far more certain.

Caulked Lead Joints

Caulked lead joints, or leaded joints, are usually made to join one section of cast-iron soil pipe to another. But sometimes the joint is used to connect one length of galvanized pipe to another, especially when the two pipes are not in perfect alignment and they are used for venting or waste.

To make a leaded joint, the spigot is inserted into the hub, and the space between is half filled with oakum, a kind of tarred rope. The balance of the space is filled with lead, which is "caulked"—that is, pounded with a special chisel until the lead expands and is permanently locked between the spigot and the hub.

Packing the oakum. The first step in packing the oakum is to make certain the spigot is centered within the hub and rests on the bottom. If you are joining large-diameter pipe with about an inch of space between spigot and hub, cut off a couple of feet of oakum and wrap it around the pipe. If you are joining standard 4-inch soil pipe, the oakum as it comes out of the box is too thick. Unravel the oakum and wrap the thinner strands several times around the pipe. Wrap enough oakum into the pocket to completely fill the space. Then, using a yarning iron—a special offset chisel with a blunt point—hammer the oakum into the space between spigot and hub. About half the space should be filled with tightly compressed oakum. Oakum alone, properly compressed into the packing space, can easily hold considerable water pressure, so pack it as tightly as you can.

Keep packing and pounding until you have a solid bed about 1 inch below the rim of the hub. This will about half fill the pocket. In some localities the code specifies the height of the lead caulking—usually ¾ to 1-inch. But it's the oakum that seals; the lead just holds it in place.

If you are joining two vertical pairs of pipes you will have no trouble centering one within the other. If the pipes are at an angle or horizontal, use wood chips to hold the inner pipe centered.

The accompanying drawing shows the irons used when there is sufficient working space and the special irons available for working in a corner or next to the ceiling.

Pouring the lead. After you have packed the oakum, make certain no loose strands are projecting. Should the lead be poured

Making a Lead-Caulked Joint

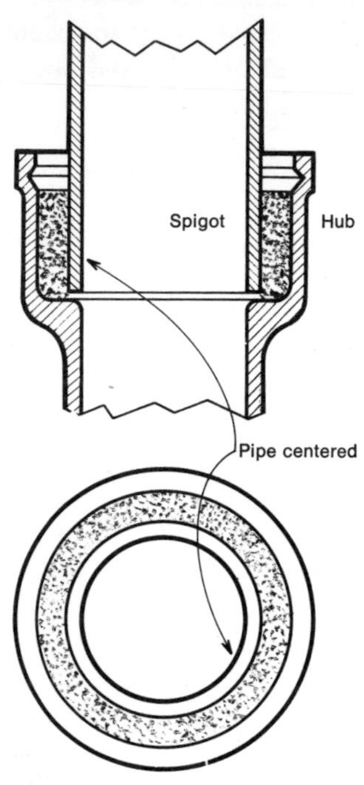

1. Center the spigot of one cast-iron pipe inside the hub of the other.

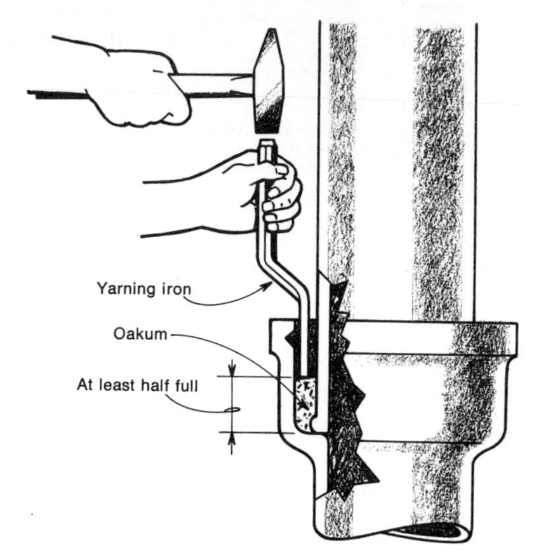

2. Use a yarning iron to pack oakum into the pocket between the spigot and the hub.

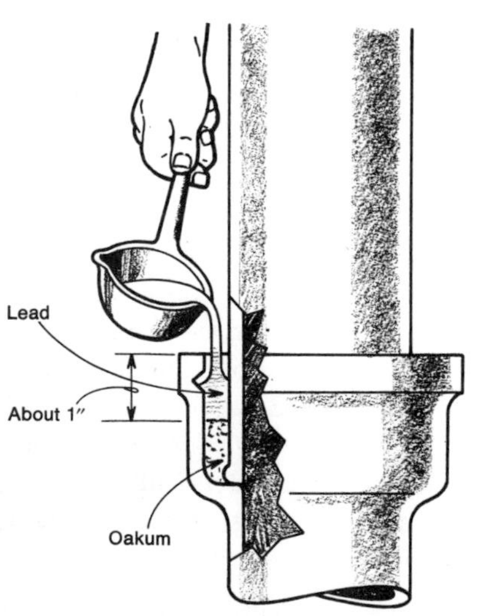

3. When the oakum is packed into the pocket, pour molten lead into the joint. This is the arrangement used with a simple vertical joint.

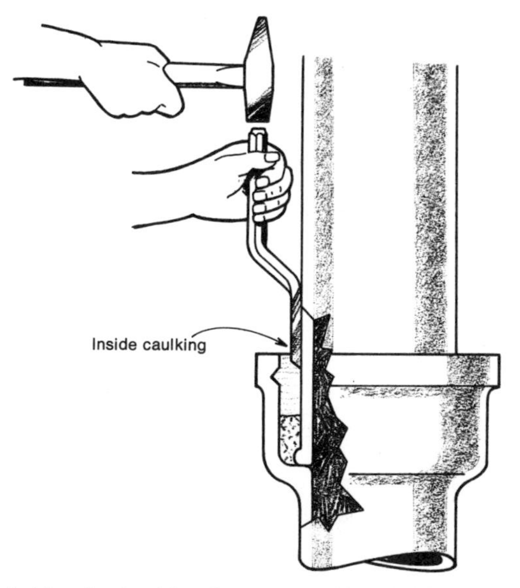

4. After the lead has been poured into the joint, it is caulked—expanded with a caulking iron. An inside iron—as shown—is used first.

with a strand of oakum reaching through, the oakum will be burned to charcoal, which can be easily pushed out by water pressure. Inspect the joint carefully to make certain there is no moisture inside. It is good practice to pour the lead as soon as the oakum is in place, before it picks up moisture.

Table 14

Oakum and Lead
Needed for Caulked Joints

Material	Size of Pipe (inches) 2	3	4	5	6
Oakum (feet)	3	4½	5	6½	7½
Lead (lbs.)	1½	2¼	3	3¾	4½

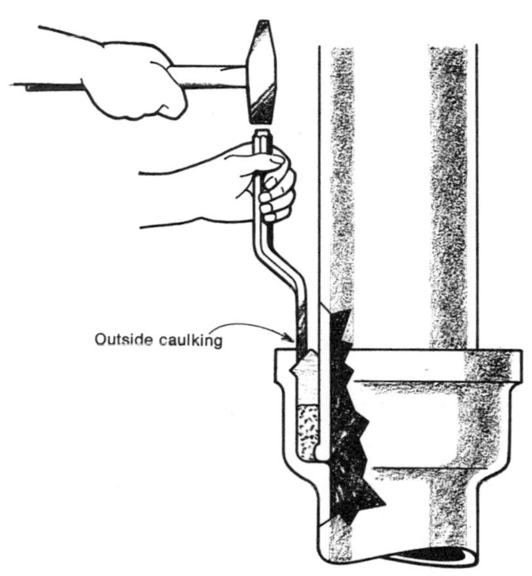

5. Now the outside circumference is caulked with an outside caulking iron. Notice difference in tip angle.

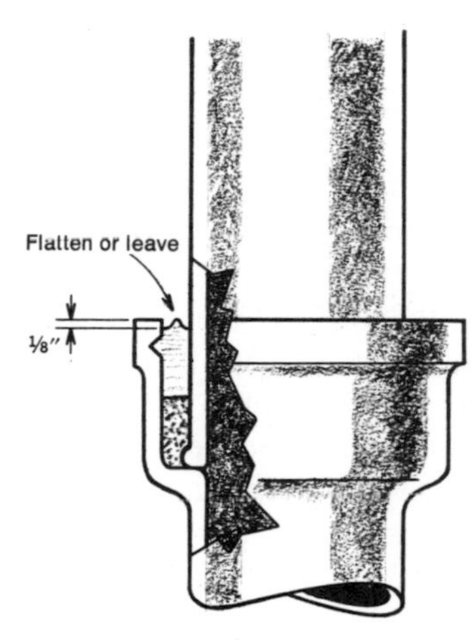

6. Caulking leaves a ridge or ring on the surface of the lead. You can trim this, flatten it or leave it alone.

Moisture in the joint can be a hazard. The hot lead will turn the moisture into steam, which will explode the lead out of the joint. Wear leather gloves with sleeves—gauntlets—and always hold the ladle at arms length and to one side.

Should you get splattered, immediately immerse the injured part in cold water. Don't apply grease or oil or any unguent. Keep the injured member in ice water and have someone call a doctor. Immediate application of ice water can prevent third-degree burns. Should you merely get a small splatter, don't put off the ice-water treatment. Immediate application of cold water can prevent a blister from forming.

If you have any doubt about the dryness of a joint, warm the outside of pipe with a torch, but be careful not to char the oakum.

You will need something less than 1 pound of molten lead for each inch of pipe diameter. Table 14 gives the approximate weight of lead needed for various sizes of pipe. This is the melted weight; figure on losing lead to dross, the pot, the ladle and drips.

Lead is melted in a plumber's furnace, which consists of a cast-iron pot on a gasoline or propane-fired stove. The complete furnace, with a long-handled iron ladle, can be rented.

Place clean bar lead in the pot, preferably while it is still cold. Allow the ladle to dry out on the pot. As there may be an oxide layer on both the lead in the pot and the added bar of lead, use the dry ladle to stir the two gently. When the lead melts, use the ladle to skim the dross—whatever collects atop the molten lead—to one side. When the lead is fully liquid, you are ready to pour.

Pouring is easy when the joint is in a vertical position and easily reached. Just pour the molten lead slowly and carefully into the space between the spigot and the hub until the surface of the lead is slightly higher than the rim of the hub. The lead will curve. Fill the entire "pocket." If more than one ladleful is needed, don't wait until the first pour hardens before adding more lead. That is why plumbers always keep the lead pot full.

Caulking. Some plumbers like to caulk the poured lead before it is cold. Some wait until it is cold. There seems to be no evidence that one technique is better than the other. However, the inner edge of the lead is always caulked first, and this is done with an inside caulking iron. The tip of the iron is held against the lead while one side is pressed against the spigot and the iron rapped with a hammer. When the oakum has been properly packed, driving the lead down about ⅛ inch seals the joint. (If the oakum hasn't been properly packed, the lead will be driven more deeply into the socket.) When the inside edge of the lead has been caulked, an outside iron is used to caulk the outside edge. The purpose of

Using a yarning iron on the job. A transition fitting, which will connect a copper drainpipe to a cast-iron fitting, is sealed with oakum. After the initial compression, the plumber will use a hammer on the tool.

this dual caulking is to force the lead to expand and lock itself against the surface of the pipe.

When both the inner and outer edges have been caulked, there will be a ridge of lead, perhaps ⅛ inch or more in height, between the inner and outer caulkings. If you wish, you can flatten this ridge with a hammer, but it isn't necessary.

If the joint is pitched or in a horizontal position a "joint runner" is needed. This is a fat asbestos rope with some type of clamp.

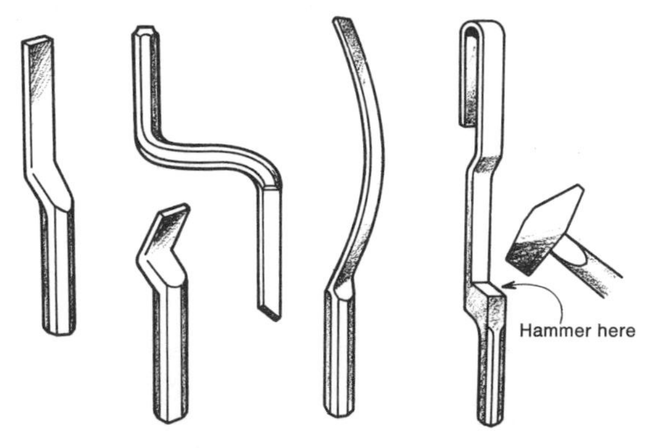

A number of different yarning tools and caulking irons that are used for making lead-caulked joints. Tool at right is used near the ceiling, where there is no room to swing a hammer.

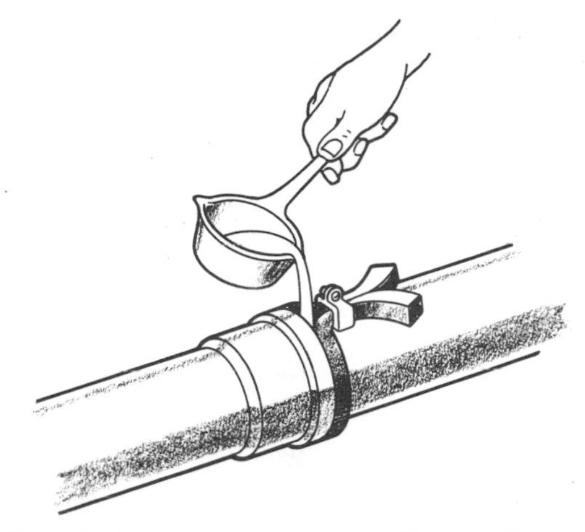

When the joint is in a horizontal position an asbestos joint runner is needed. The runner is clamped around the pipe and pushed up against the hub, forming a small opening through which hot lead is poured into the pocket.

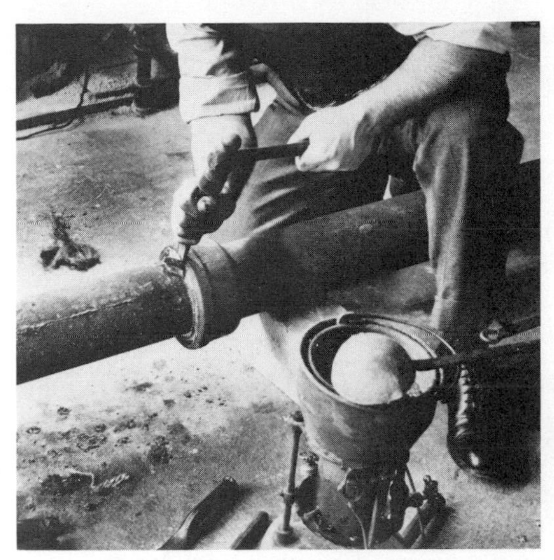

Trimming the wedge of lead with the aid of a joint runner after a horizontal joint is poured.

Using a joint runner to pour a horizontal joint on the job. *Courtesy Lead Industries Assoc.*

The rope is wrapped around the pipe, clamped into a circle, then pushed up against the hub, clamped ends upwards. Next, the molten lead is poured between the clamped ends (the thickness of the asbestos rope leaves an opening there). When the lead has firmly solidified, the runner is removed.

In using a joint runner in any position it is well to have the lead a little more liquid than usual, and it is wise to have considerably more lead in the pot than needed for the joint itself, as some of the lead remains between the runner ends.

Horizontal leaded joints are caulked in the same way as vertical joints. The lump of lead that protrudes after the runner has been removed is chiseled off and caulking done as usual.

Should the hub crack during any of these operations, the entire section of pipe has to to be replaced. You may be able to seal the crack with "cement," but the plumbing inspector will never pass it.

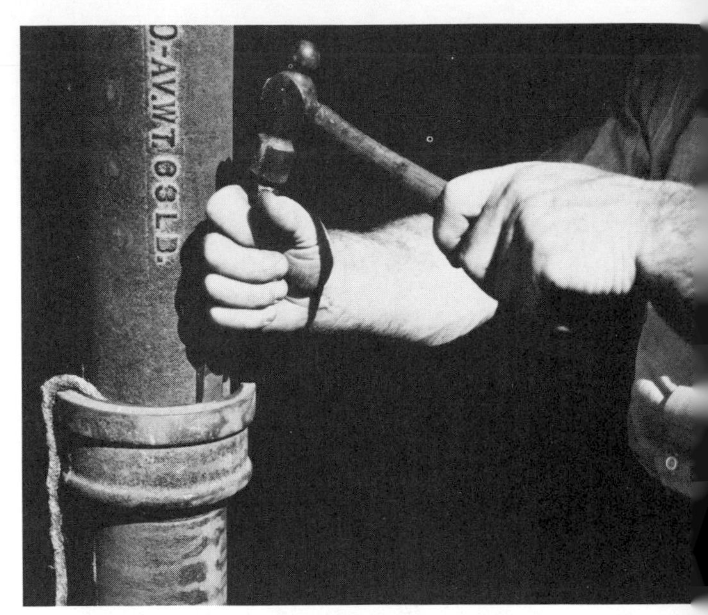

Lead wool is also used to make a caulked joint. The wool is worked into a rope and pressed into the hub (left). The lead wool is caulked with an iron, like poured lead (right).

Nonpoured Lead Caulked Joints

Lead caulked joints can be made without molten lead. In its place lead wool is used. Lead wool is ordinary lead that has been shredded and looks like fine "snow" that is used for decoration around Christmas time. It costs about twice as much per pound as solid lead and does take more work, but the resultant joint is just as good and is acceptable to most plumbing departments.

The lead wool is worked into a thick strand, much like rope. It is not pulled apart into little bunches. The strands are wrapped around the pipe and pushed down into the hub. It takes a little practice to estimate how much lead wood is needed for a joint, as it is of course bulky. Twist the strands into as tight a rope as you can and insert about 1½ the thickness of metal needed. Then pound it down as described for poured lead.

Threaded Cast-iron Pipejoints

Whenever a galvanized iron pipe has to be joined to a cast-iron pipe fitting the fitting is commonly provided with one or more female threads. The male-threaded end of the galvanized pipe is given a generous coating of pipe dope and screwed home.

Lead Caulked Galvanized Pipe Joints

Galvanized and black steel pipe can be strongly joined by the same lead caulking materials and methods used for cast-iron soil pipe.

In place of the hub that is normal to cast-iron soil pipe, a reducer is screwed onto the pipe, small end first, leaving a "hub" exposed. The following pipe is inserted into the hub, caulked with oakum, which is followed by poured lead or lead wool and then caulked in the standard manner.

Lead caulked galvanized joints are only used for vent and drain lines. They are not used for water lines.

The only problem in joints of this type lies in finding a reducer with a sufficiently large end to accept the following pipe and still leave room for the yarning and caulking tools.

Gasket Joints

Cast-iron soil pipe can be joined by means of a neoprene gasket made for the purpose. The gasket is a preformed, length of ribbed tubing. It is lubricated inside and out with soapy water and positioned within the hub of the cast-iron soil pipe. The spigot of the succeeding length of pipe is forced into the center of the tube gasket. This completes the joint.

A gasket joint has an advantage over a lead caulked joint in being slightly flexible. Unfortunately, gasket joints are not acceptable to all plumbing departments.

No-hub Joints

These joints are made with a special fitting that clamps the pipe ends together. No-hub joints can be used with any thick-wall metal pipe: cast-iron, galvanized iron, copper and brass.

The fitting consists of a neoprene rubber sleeve, a flexible sheet of stainless steel and two stainless-steel clamps similar to those used on automobile water hoses.

The pipe ends are cut squarely across. (Remove the hub on cast-iron pipe.) The neoprene sleeve is slipped over the pipe ends. The pipe ends are brought together. The metal sleeve is wrapped around the neoprene sleeve. The clamps are then used to tighten the metal sleeve in place.

This joint is simple, easy to make, easy to disassemble and even more flexible than the gasket joint, allowing the assembled sections of pipe to be bent a few degrees at the joint.

Unfortunately, some plumbing departments do not permit No-hub joints.

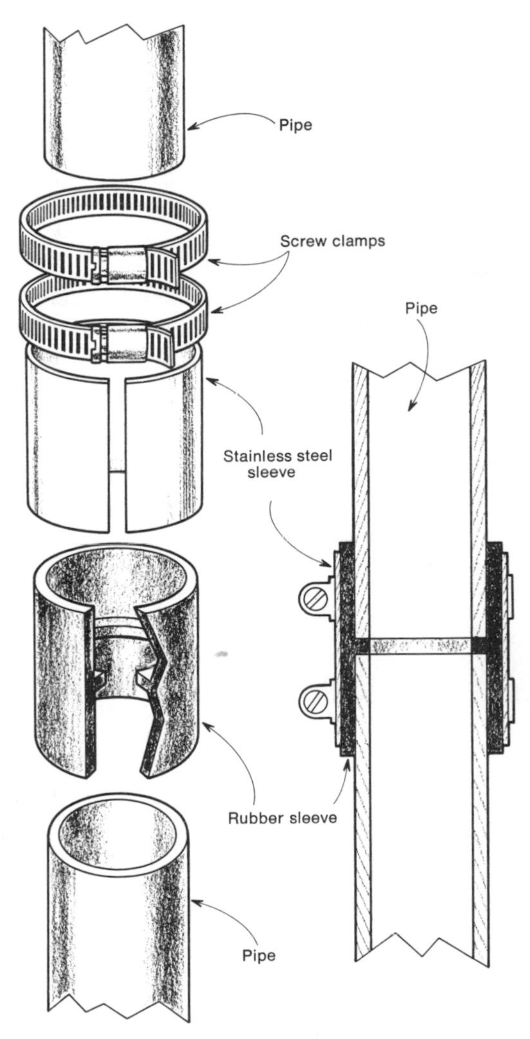

Exploded view of No-hub joint (left), showing rubber and stainless-steel sleeves with screw clamps that hold them in place. Drawing at right shows how slit stainless-steel sleeve slides over rubber sleeve and is tightened by clamps.

LEAD PIPE

Lead pipe is easily cut with a hacksaw. Lacking a hacksaw, you can also cut lead pipe with a sharp knife.

When two sections of lead pipe are joined, plumbers use what is known as a wiped lead joint. This is the most difficult of all plumbing joints to make. It takes practice and a certain amount of fortitude to work with molten lead.

Preparing a horizontal joint. Taper one pipe for a distance of about ⅜ inch with a clean file or rasp. Expand the other end a bit by lightly tapping it with a ball-peen hammer. When the tapping and expanding are completed, the tapered end should enter the expanded end for a distance of about ⅜ inch. Smooth the lip that was raised on the expanded end and file the ends of both pieces of pipe square.

Next, apply 2 inches of plumber's soil with a brush. This is a mixture of lampblack and glue dissolved in water. It is available in prepared form. Following, the soil is removed from the pipe ends for a distance of about 1 inch. This may be done with a plumber's shave hook or clean steel wool. This will be the extent of the joint—1 inch on each pipe end. Following, coat the bright metal with soldering flux, and press the two ends tightly together. Use concrete blocks or brick to hold the pipe sections in a straight line.

Melting the lead. The lead, actually solder, is a mixture of 37 percent tin and 63 percent lead that is brought to a point somewhere between its solidus and its liquidus temperature. As it is an alloy, it doesn't suddenly melt, but first becomes slushy. When the molten solder will just scorch a piece of paper, it is the correct temperature.

Making a Wiped Lead Joint

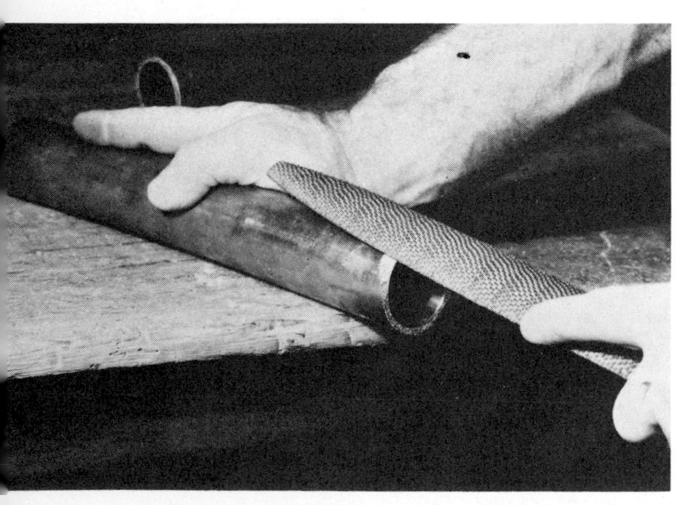

1. One pipe end is filed to a bevel; the other end is expanded somewhat with the aid of a small hammer.

2. The joint is assembled and the area not to be joined by lead is painted with black soil. The area that is to be joined by lead is cleaned with a plumber's shave hook or steel wool.

Pouring the lead. In addition to substantial gloves, you will need one or two pads of herringbone ticking folded into thick squares about 2¾ by 3 inches in size. Run wax onto the pad so the solder won't stick. You will also need a ladle with which to pour the solder.

Dry the ladle by holding it bottom side down on the molten solder. Fill the ladle with solder and, placing one hand with the pad in its palm under the joint, slowly pour the solder over the joint. The purpose of holding the pad under the joint is to catch all the molten solder that rolls off the pipe; but don't pour so much it runs off your pad and onto your glove.

Some mechanics let the first pouring of solder run right to the ground. The reason is that the purpose of the first and subsequent pourings is to heat the lead pipe. If you recall the soldering process, you will remember that it is necessary for the base metal, in this case the lead pipe, to be hot enough to melt the solder.

Unless the base or parent metal is sufficiently hot, the solder will not form a strong bond. So the first few pourings can be warm-ups. Afterwards, it is necessary to catch the overflow with the pad and bring it to the top of the joint, from where it is pushed around and under the pipe joint with the pad. This is where some mechanics find a second pad useful. Once enough slushy solder is on the pipe, and the pipe is sufficiently hot, the joint is shaped and the edges feathered down tightly.

Vertical wiped joints. The same preparations are made as for horizontal joints. The difference lies in the way the molten solder is applied. It can't be poured sideways; it must be carefully splashed on. When sufficient solder has adhered it is wiped into shape. The top edge is wiped first because it cools first.

3. The hot lead is poured over the joint, caught with the wiping pad and shaped as desired. This is how the joint looks as you begin wiping it. *Courtesy Lead Industries Assoc.*

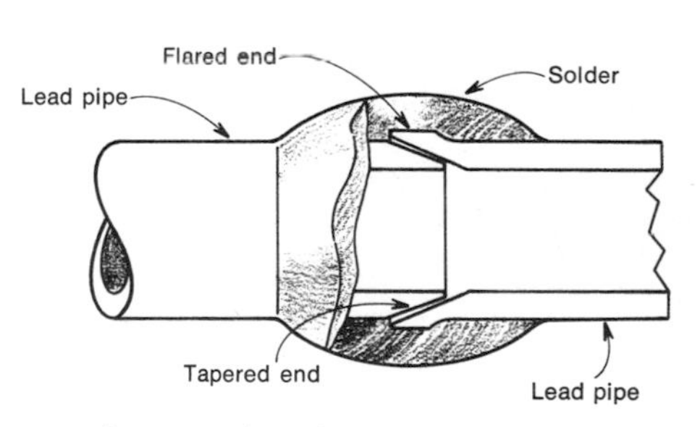

Cross section of a wiped lead joint.

Other joints. Tees and Wyes are made the same way. The branch is tapered into a spigot, while the main run is opened to form the hub. When there is sufficient solder and the joint-to-be is sufficiently hot, it is wiped just like the others. The same system is used

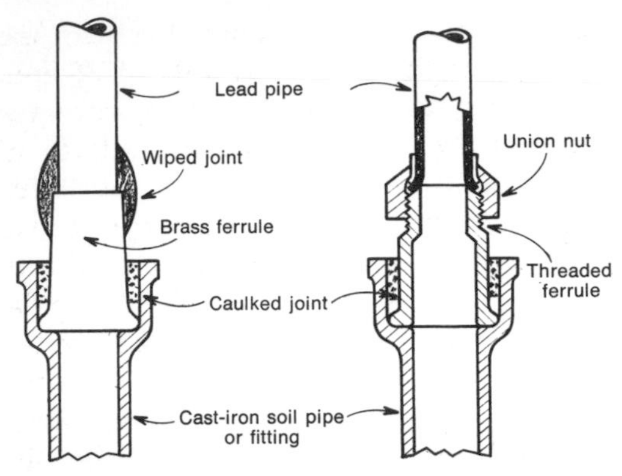

Left, how a brass ferrule may be joined to a lead pipe by a wiped lead joint. Right, how a threaded brass ferrule may be joined to a lead pipe by means of a large flare joint. Note that in both cases the brass ferrule is lead caulked to the cast-iron pipe.

Filing a brass ferrule preparatory to making a lead wiped joint.

Applying soil to a brass ferrule to limit the distance the lead will go when making a wiped joint. *Courtesy Lead Industries Assoc.*

to join lead pipe to a brass ferrule, which is then lead-caulked into a cast-iron hub. The ferrule is not shaved, but should be filed or sandpapered bright. If you wish to be certain of a good lead-to-brass wiped lead joint, tin the brass first. Clean it, flux it, heat it sufficiently to enable you to melt solid-wire solder on it to give it a thin coating of solder. No harm if the solder covers more than the wipe-joint solder will.

VITRIFIED CLAY PIPE

Support the pipe to be cut on a piece of wood or a pile of earth, the bell end hanging clear. Draw the line of cut with chalk. Use a cold chisel, preferably sharper than that used for cast-iron pipe. Tap it lightly against the line, the first time around to break the glaze. Then tap the line lightly until the pipe separates. The trick to cutting clay pipe without breaking it is patience and proper support. Each tap of the chisel crushes a small area of clay. If you hit too hard, the chisel opens a big crack that usually wanders from the line. With experience you will soon learn just how hard you can rap the chisel without breaking the pipe.

To use the soil-pipe cutter, wrap the cutting chain around the pipe and tighten it until the pipe cracks.

To join the spigot at the end of one length of vitrified clay pipe to the bell of another length of clay pipe, first clean the bell and spigot of foreign matter. The spigot is centered in the bell and sufficient oakum is "yarned" into place to hold the spigot in the center and form an oakum ring about half inch high. The oakum is packed very lightly, to avoid cracking the clay pipe. The oakum makes the joint watertight and prevents the joint sealent from getting inside. This eliminates the need of a swab, which

Cutting clay pipe with hammer and cold chisel.

is an old-fashioned mop that is dragged through the joined clay pipe to remove any sealent remaining inside.

Mortar cement, asphalt, molten sulphur and sand and commercial "plastic" preparations may be used to seal the joint. To use mortar cement, mix 3 parts Portland cement with 1 part mason's lime. Or purchase mortar cement, a prepared lime and cement mixture. Either way, mix 2 parts clean sand with 1 part dry mortar cement. Add sufficient water to make a completely wet mixture that can be piled into a little hill about 6 inches high and 6 inches wide at its base.

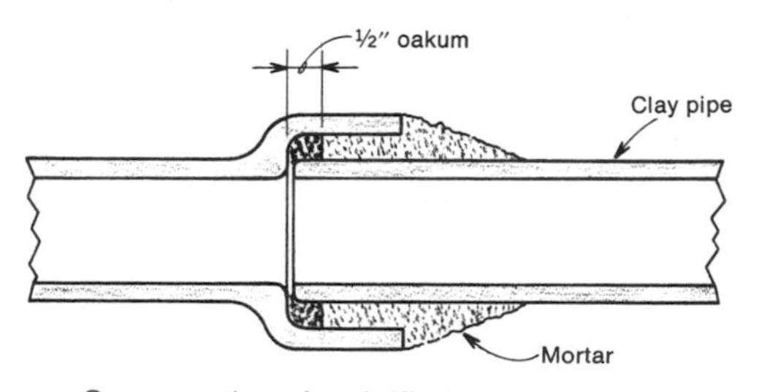

Cross-section of a vitrified clay pipe joint

The mixture should be wet, but firm enough to be molded. You'll find it easier if you add water to the cement, rather than the other way around.

Trowel the mortar into the joint and then "butter" more mortar on for a distance of 2 to 3 inches. It is not necessary to trowel the cement up and over the hub.

Pipe so joined should be in a perfectly straight line for best results. If necessary, a little angle can be tolerated. But don't tilt the spigot enough to eliminate cement from between the spigot and bell or the joint will leak.

Some plumbers do not bother to use oakum before applying the mortar. Whether it is faster to omit the oakum and swab the pipe clean than use oakum and not swab is difficult to say. However, I prefer to use oakum.

The commercial plastic preparations used to seal clay pipe joint are applied the same way, with or without oakum. While the plastic preparations have the advantage of slight flexibility—cemented joints have no flexibility when hard—cemented joints are easier to make when you have to reach down under the pipe and apply the sealant by hand. Mortar cement is readily washed off; asphalt and similar compounds are not.

Mechanically sealed joints. Vitrified clay pipe is now available with "speed seals" made of permanent polyvinyl chloride ring gaskets affixed around the spigot end of the pipe by the manufacturer. The joint is lubricated with a solution of liquid soap; then the spigot is pressed into the hub. That is all there is to it. The resultant joint is not only tight and permanent, it is flexible and permits the pipe to be bent a little, either during installation or in later years should the earth shift and the pipe sag. Cemented lines open under such conditions.

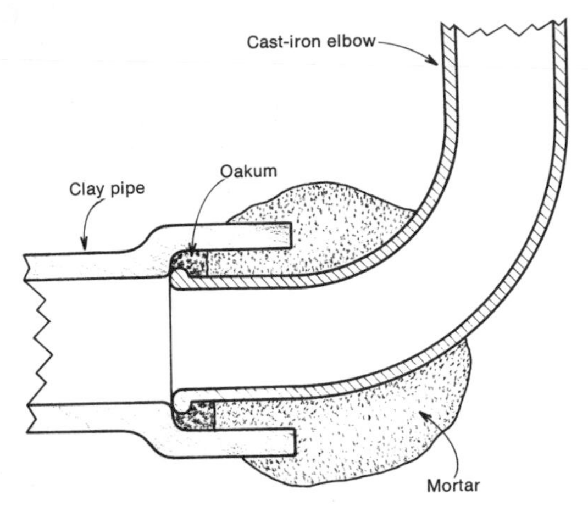

Cross section of a vitrified clay pipe joined to a cast-iron soil pipe. Note how mortar is piled up to provide additional strength around joint.

Using a hacksaw to cut asbestos-cement pipe.

Clay to cast-iron joints. It is common practice in many parts of the country to run vitrified clay pipe from the cast-iron sewer pipe leaving the house on to the city sewer line. The joint between the iron and clay pipe is always an iron spigot into a clay bell.

Such joints should always be caulked with a ring of oakum before the cement or plastic sealant is put into place. Don't tamp the oakum hard.

In some towns the city will run a 6- or 8-inch sewer line stub-out to the property line. (The city engineer will tell you exactly where it is.) As most house sewer lines are 4 inches in internal diameter, a reducer is used. This is joined to the two pipes exactly as described for regular joints. Take care to keep the pipes centered.

ASBESTOS-CEMENT PIPE

Pipes of asbestos cement, frequently called Transite, as that is one manufacturer's name, have ends machined to take couplings of the same material. The coupling is hand pressed on the machined end. Rubber rings inside the coupling make a tight, waterproof and rootproof seal.

To press the coupling on the pipe end, you can use brute strength or you can place a thick block of wood over the coupling and gently tap it in place with a hammer. You can also use a short piece of wood as a stop on which you stand, and a length of 2-by-4 as a lever to press the coupling in place. The end of the 2-by-4 rests against the stop under your foot.

Driving an asbestos-cement coupling onto the end of a length of asbestos-cement pipe.

Asbestos pipe can be cut with an ordinary carpenter's handsaw. But to join a cut end to a machined end you need a special coupling made for the purpose. There are also couplings made to join two "cut" ends.

You can join asbestos to clay vitrified or to cast iron by inserting the asbestos within the hub, caulking it with oakum and following up with mortar cement or one of the compounds made for the job.

BITUMINOUS PIPE

Bituminous pipe, often called Orangeburg, as that is a leading manufacturer's name, can be cut with a handsaw.

Pipe ends are chamfered to accept couplings and fittings. These are simply tapped in place, with a block of wood placed on the fittings to absorb the shock.

Pipe ends cut in the field can be chamfered with a special tool made for the purpose, or they can be joined to other "cut" ends with special fittings and joining compound available from the pipe suppliers. There are also special fittings to permit bituminous pipe to be joined to threaded pipe. Connections to the hub of cast iron or vitrified clay pipe are made by first caulking the joint with a layer of oakum and following it up with mortar cement or a sealing compound. Care has to be used when caulking not to crush the fiber pipe. Follow the same procedure explained for caulking vitrified clay to cast-iron pipes.

Using a crosscut saw to cut bituminous pipe.

5

Repairing and Installing Faucets and Valves

The handle you turn every time you want to control the flow of water into your sink or tub is attached to a device called a faucet, which is a special kind of valve. It is a valve made to terminate a run of pipe; very often it has a spout attached. It is made to take abuse and to be easily repaired. In a busy kitchen it may be opened and closed a hundred times a day.

Valves and faucets are very much alike, and what applies to faucets usually applies to valves. A valve is a faucet that is connected to the middle of a run of pipe. A valve can accept a pipe at both ends; a faucet only at one end. Faucets and valves are less complicated than the number of different designs may lead you to believe.

COMPRESSION FAUCETS

A compression faucet is the faucet found at the end of most of the hot- and cold-water pipes in our homes, offices and factories. It is an old standby, still selected for most systems.

It is called a compression faucet because when you turn the handle to shut off the flow of water, the stem moves down into the body of the faucet and a washer is compressed against the opening that passes water from one end of the faucet to the other.

The accompanying illustration shows the various parts of a compression faucet and their relation to one another. Note that the flow of water is up through the hole in the bottom of the faucet, through the hole in the removable seat and out the spout. When the threaded spindle is turned, the washer on its end moves down and seals itself against the top of the seat, stopping all flow of water. Note too that when the faucet is open and the spindle is in its raised position the water can slip past the threaded spindle, up along the stem until it reaches the packing which stops farther movement.

There are several major points of wear in a faucet of this type. First and foremost the washer wears as it is repeatedly tightened against the valve seat. Secondly, the packing wears as the spindle is turned.

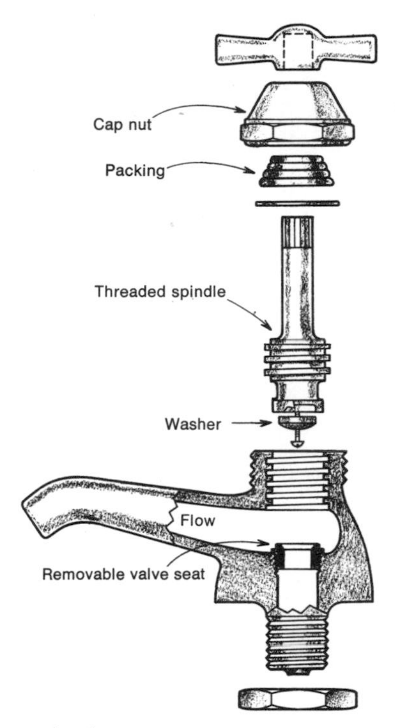

Major parts of a compression faucet. When the threaded spindle is screwed down, the washer is compressed against the valve seat and water cannot flow through.

Third the seat wears as the washer rubs against it and as the water flows over its edge. Next the spindle itself wears along its thread and along the area in contact with the packing.

Leaks at the Washer

A faucet in proper working condition will shut off the flow of water completely with little more than light finger pressure on its handle. A faucet in proper working condition does not leak water at its stem (spindle) when it is opened.

Leakage through the faucet spout is due to wear at the washer and/or seat. Leakage past the stem is due to a worn stem and packing. These are two independent problems. A leak at the washer does not affect a leak at the stem or vice versa.

When heavy pressure is required in order to completely close a faucet, its washer and/or seat will soon become defective and need to be replaced or repaired. If a repair is not made, the washer will be chewed up very quickly and the water will leak out badly no matter how much pressure is applied.

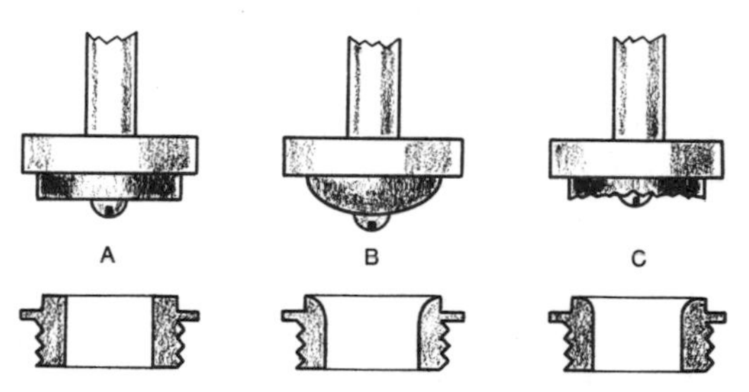

Three conditions you may find upon removing a faucet stem and exposing the valve seat. (A) The surface of the valve seat is square with its body. This seat normally takes the flat washer shown. (B) The surface of the valve seat is curved. Normally a hemispherical valve is used here. (C) The washer is chewed up and there is no definite shape to the valve seat. All you can do is try one type and hope for the best. If it doesn't work, try the other.

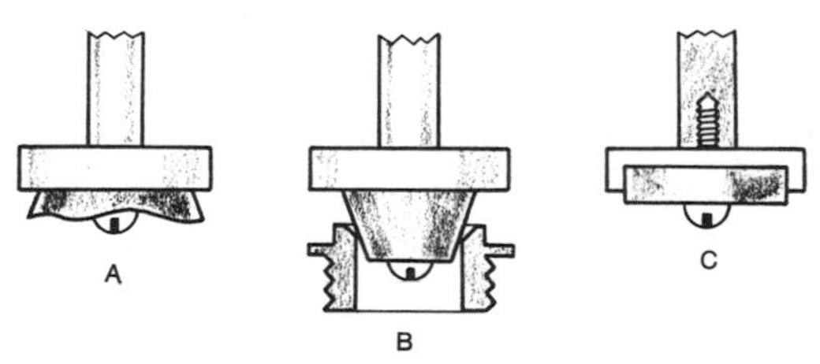

(A) When the washer is too large and overtightened, it wrinkles and leaks. (B) When a hemispherical washer is too small it gets caught in the valve seat and is unscrewed. (C) The correct size washer is a fraction smaller than the allotted space and its screw bottoms with a slight pressure on the washer.

A few of the many types and sizes of washers used today. The sandwich type and the washer with the split leg are noted for durability.

When light or heavy pressure does not prevent a faucet from dripping, the washer and/or seat are defective and must be replaced or repaired. In less time than one might imagine, the drip will wear a channel through the washer and seat and the small drip will shortly turn into a bad leak. If the repair is made quickly enough, it is very likely that only the washer will need to be replaced.

Removing the handle. To the uninitiated, removing the faucet handle may appear simple, but it can be the most difficult part of the entire job of replacing a washer. When the stem of the faucet is long, you can often remove the cap nut and unscrew the spindle to expose the washer without removing the handle. If the stem is short, and the handle completely covers the stem and cap nut, it is necessary to remove the handle before putting the wrench to the cap nut.

If the screw holding the handle in place on the stem is visible, simply back it off and out. If it is not visible, but a circle of metal covers the top of the handle, pry this circle up with a sharp knife. The screw is underneath.

If the handle comes up and off easily, all is well. But if it is corroded, try prying it up with two screwdrivers at the same time, one on each side. If that fails, purchase a handle-removing tool.

When the handle is removed, clean the splines on the stem and inside the handle so they'll slide together easily when you reassemble the faucet and handle.

Removing the cap nut. The only precaution necessary here is that the nut be heavily wrapped with adhesive tape or it will be scratched. Even a crescent wrench will leave marks on a shiny fixture. A pipe wrench will chew ridges into bare metal. If it is a thin, under-the-handle cap nut, watch for a thin plastic washer. If lost, it is a nuisance to replace.

Replacing the washer. With the cap nut off, unscrew the spindle—to the left for right-hand faucets and to the right for left-hand faucets. The washer, or the remains of it, is in a shallow, flat-bottomed cup at the bottom of the stem. If this cup is missing, or if a portion of the rim has broken off, the stem must be replaced. A new washer may hold for a while, but it will not last long.

Remove the screw holding the washer and the washer itself. If the bottom of the cup is rough, or parts of the old washer adhere, it must be smoothed and cleaned. Then insert a new washer and replace the screw.

Again, this is not as simple as it may appear. Washer shape, size and screw pressure are critical. There are two washer shapes. Both are circular in outline, but one is flat on both sides. The other is flat on one side and round on the other. If the original shape of the washer is clearly distinguishable and the faucet worked well up to this point, replace it with a washer of the same shape. If the washer is chewed beyond iden-

Replacing a Seat in a Modern Faucet

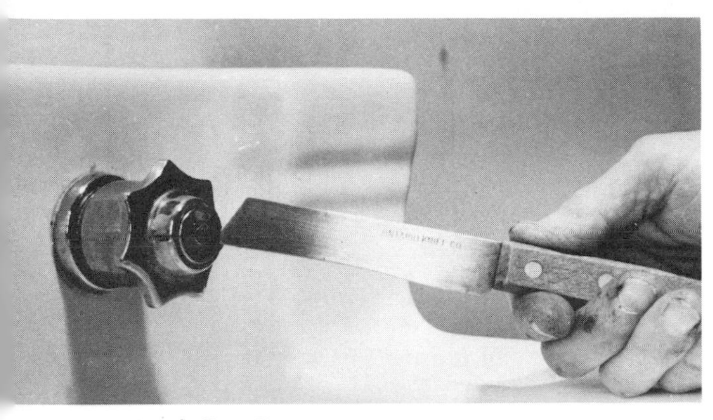

1. Pry off the cap covering the screw that holds the handle in place.

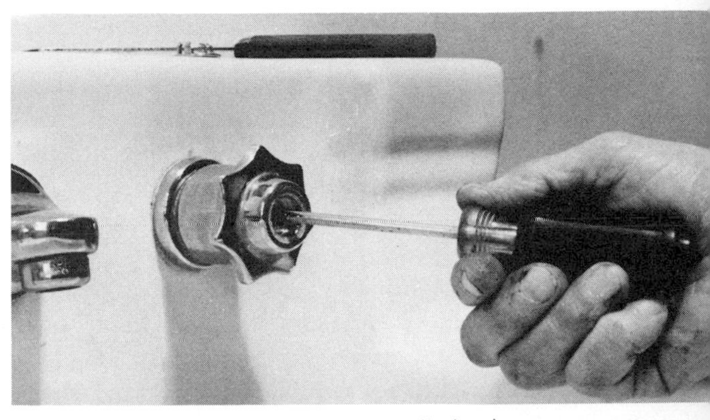

2. Remove the screw that holds the handle in place.

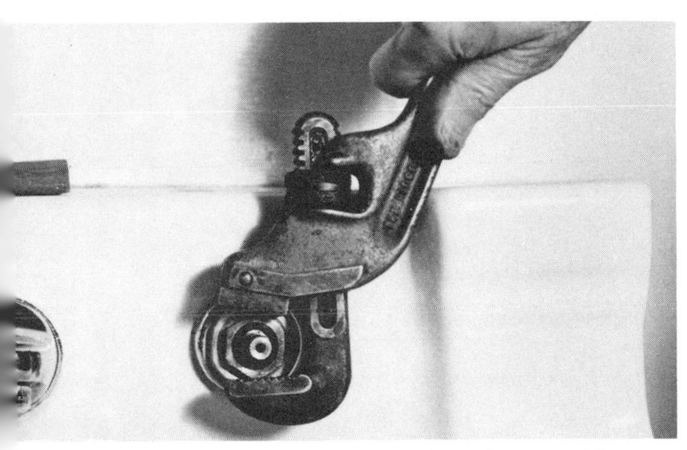

3. Use a wrench to loosen the valve assembly.

4. Remove the valve assembly. This is a diaphragm assembly made by American Standard.

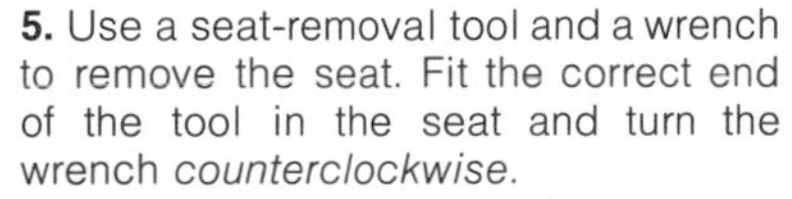

5. Use a seat-removal tool and a wrench to remove the seat. Fit the correct end of the tool in the seat and turn the wrench *counterclockwise.*

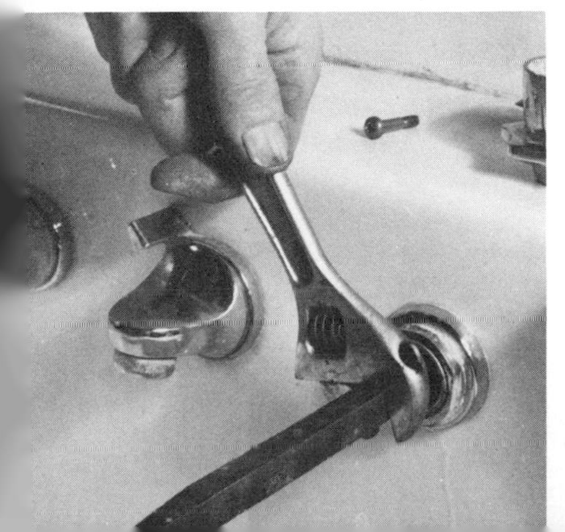

6. The old seat, on the end of the tool, should be matched with the new seat. The one shown here is the wrong size and will not fit.

7. When you have found the right seat, slip it on the end of the tool and carefully screw it in place. Then reassemble the rest of the faucet.

tification, you have the choice of disassembling a similar faucet to determine whether a flat or curved washer is required, or attempting to identify the shape by the conformation of the valve seat. Both are difficult. It is easier to try one shape and then the other. If a flat washer is chosen, be certain that the rough side goes face down against the metal.

Do not use old-style washers of paper composition. They do not last long. Use rubber washers. A box of assorted shapes and sizes, along with screws, can be purchased at hardware stores.

Select a washer with a diameter that is slightly smaller than the space provided. There must be a hair's breadth of play or space between the edge of the washer and the inner edge of the rim of the metal cup.

Checking the seat. When the faucet handle is turned to the closed position, the washer is pressed against the seat, forming a watertight seal. If the surface of the seat is rough, if it is traversed by fine cracks (caused by a constant drip), a watertight seal is impossible. Should the handle be turned with enough force to stop the drip of water through the cracks, the washer will be quickly destroyed.

The seat of a faucet in normal use is shiny. Channels from the center of the seat across its rim appear darker because of corrosion. Most of the time a visual inspection with the aid of a flashlight is all that is necessary. Sometimes, when a new washer has been installed and the faucet persists in dripping, it is necessary to remove the seat for closer inspection.

Removing the seat. The seat is easily removed with a seat-removal tool, a steel rod with one end square, the other end octagonal. The correct end is fitted into the center of the seat, tapped home, and turned counterclockwise. If you cannot get a good "bite" on the seat, see if the tool is striking the bottom of the inside of the faucet. If it is, use a hacksaw to remove a little of the end of the tool.

The old seat is used as a guide to selecting a new seat. It is possible to select a new seat by giving the plumbing shop the make and model number of the faucet (or valve), but it's simpler if you use the faucet make and model number as a starting point and use the old seat itself as the final guide.

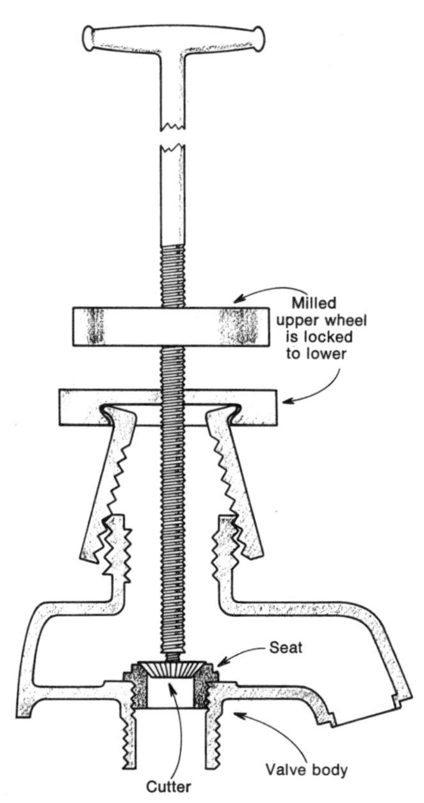

Cross section of a professional valve-seat refacing tool. To bring the cutter against the seat, the two wheels are separated and the handle turned a fraction while the lower wheel is held fast.

Plumbing supply shops stocking replacement seats usually have a "test board," a plate of metal with twenty-four threaded holes of various sizes. The old seat is tried in as many holes as necessary until the correct one is found. The seat must thread easily. If it must be forced or if it shakes after being tightened all the way it is no good. *Do not force a replacement seat into the faucet.*

The test board will enable you to get the diameter and thread size of the old seat, and to make certain (by trying) that the new seat is the same diameter and thread.

If the shop doesn't have a test board, check thread replacement by placing one seat atop the other; there should be no spaces between the threads. Also check diameter by mating.

You may find that you can duplicate the diameter and thread, but the shoulder and top of the new seat differ. The only way to determine whether or not the difference in the top and shoulder of the new seat affects the operation of the faucet is to try it.

Replacing the seat. This is accomplished by forcing the new seat up a bit on the seat-removal tool. In some instances you may want to use a little putty on the tool to make the seat stay on. Always use pipe dope on the threads—it helps to seal the joint. The seat is then lowered into the faucet and screwed into place. To make certain the new seat is not cross threaded, turn it forward and backward half a turn or so to make certain it is started correctly.

If the top of the new seat is different from the old, you will have to use a washer to correct the problem. If you are not certain which is correct, try the curved washer first. There is about 3⁄16-inch difference between the height of the flat and curved washers. If you are short of space because the new seat is too high, try using a flat washer. If the new seat is much lower than the old one, try a curved washer.

Refacing the seat. When the old seat cannot be removed, or when the faucet or valve does not have a removable seat, you have the choice of replacing the faucet or refacing the seat. Refacing the faucet seat is a machining job. Sometimes it can be correctly done with a professional-grade refacing tool, but more often than not the job is unsatisfactory. In any event it is worth a try. Generally you will get better results if you use a "cutter" that will reface the seat to accept a hemispherical washer.

Bear in mind that refacing reduces the height of the seat. Sometimes, the seat may be properly refaced (it does not drip when the faucet is reassembled), but it is so much lower that the handle strikes the cap nut or sink top.

Remove the cap nut and valve spindle from the faucet. Select a cutter of suitable size and screw it onto the end of the refacing tool. Turn the handle to the left while the wheel attached to the cone is prevented from turning. This backs the tool spindle out of the cone. Screw the cone down on the exposed thread. As the cap thread is coarse and the cone thread is fine, the cone cannot be tightened too firmly. Now turn the handle until the cutter's face just touches the top of the valve seat. Then turn the second wheel until it comes down the shaft and contacts the first wheel. Hold the first wheel firm and tighten the second wheel. The result of locking the two wheels together is to "set" the cutter against the valve seat. The handle can now be turned right or left by virtue of the slip joint without altering the distance between the cutter and the seat. In other words, we have locked the tool onto the valve body and are now bringing a hand-operated miller against the surface to be made smooth.

Turn the handle to the right until there is no resistance. Then separate the two wheels. Hold the lower one fast and give the handle a fraction of a turn into the work. Tighten the upper wheel against the lower. Again turn the handle to the right. In this way, the cutter removes a layer of metal.

If the threaded cone on the refacer mated perfectly with the threads on the cap nut, a satisfactory milling job could be done on the valve-seat face. But they don't mate properly. The tool's manufacturer provides a cone and an adapter to permit the device to fit many of the hundreds of different faucets and valves now in use. The tool's spindle is often a fraction of an inch off center, and sometimes tilted a few seconds of arc in the bargain. Even pressure cannot be brought to bear on the cutter, and when the cutter is dull and too much pressure is used, the entire device pops free. However, when it is impractical or impossible to replace the valve or its seat, the refacing tool is worth a try.

Alternatives to refacing. When the seat is defective and cannot be replaced for one reason or another, there are two alternatives known to this author. One is a combination drill and tap. This is a T-shaped device terminating in a special screw of steel. You simply slip it into the faucet and screw it down into the old seat, working it back and forth a little as you go. It drills a hole and threads it for a replacement seat. It is fine if you can get one that fits your faucet.

The other alternative is a special washer sold under the tradename Miracle Washer. It is an inverted, shallow cap of rubber backed by a shallow metal cup. The miracle washer goes over the seat and makes contact with the metal underneath. In theory, the seat is not used.

Repairing Leaks at Packed Stems

Despite its wax or graphite lubrication, the material wrapped around a stem, the packing, wears a little each time the valve or faucet handle is turned. Turned a sufficient number of times the liquid, gas or steam within the faucet body leaks out through the packing and past the stem.

When this happens the packing nut or the cap nut is given a tightening turn—just a little more than is necessary to stop the leak. That does it until the packing wears a little more, then another partial turn of the nut is needed.

Should the packing nut be made tighter than necessary, the rate of packing wear will be increased; there is more friction with increased pressure. Should the packing nut be too tight, it will be very difficult or even impossible to turn the stem and operate the faucet.

In time, tightening the packing nut will fail to stop the stem leak. This indicates the packing is completely worn. The packing nut will hit the shoulder of the faucet body. Sometimes the packing, especially if it is not commercial packing but some emergency concoction of household string rubbed in candle wax, has deteriorated to where it cannot be moderately compressed and still hold water. Tightening stops the leak, but turning is impossible.

Replacing the packing is easy. Close the faucet and back off the packing nut. Use a piece of wire to remove the remains of the old packing. Wrap the new packing around the stem and push it down into the space between the stem and the faucet body—called the stuffing box. Just enough packing

Commercial thread packing and a preformed washer packing. The thread is normally too thick to use in a faucet and must be unraveled.

is added to make it difficult, but not impossible, to force the packing nut down far enough to engage the threads on the faucet body.

Commercial "thread" packing is too thick for most of the faucets encountered in the home. It should be unraveled and a thread thin enough to go around the valve stem several times before filling the space should be used. Do not use one short piece of packing that is so thick it cannot be overlapped. A leak can occur between the butting ends.

If the valve used washer-type packing, secure some from a plumbing supply house or adapt the thread packing to the job. If you have washer-type packing, and it fits, it can usually be used in place of the string. And if you have no commercial packing, you can make do with cotton thread rubbed generously in candle wax.

Incidentally, when the old packing is out and the valve stem is exposed, examine it for burrs and straightness. If it has been roughened or bent, the packing will be short lived. When the stem is badly worn or bent, no amount of packing will prevent it from leaking. You will find that worn stems can be tilted from side to side; that the portion making contact with the packing is worn and slightly grooved. Worn stems must be replaced. Many of the plumbing supply

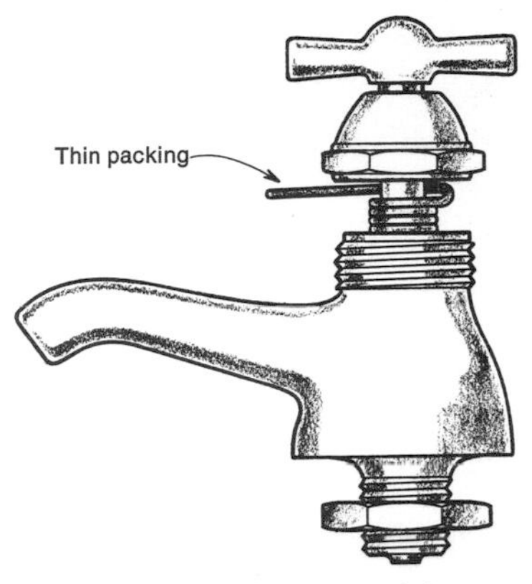

If you can raise the packing nut high enough, you don't have to remove the faucet handle to replace the packing. Use packing thin enough to go around the stem several times. Wind in the same direction that the stem revolves when closing the faucet.

houses carry an assortment of replacement stems. If the valve itself can be easily replaced, it is advisable to replace the entire valve or faucet, rather than the stem alone. The reason is that the valve body, which carries the coarse threads on which the stem rides, also wears. While the new stem will not "tilt" as much as the old stem, it still will tilt because the threads and cap nut are worn.

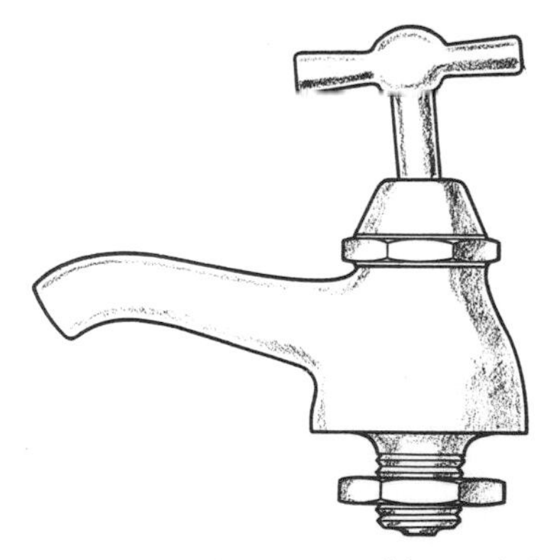

When you can tilt the faucet stem this way in the open position, it is best to replace the entire faucet. If that is not practical, replace the stem. Tilting faucets always leak at their stems.

REPAIRING DUAL FAUCETS

The earlier individual faucets found on wash basins and sinks of yesteryear have been largely replaced by faucets in tandem. Water passing through each of the faucets meet and comingle in a single, swivel-mounted spout. The mechanics of the individual faucets remain the same. The advantage is a single casting for the body of both faucets and the convenience of easily mixing hot and cold water to secure luke-warm water.

The presence of two valves feeding one supply pipe leads to sometimes confusing problems. For example, water may leak from the cold-water faucet stem when the hot-water faucet is opened. This is caused by a leaking cold-water stem. As pressure from the open hot-water line can build up in the closed cold-water line, water can emerge from the cold-water stem.

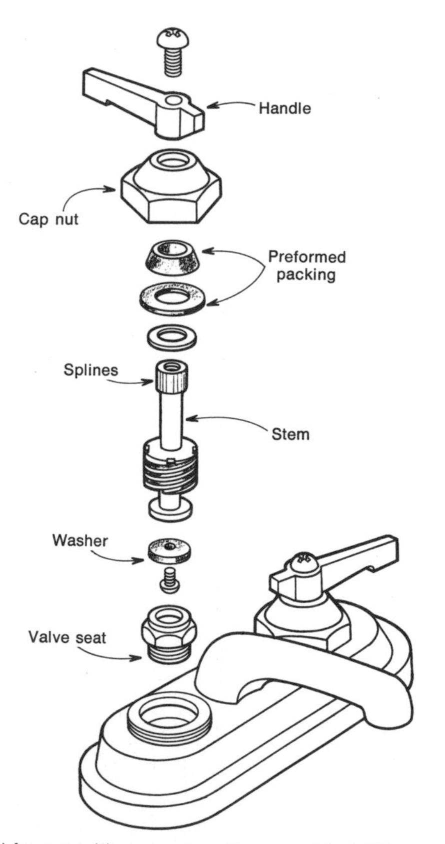

Dual faucet with one valve disassembled. Disassembly begins with removal of screw holding the handle, followed by the removal of the cap nut and then the rest of the parts.

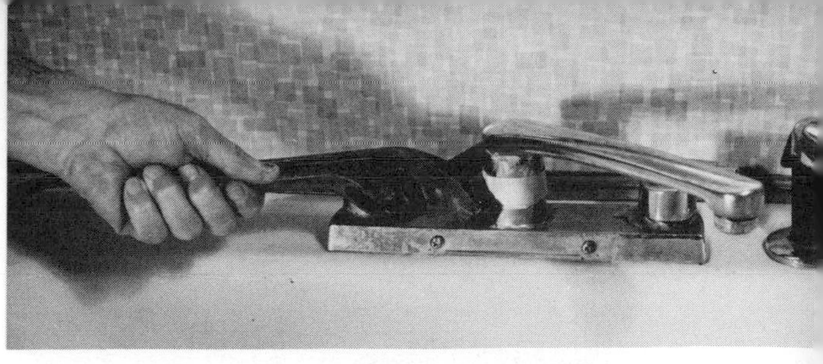

To disassemble a swinging, dual-faucet spout, put some tape on the large nut beneath the spout and loosen it with a stillson wrench (above). With the spout removed, you can now remove and then replace the O-rings (below), which must be the correct size or the joint will leak.

Leakage at the swivel spout joint is usually due to worn O-rings at the joint. Tightening the nut holding the swivel spout in place usually does not help. Wrap the spout nut in tape of some kind to protect it. Back the nut off and gently lift the spout up. The one or two O-rings visible there need to be replaced. Their size is critical. Replacement rings must remain in place with very little tension and protrude just a hair's breadth beyond the metal.

To determine whether the drip from the central spout is due to the hot or cold faucet, hold your finger under the drip. If that isn't positive, try shutting off the water to one faucet under the sink or wash basin.

FULLER FAUCETS

Fuller faucets are used on low-pressure water lines where fast action is required. To open, the handle is given a quarter turn in either direction. The stem or spindle has an eccentric pin at its end. This engages a rod terminating in an acorn-shaped lump of rubber that serves as the valve. When the valve is pulled forward towards the spout, the aperture leading to the spout is covered. When the acorn plug is moved rearward, the aperture is opened and the water flows.

To service, the entire faucet is unscrewed from the supply line (remember to shut off the water and drain the pipes), then the valve cap is unscrewed from the body of the faucet. This makes it possible to unhook the valve from the eccentric.

Generally, a leak indicates a worn valve ball. But sometimes the eccentric pin or its bearing has worn so there is too much play. Then even a new valve will not be pulled up with sufficient pressure to stop the leaking. If so, replace the entire valve. If the valve seat is grooved, it alone can be replaced.

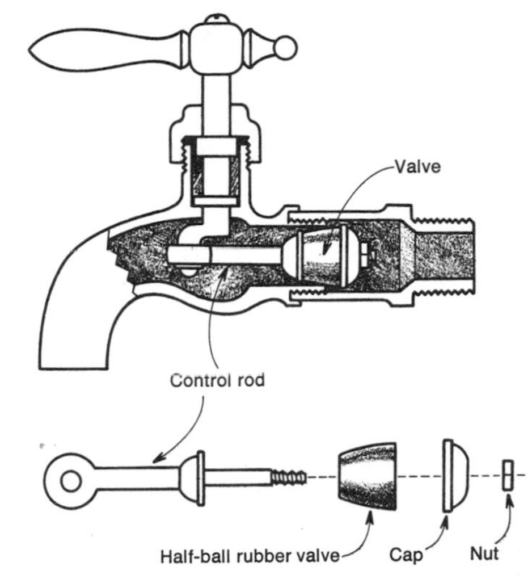

Major parts of a Fuller faucet. Recognized by the handle, which can be turned both ways, the faucet leaks when the rubber valve becomes worn or the hole in the end of the control rod becomes enlarged.

SHOWER FAUCETS

Shower faucets differ from the faucets just discussed in that they have long stems, inserted cap nuts and long bonnets. They are serviced exactly like the other faucets. Some are mounted inside the bathroom wall and cannot be reached by ordinary plumbing tools. You need automotive-type deep-socket wrenches.

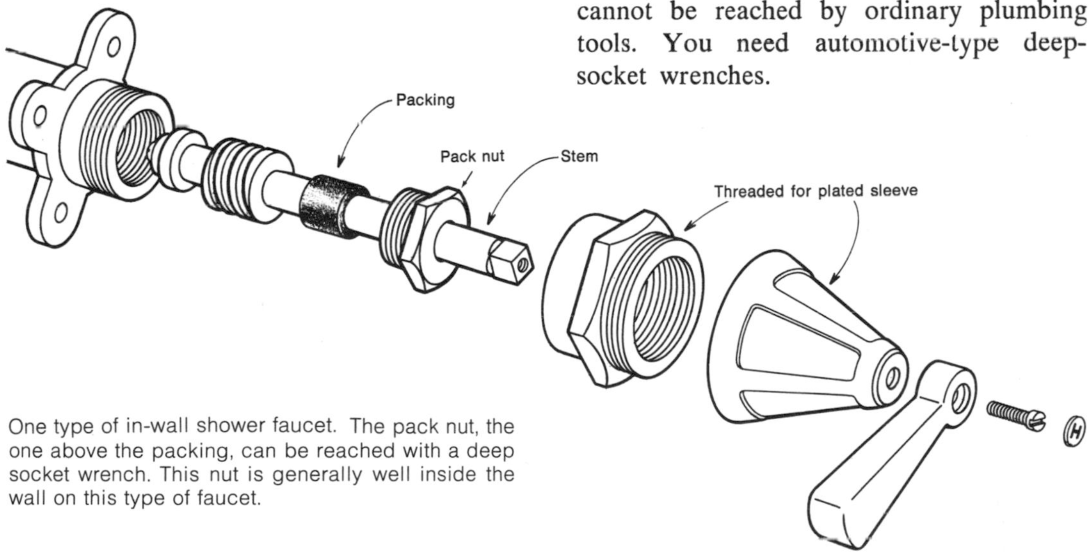

One type of in-wall shower faucet. The pack nut, the one above the packing, can be reached with a deep socket wrench. This nut is generally well inside the wall on this type of faucet.

When you plan to remove a shower-faucet stem for inspection and possible washer replacement, the water line leading to that faucet must be shut off, which is what is done with all water-filled lines. However, when you remove the bonnet, the top end of the faucet is within the wall. All the drip remaining in the pipe above your opened valve will come out and run down inside the wall. There is no easy way to remove the water. To prevent this, open all the faucets in the lines *above* the faucet you are going to disassemble and wait long enough for all the water to drip out before you disassemble.

To disassemble a shower faucet, remove the handle, then the escutchen or flange. If there is a plated tube, unscrew it. Use a deep-socket wrench on the packing nut, and turn it with a small pipe wrench. To get at the bonnet, a larger size deep-socket wrench is needed. One ¾-inch and one 13/16-inch deep-socket wrench will generally handle most shower faucets.

Should you need to remove the valve seat, do so very carefully. If you strip the hole in the removable valve seat you are in trouble. To remove the dual faucet you have to knock a hole in the wall. If your wall is tiled, it can be a messy, time-consuming job. If the plaster around the hole in the wall disintegrates, as it sometimes does if the stem has been dripping for a long time, rebuild it with plaster. Slip a paper tube from a roll of toilet paper over the stem to protect the packing nut from plaster.

FREEZEPROOF FAUCETS

If you live in a cold climate you know that each winter you must close the sill cock—the faucet to which you connect a garden hose—and then close the stop valve in the line leading to the sill cock. After which you open the sill cock so that the cock is drained. If you do not do this, a hard frost can freeze the water in the sill cock and it will burst.

Freezeproof faucets eliminate this fuss. A freezeproof faucet is constructed like a fire hydrant. Although it appears to be a standard sill cock or faucet from outside the building, the standard compression faucet is at the end of the device, far enough inside the house to be clear of frost.

Freezeproof faucets are repaired in exactly the same way as the packed-stem compression-type faucets discussed previously.

O-RING FAUCETS

Modern faucets do not depend on string or washer packing for sealing their stems. Instead they utilize two small rubber O-rings fitted into slots on the stem, which in turn fits into a polished tube. The seal is affected by the contact between the O-rings and the tube. An O-ring faucet can be distinguished by its short stem. Generally, you cannot see the stem, it is hidden by the handle. If you cannot see the screw holding the handle to the stem, pry up the tab on top. It hides the screw.

O-ring faucets rarely suffer from worn O-rings. They appear to last much longer than ordinary string packing. However, the stems and stem threads are so short that when the metal parts wear, the stems cock, that is, they tilt when the faucet is open. Thus, even though the O-rings themselves do not permit water to pass, the faucet leaks at the stem. To repair such a leak, take the faucet apart and try the stem in its little tube. If you can feel the O-rings mak-

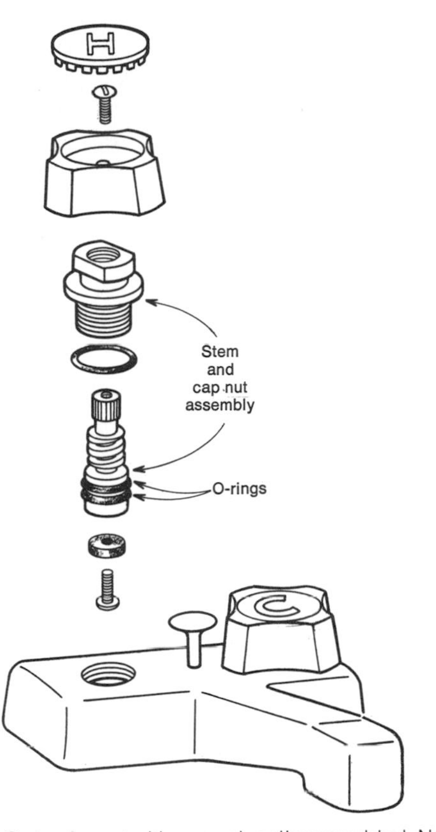

Dual O-ring faucet with one valve disassembled. Note the two O-rings just below the screw thread on the stem. If the faucet leaks around the stem these rings are worn. If rings of the correct size cannot stop the leak, the entire valve assembly needs to be replaced.

ing contact, they are satisfactory. Leakage at the stem cannot be cured by new O-rings. The only cure is a new stem. Fortunately, individual stems are for sale.

DIAPHRAGM FAUCETS

In an effort to avoid the problem of replacing washers and eventually seats and packing, a new type of valve has been developed. It uses a rubber diaphragm in place of the

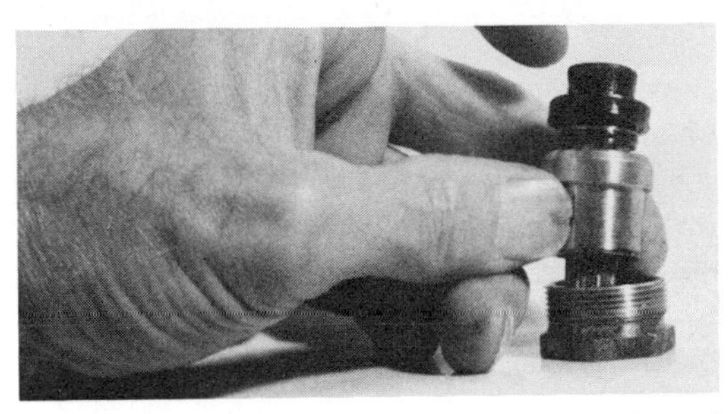

Close-up of a diaphragm valve assembly. Hat-shaped rubber cap under forefinger serves as both washer and O-ring.

washer and packing. The diaphragm, of rubber, rides on the lower end of the stem. Its sides contact the walls of the valve, sealing the stem. When the stem is screwed down, the central portion of the diaphragm rests over a stainless-steel valve seat. It is still a type of compression valve, but different.

To service this type of faucet, remove the handle and unscrew the insert holding the stem. The diaphragm can then be easily removed and replaced. Note: The new diaphragm must be exactly the same. Keep track of the metal ring washers that accompany the diaphragm. They must be replaced in the same position. They help keep the diaphragm from sticking to the stem and turning with it.

Diaphragm valve parts. Left to right, the stem with a small O-ring in place, cap nut, assembly body into which stem screws, rubber cap and metal washer that fits under it.

DIAL-EASE FAUCETS

The Crane Company has developed still another alternative to the conventional washer-and-seat faucet. They call it the Dial-Ease faucet. It differs principally in that its stem moves upwards to close. The faucet itself is similar to a conventional faucet, but its handle comes up when you close it.

To service, remove the inner valve assembly and replace damaged parts.

Cutaway view of the Crane Dial-Ease faucet. Faucet is shown in open position. *Courtesy Crane Co.*

Valve assembly of the Crane Dial-Ease faucet. Unit is shown in open position. *Courtesy Crane Co.*

SINGLE-LEVER FAUCETS

In response to our continued search for creature ease, inventors have responded with a single-lever mixing faucet, which manufacturers have marketed with promises of lifetime, troublefree service. Nevertheless, repair kits are already available at some plumbing shops.

There are various designs. Generally, pulling the single level toward you opens

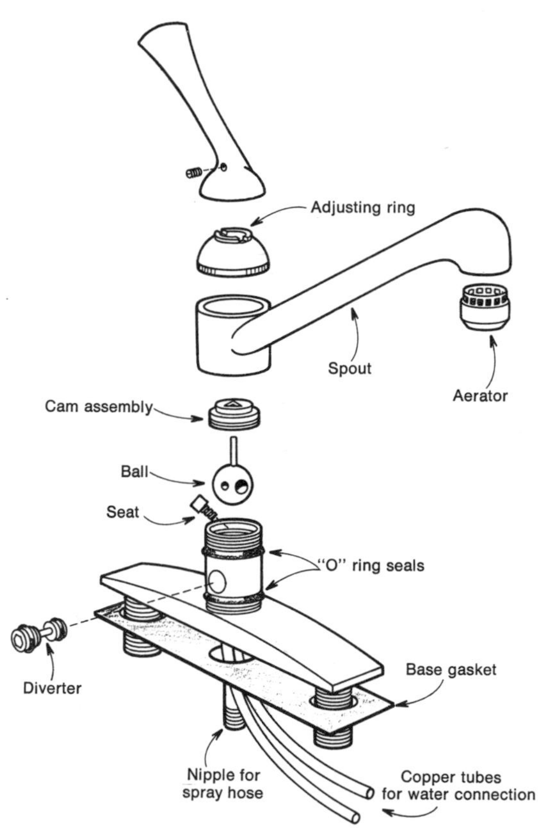

Disassembled single-lever faucet utilizing a rotating ball and pressure seats as valves. Biggest enemy of these faucets is the child who slams the handle to and fro. To disassemble this unit, loosen the setscrew in the handle, unscrew the cap and then remove the cam assembly to get at the seats, which wear in time.

Disassembling a Single-Lever Faucet

1. Lift the lever and loosen the setscrew underneath with an allen wrench.

2. Pull the lever up and off, unscrew the metal cap above the spout. Lift it up and off.

3. Loosen and remove the two brass screws holding the ceramic plug in place. Lift the plug up, as shown, to inspect the rubber circles which serve as washers. Replace if necessary. If brass lower section is worn, bring it to a machine shop and have it refaced.

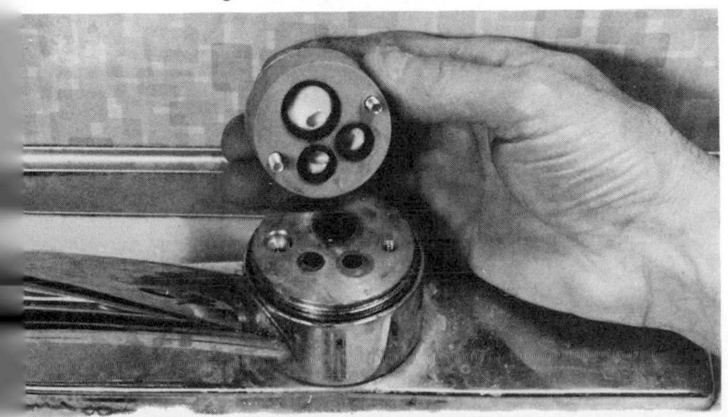

both valves; swinging the handle to the right or left favors the hot or cold line. The middle position provides a mixture.

The claim to longevity is based on the fact that the single-lever faucet has no washers under compression, but it does have O-rings, metal and composition valves. They wear, and in time must be replaced to prevent the faucet from leaking.

Kits for various makes of single-lever faucets are available at local plumbing supply shops. The kits contain the necessary pads and springs and other wearing parts along with specific instructions.

Generally, you will find a setscrew in the rear of the collar at the base of the control lever. Loosening this screw usually permits the collar to be removed by turning it counterclockwise. Lift it off gingerly so as not to lose any parts or forget where they belong.

SINK SPRAYS

Although the mechanism of a sink spray is rightfully classified as an automatic valve, it is used with a mixing faucet, so perhaps it is best to discuss it here.

A sink spray is a long hose, terminating in a nozzle with a pushbutton. It is used for washing dishes. With the faucet on, pushing the button on the nozzle handle diverts water from the faucet spout to the nozzle, enabling you to spray water directly on the dishes.

There are two types of sink sprays. One simply sprays water. The other has a brush and soap reservoir in the nozzle's handle which is controlled by a pushbutton.

Both types of nozzles connect to a mixing faucet by means of a flexible vinyl hose that passes through a grommet either in the sink itself or in the sink's cabinet and is attached to a tubular projection below the center of the faucet's body. Single faucets never have this fitting, and all mixing faucets do not have this fitting, and all sinks do not have provisions for this fitting. But most of the new mixing faucets, including the single-lever types, do.

The tubular projection that accepts the end of the nozzle hose communicates with the interior of the faucet. A diverter valve atop the inside end of the hose connection seals it closed. When the faucet is opened, water pressure is exerted on the diverter valve. However, as the spray tube is filled with water (from previous use), and as its further end is closed, the diverter valve remains closed and all the water leaves the faucet through its spout. When the button on the spray nozzle handle is depressed, back pressure on the diverter valve is relieved, and most of the water is diverted from the faucet spout to the spray nozzle.

Most spray problems arise from the collection of debris in one or both aerators. When they are clogged, the pressure differential on the diverter valve changes and it doesn't work correctly. To clean an aerator, simply unscrew it and take it carefully apart, *making notes on the order of the parts.* If you forget the sequence, the aerator won't work, no matter how clean you get the screens. The device mixes air with the stream of water, and unless the openings are positioned correctly it doesn't do its job.

With the aerators thoroughly cleaned, the next step is to clean the diverter valve. Ordinarily this valve is good for the life of the mixing faucet, but it can collect grit. All water supplies contain grit, especially during municipal repairs.

The diverter valve is positioned within the faucet body directly beneath the central spout. To get at it you have to remove the spout. Generally, you will find some sort of a ribbed ring or large nut between the spout and the body of the double faucet. Wrap tape around the nut before you apply your wrench, or you will mar it for life. Loosen the nut and gently lift the spout straight up. Use a flashlight to look down the hole. You should see a little metal tip projecting towards you. It will be attached to one end of the diverter valve.

Carefully rinse the valve under another faucet. If any mineral deposits have accumulated, try removing them with a toothpick or something equally soft. If all appears to be well, replace exactly as it was originally installed.

If, after having cleaned both aerators, checked that connecting hose is not kinked, cleaned and replaced the diverter valve, the spray still doesn't work properly, bring the old diverter valve to a plumbing shop and have it replaced.

REPLACING FAUCETS

Replacing a faucet can be easy or difficult. The methods you use and the problems you encounter will depend on the type of faucet, the amount of room for working, and the condition of the pipes and fittings.

Old-timers. These are single-unit faucets that are screwed directly onto the end of a threaded pipe that projects right through the front of the sink or tub. If you look to the rear of these faucets you will see the ¾-inch pipe coming through. The faucet itself has a thick flange resting on the sink or tub.

To remove such a faucet, wrap some tape around it to prevent marring and back it off with a large wrench. To replace, apply pipe dope to the male thread and screw the new faucet on.

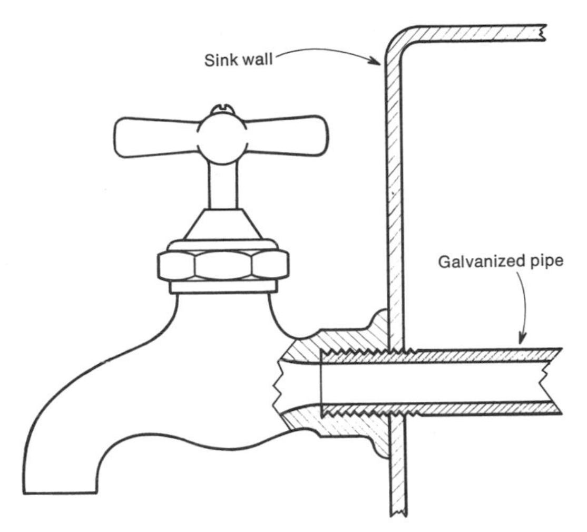

An old-fashioned faucet is screwed to the end of a threaded pipe that projects through the fixture or wall. Use a wrench to unscrew the faucet, which should be protected by tape or a rag.

Clean the old thread and give it a fresh coat of pipe dope. Then screw the new faucet in place, starting by hand and tightening with a wrench.

Single-unit faucets. These are like the old-timers in that each faucet is an individual valve. However, they differ in that they are not held in place by the pipe alone but by a large nut that is screwed to the underside of the faucet body. The feed pipe has a preformed tip which fits into the end of the faucet and is held there by a second nut that goes up over the threaded, hidden end of the faucet.

To remove this type of faucet, back off the small nut. Then back off the large nut on the underside of the lavatory or tub. The faucet can then be lifted up and out.

To replace this faucet simply reverse the steps, putting a little pipe dope on the coarse thread that carries the large nut. It will make it easier to remove in the future.

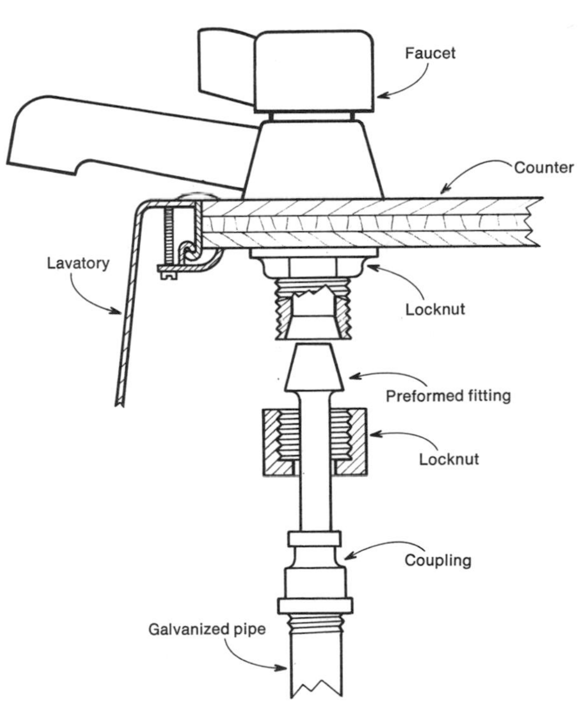

The modern single-unit faucet mounted on a lavatory or tub accepts a preformed pipe fitting, called a Speedee. To remove this faucet, back off the fitting nut, then the nut under the faucet.

The preformed tip makes a dry, clean connection.

The problem here is that the replacement faucet must be exactly the same size as the old unit. If it is not, it will not make proper contact with the feed pipe. If the feed pipe is too long, the faucet will end up above the sink, and if the feed pipe is short, it may not reach. Of course, you can always replace the feed pipe or cut and thread as necessary. If you find the preformed tip leaks when you tighten it, dry it and try some automotive gasket cement on the joint.

Also be advised that some of those single faucets have ridges on their bodies that permit them to fit only certain sinks (ridges keep the faucets from turning), so you may not know whether or not you have the correct replacement until you actually remove the old faucet.

Replacing dual faucets. Modern lavatories and sinks usually have dual faucets. Faucets on kitchen sinks have an exposed joining pipe that rests atop the sink. Duals used on lavatories have a hidden joining pipe.

Dual faucets on kitchen sinks are removed

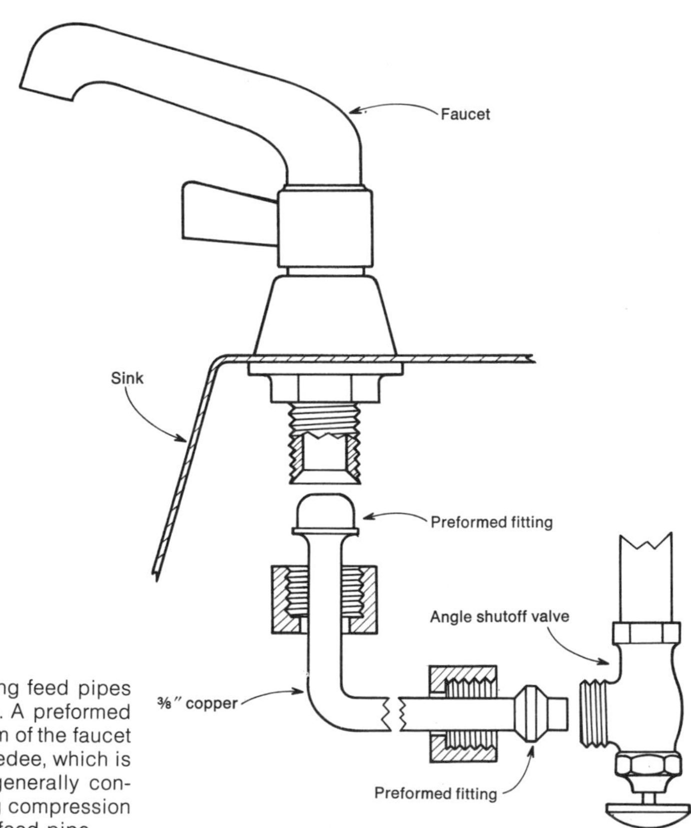

The most common method of connecting feed pipes to dual faucets on lavatories and sinks. A preformed fitting is held in place against the bottom of the faucet pipe by a nut. The lower end of the Speedee, which is a length of ³/₈-inch copper tubing, is generally connected to the stop valve by a metal-ring compression fitting. Sometimes it is soldered to the feed pipe.

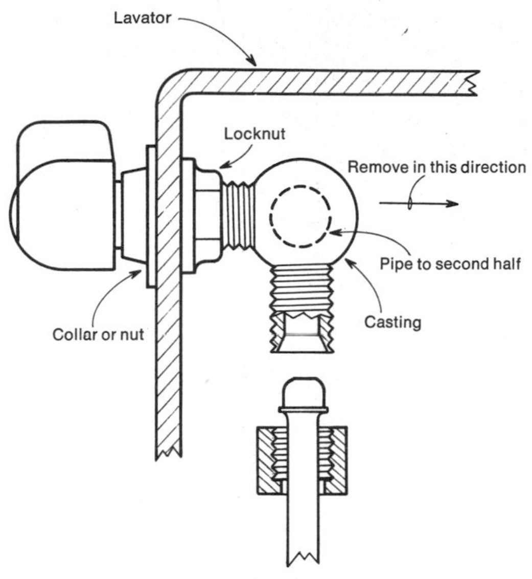

One of the methods used to mount a dual faucet on a lavatory. A collar (with internal threads) or a nut, hidden by the faucet handle, threads onto the faucet casting, which projects through the wall of the lavatory (above). The central spout screws onto the center of the casting (below).

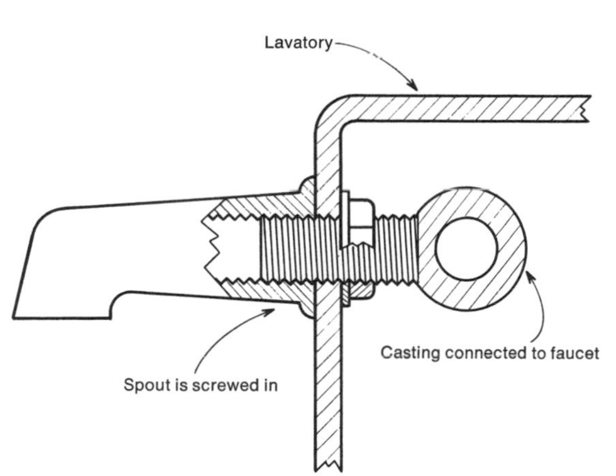

by first disconnecting both the upper and lower ends of both feed tubes. Note their original position and remove. Now you will have better access to the nuts holding the dual faucet in place.

If your wrench can't fit into the space, you will have to get a basin wrench. If the nuts cannot be turned, do not *drive them* with a chisel and hammer; you will damage the porcelain. The only alternative is to cut them with a hacksaw. Sometimes, however, they are rusted so badly they crumble under the wrench.

To replace a dual faucet you must make certain you have the correct number of holes, the correct distance apart, in the sink. Some faucets have a central projection even though they do not have a spray connection. And before you position the new faucet, load its base with plumber's putty so that it seals to the top of the sink. You can scrape the excess away later.

If you have replaced the faucet with the same model, the old feed pipes will fit. Should the replacement differ as to the distance it enters the sink, you will have to alter the feed pipes. Generally, you'll save

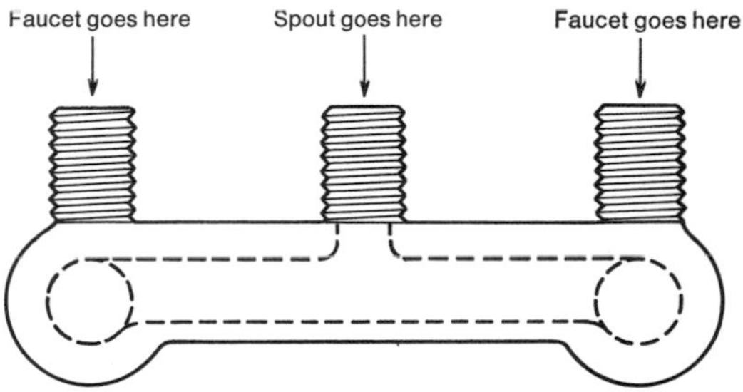

Top view of the body of a dual faucet often used on modern lavatories.

time and temper by purchasing new feed pipes rather than attempting to alter the old.

Called Speedees, the new ⅜-inch plated copper tubes with the preformed ends generally come in 24- and 36-inch lengths. The original nut that held the original feed pipe can be used at the top; but it is advisable to use a new metal compression ring and a new compression nut at the bottom.

Take care when bending the new feed pipes. Do not use the upper or lower joint as a fulcrum. Instead, bend an offset into the pipe with your hands alone, and then, by matching the pipe to the distance between terminations, cut the pipe with a hacksaw. Remember, the more bend you leave yourself in the feed pipe before you cut it, the easier it will be to work it into place.

Dual lavatory and tub faucets are usually removed from the rear. Disconnect both feed pipes leading to the faucet, at both ends, and remove them, to provide more work space. Also remove the mechanical linkage, if there is any, from the pop-up valve handle to the valve itself (it opens and closes the drain). Cover the central spout with tape and remove with a wrench. Next, remove the individual faucet handles and escutcheons, if any. This should reveal a large nut flush against the porcelain at each faucet projection. If there isn't a large nut, the large plated collar may be acting as a nut. In either case, back it off with a wrench. The dual faucet is now free to be pushed to the rear of the fixture and removed.

Replacement follows the same procedure in reverse. If you find any plumber's putty under the nuts and collars or behind the central spout, place a quantity of plumber's putty in similar locations on the new faucet pair. There are no particular precautions necessary during replacement except to remember not to overtighten the faucet against the porcelain. It cracks easily.

The preceding directions will cover most of the faucets, single and dual, that you may encounter. However, as there are hundreds of different designs, we can rest assured that some will be different. Therefore, don't remove any faucet until you are certain you can find a replacement, and don't be surprised if your faucet doesn't come off exactly as discussed.

Details of pop-up valves, their repair and adjustment, are discussed near the end of this chapter.

A faucet is a valve that is connected to the end of a run of pipe instead of in the middle. This is the major difference between them. A valve can thus accept pipe at both ends; a faucet normally can't.

GLOBE VALVES

Several different kinds of valves are used in a home plumbing system. The most common is called a globe valve because the lower portion of its body is generally globe shaped. It is a form of compression valve.

As globe valves are identical to the faucets we just discussed, they are serviced exactly the same. The only difference is that valves rarely have replaceable seats. When the seats are too far worn to be refaced, the entire valve has to be replaced.

If the valve is fastened by screw threads, the joining pipe has to be unscrewed and then the valve is unscrewed. If the valve is sweated in place, the upper half of the valve is removed by unscrewing. Drain the water in the line. Apply heat to the sweat joints. When the solder is liquid, the valve is pulled free of the pipe with the help of a pair of pliers.

The new valve is installed by screwing it home or soldering it in place. There are just two things to keep in mind. Whenever possible always connect the valve so the water enters from the bottom (there is often a directional arrow on the valve's side). Whenever possible install the valve on its side. This helps drain the valve when you want to drain the system of water for the winter.

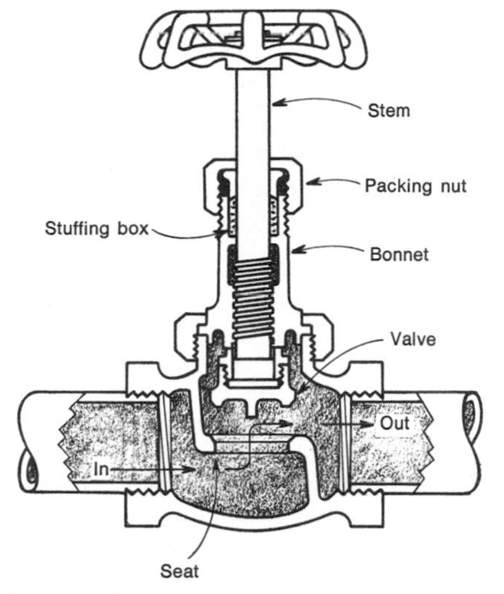

Interior view of a large globe valve. All globe valves have the same parts arranged in the same way.

GATE VALVES

A gate valve incorporates a "gate" which is lowered, by turning the valve handle, into a position between the valves inlet and outlet ports. The sides of the gate are tapered to mate tightly with two valve seats, one on either side. When the control wheel or handle is turned—and it must be turned several times—the gate is lifted clear of both ports.

Gate valves are used to shut off the flow of a liquid or a gas. They are not used to vary the flow, so they are not operated in the half-open position. The gate itself is a finely machined double-faced piece of metal. It seals by metal-to-metal contact. As there is almost no resilience, water flowing forcefully over the gate can easily scour it and the seat. The valve would then leak, necessitating complete replacement of the valve or remachining of the valve surfaces.

Gate valves are used in main pipe runs where there is no need for partial flow. Their advantage over globe valves is reduced friction. Water passing through a globe valve has to make two turns and pass through an orifice, generally smaller than the feed-pipe diameter. Passage through an open gate valve is almost free and clear.

Gate valves are found in the main line of a hot-water heating system, where the valve is closed in the summer and opened —all the way—in the winter. Gate valves are sometimes used in individual feed pipes leading to steam radiators.

A gate valve can usually be identified by its unusually large bonnet. Also, a gate valve usually must be turned a few times before you can feel anything happening inside. Globe valves respond faster. But once a gate valve is disconnected from its pipe, it is easy to peer inside and see the gate.

If there is a leak at the stem, tighten the cap nut and/or replace the packing. A gate valve's stem is identical in this respect to that of a globe valve.

If you cannot shut the valve completely, the gate and/or the gate seats are worn. There is no way to reface the seats and gate with hand tools. The entire valve must be replaced. However, it is worth your time to take the valve apart to make certain dirt isn't keeping it from closing completely.

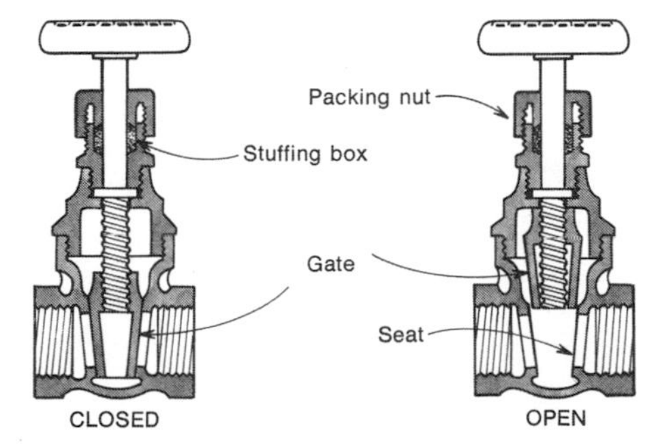

Interior view of large gate valve. Valve takes its name from the "gate" that moves up and down. You can usually recognize a gate valve by its large bonnet.

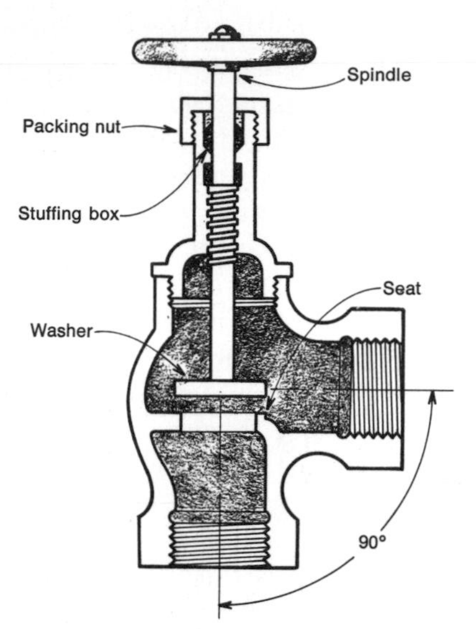

Interior of an angle valve. Actually it is a globe valve with its inlet and outlet connections at a right angle. Using this valve eliminates an elbow and nipple and simplifies piping.

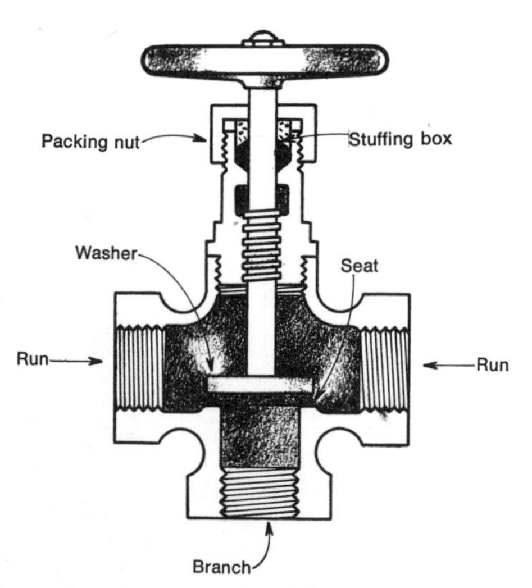

Interior view of a cross valve. This is used in place of a standard valve nipple and Tee connection. It saves labor, space and material.

ANGLE AND CROSS VALVES

An angle valve, sometimes called an angle stop cock or an angle shut-off valve, is often used to connect a pipe emerging from a wall to an overhead faucet. You see angle valves beneath lavatories, toilet tanks and some sinks. They are simply compression valves and are installed and serviced exactly like the ordinary faucets discussed some pages back. With an angle valve, it is unnecessary to use a Tee fitting and a nipple. Whenever you want to tap onto a line and control the flow of water into the tap a cross valve is used. It saves space by eliminating a Tee and a nipple.

GROUND-KEY VALVES

Sometimes called a cock, sometimes called a petcock and sometimes a plug valve, a ground-key valve has a metal plug tapered to fit into a hole in the body of the valve. The plug (key) has a hole in it. When it is turned so that the hole is in line with two ports, liquid or gas can flow from one pipe to the other. A partial turn produces a partial flow.

Ground-key valves are to be found in water-service lines, gas stoves and gas mains, at the meter. When you don't pay your gas bill, the company man comes, closes the valve, and locks it through a hole in the handle.

The same type of valve is sometimes installed in the water line of a summer home. It can be locked to prevent someone from accidentally turning on the water in the winter.

Ground-key valves normally require no maintenance. If they leak, check the spring; perhaps the plug is not held tightly enough. If that is satisfactory, take the spring off, re-

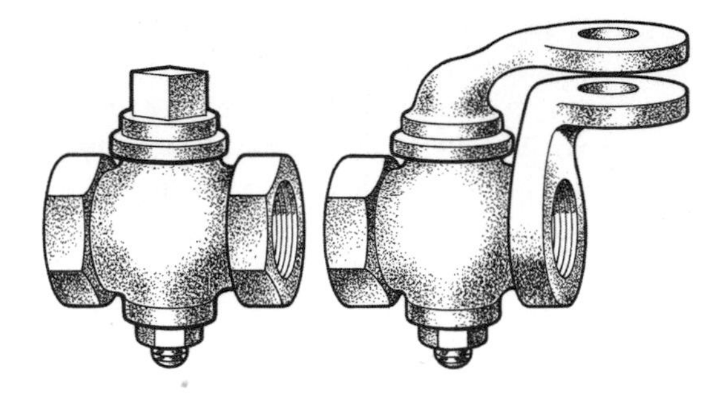

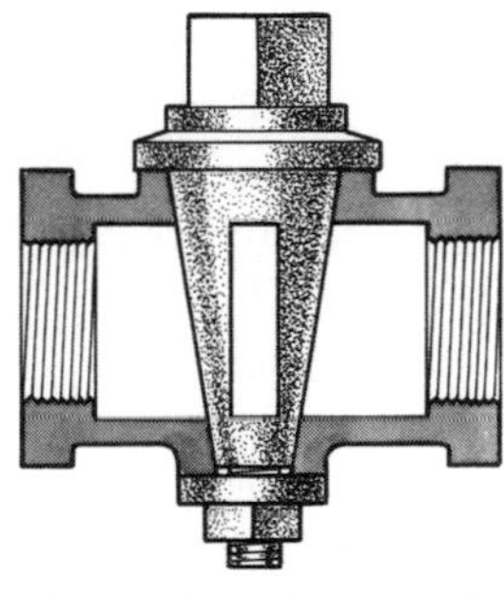

Top, two ground-key valves. Valve at left is turned with a wrench. Valve at right can be locked closed with a padlock. Lower drawing shows cross section of ground-key valve with valve in closed position.

move the plug, wipe it and its seat, the hole, very carefully with a perfectly clean cloth and try again. If that doesn't fix it, use a fine polish, a jeweler's rouge or the like. Clean afterwards and try again.

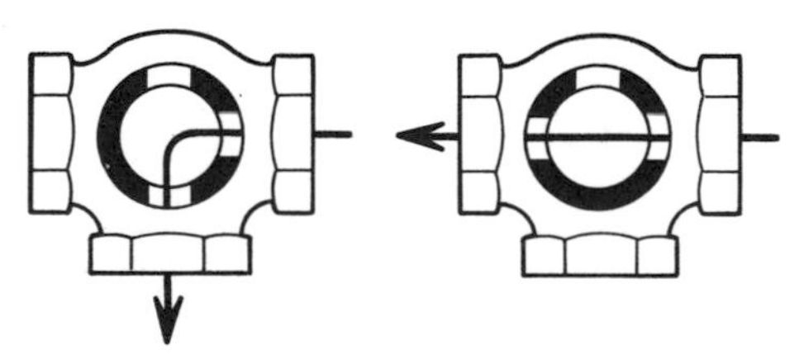

Top view of a 3-port ground-key valve. Turning ground key (center piece) stops flow or directs flow to either of the two connecting pipes.

LAVATORY POP-UP VALVES

The valve in the center of your wash basin is usually a pop-up valve. The drawing shows how it works. The valve consists of a ball joint mounted on the side of the waste pipe. A rod extends through the ball joint. The inner end of the rod may extend through an opening in an extension attached to the casting that forms the valve itself. The outer end of the rod is attached to another rod that passes vertically up and through the faucet or the fixture, terminating in a knob. When the knob is lifted the stopper is lowered. When the knob is depressed, the stopper is lifted and the waste water passes down through the drain.

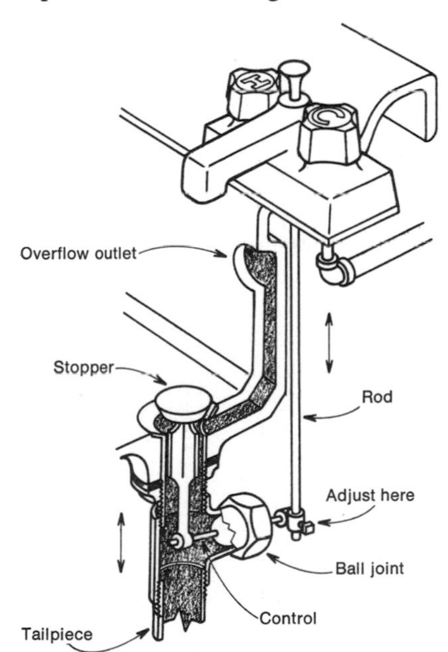

Pop-up stopper valve is controlled by a knob between faucet handles which activates a rod and ball joint. Wear at ball joint causes leaks at this point when lavatory is draining. Wear at end of control rod and bottom of stopper causes mechanical failure. Some correction can be made at the juncture of the control and rod. But if insufficient, the only recourse is to replace stopper and control. To remove this stopper, remove nut holding the ball joint and pull control rod back.

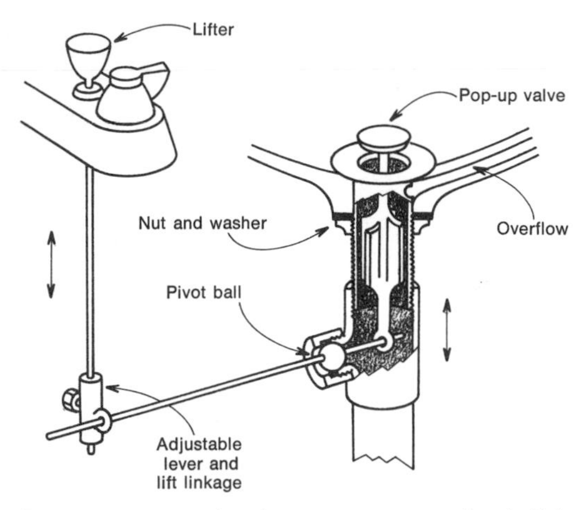

Pop-up stopper valve that rests on control rod. This type of valve can be lifted up and out.

In some designs the end of the stopper merely rests on the rod, its weight closing the drain hole when the rod support is lowered. In other designs the control rod terminates in a horizontal knob, sometimes mounted on the spout. When this knob is turned, the control rod is raised and lowered.

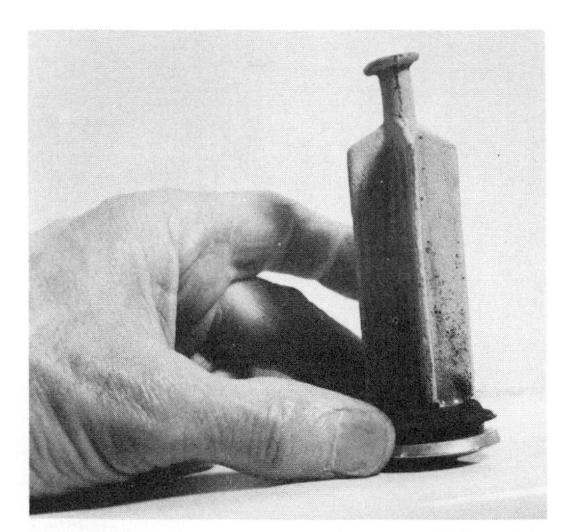

Pop-up valve showing typical tip wear. With end in this worn condition, control linkage cannot lift valve high enough for rapid sink discharge. Note ring gasket at bottom. If this is worn or ripped, valve will not seal. Try rebuilding worn tip with epoxy cement.

When the lavatory is relatively new and the valve doesn't work properly, the difficulty may generally be laid to an improper adjustment. The position of the joining device—spring, clamp(s)—holding the vertical to the horizontal rod is incorrect. Loosen same and first make certain the vertical rod is truly vertical. Then depress the vertical knob, or turn the control knob to lower the vertical rod as far as it will go. Raise the stopper about ¼ inch, lock the two rods together. Now try it again. Generally the difficulty is that the stopper is raised so high that it cannot be sufficiently closed; there just isn't enough swing in the arrangement of parts.

If the stopper is old, examine the washer on its stem to see whether this is at fault. If it is worn and the stopper still fails to hold water after the washer has been replaced, try readjusting the position of the stopper. Loosen the aforementioned connection and reduce the height of the stopper in the open position. If you still cannot get enough "swing" to pull the stopper tightly closed, the parts are too badly worn and there is too much play. You might examine the various parts and see which is so badly worn it needs to be replaced. Most likely all the joining surfaces are a little worn. My advice is to remove the stopper completely and purchase a rubber plug. Unless you are willing to spend a lot of time searching, you will not find replacement parts.

Should the ball joint leak, try tightening the holding nut a little. If that doesn't help, open the joint and replace the O-ring or packing. That should do it. Generally the ball joint never leaks no matter how worn it is until the sink is stopped; then the water level rises high enough to make trouble.

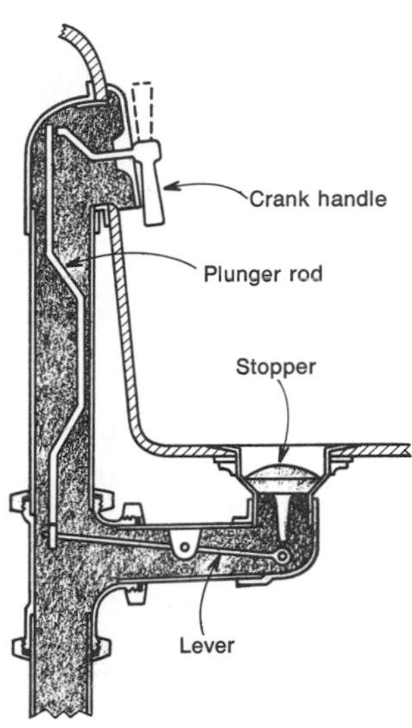

One type of bathtub pop-up valve assembly. To repair this valve, you have to get under the tub and disassemble the drainpipe.

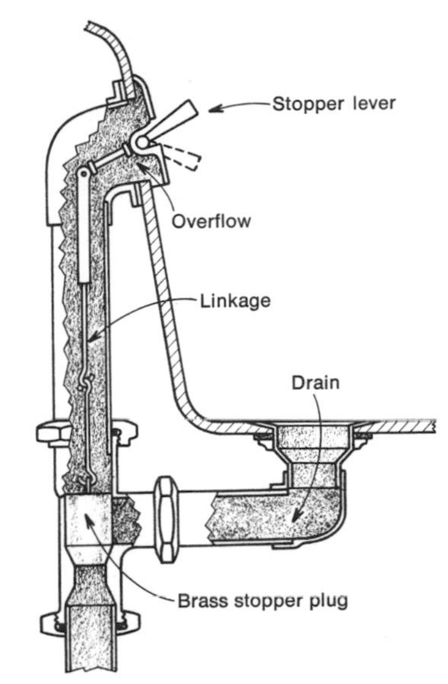

Another pop-up valve designed for bathtubs. Often the stopper lever wears at its bearings. It is made of white metal. Replace stopper lever assembly, remove and clean brass stopper plug.

POP-UP VALVES IN TUBS

Some tubs are equipped with pop-up valves that work the same as those just described. The control lever is either lifted and lowered by means of a knob, or a knob or lever is rotated, resulting in the same action. The distance between the control rod and the valve itself is no greater than that found in a lavatory, but the head of water (height) in a tub is much higher, and the quantity so much greater, that it is usual to install the pop-up valve mechanism within the overflow system. This means it is difficult to reach. Should the ball joint leak, the water will simply flow down the drain.

According to code, the back of the tub, when it has any such mechanism, must be accessible. Therefore in modern, code-satisfying homes you will find the tub backed up to a wall, the further side of which carries a trap door. Generally, you will find the trap door in the rear of the linen closet.

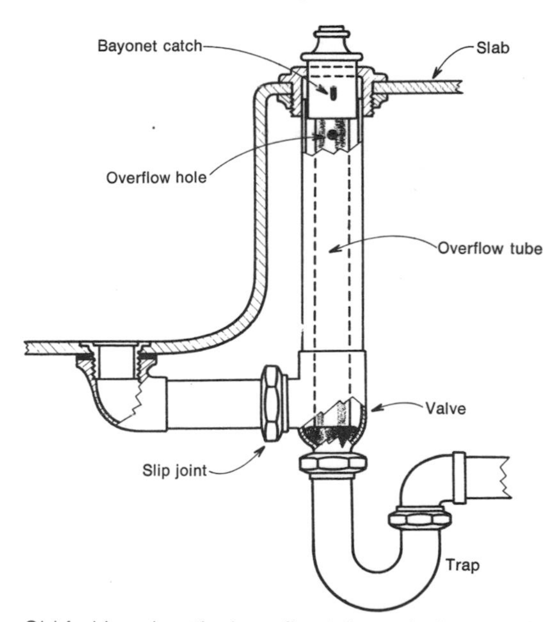

Old-fashioned, vertical overflow tube and tub-stopper assembly. Advantage of this valve is that it always works, unless something has jammed the bottom of the valve. In which case, you can remove the overflow tube and clear it. You can also use a snake at the trap through the same hole.

To get at the ball-joint mechanism, you may have to remove a metal plate, sealed in place with the aid of a gasket. When replacing, make certain the gasket is tight and intact, or you will have leaks. In fact, keep the flashlight on the joint and let the tub fill to overflowing a couple of minutes before assuming all is well.

In some tub drain-closure designs the valve is attached to a series of metal links, the end of which rests on the control lever. Should you pull one of these out, you may spend a half hour getting it back in.

In other tubs the drain is closed internally by a little piston that is lowered into the waste line. This piston is attached by linkage to a lever on the front of the tub, directly above the drain hole. There are various types, the most common has an octagonal cover and its protruding lever is lifted up and down. The cover and lever is made of white metal, which wears very fast. This is the major difficulty with this type of pop-up valve. It is a standard item, however, and easily replaced.

Disassembling a Tub Stopper-Valve Assembly

1. Pull the stopper and flexible extension out of the drain hole. If these parts are jammed with hair, stopper may not work.

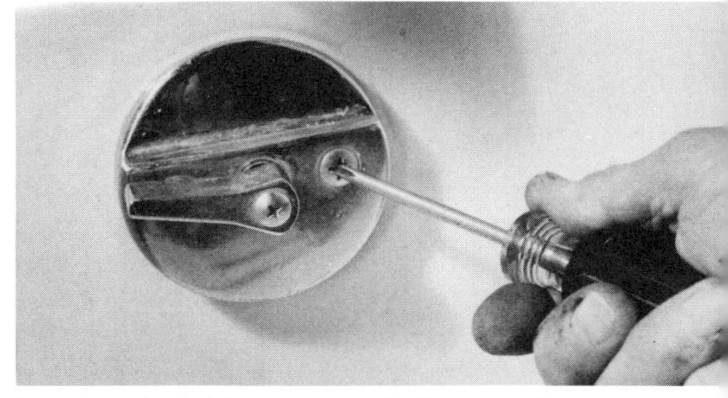

2. Loosen the holding screws and disconnect the cover plate holding the control lever.

3. Remove the rod and spring attached to the control lever. The end of the spring contacts the end of the stopper extension. To adjust the stopper, shorten or lengthen the rod by means of the adjustment screw, but if the nuts are rusted fast, expand or compress the spring.

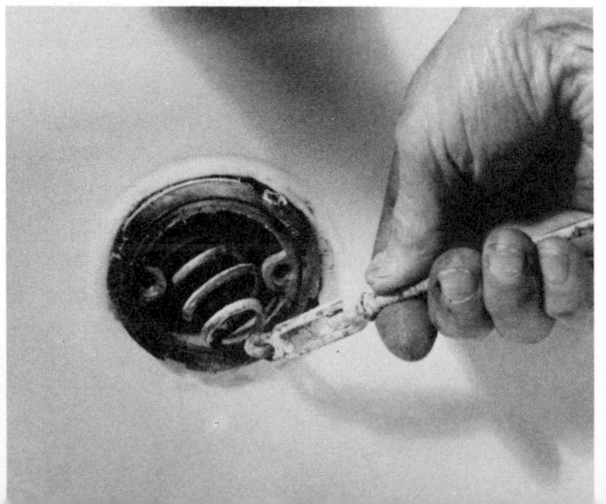

4. It takes a bit of fiddling to work the stopper and its extension back into the hole, but you can feel when the assembly is properly installed.

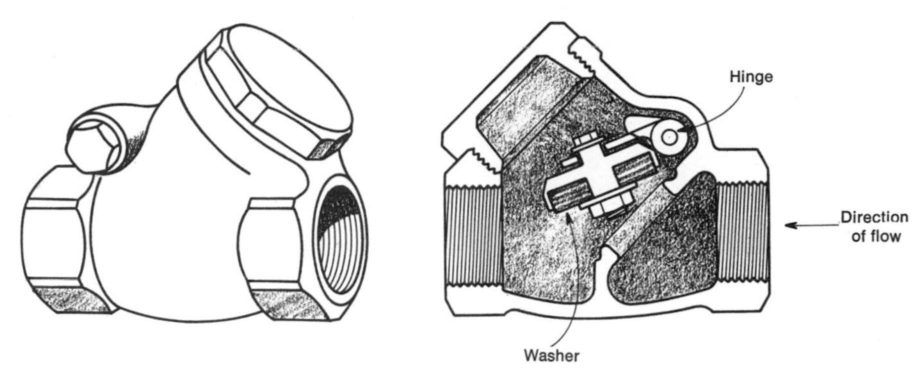

Outside and inside view of a check valve. To reach the valve itself, simply unscrew the cap.

AUTOMATIC VALVES

An automatic valve is simply a valve that works without a human assist. There are many in use in a modern home. Here are the more common types.

Float valve. This valve is usually used in a water-supply tank. When the water level drops the float drops, opening the valve, admitting water into the tank until the float rises high enough to close the valve. A common variation is used in a sump pump. When water *enters* the sump the float rises and closes a switch leading to a pump which kicks in and pumps the water out of the sump.

A most common application of the float valve is in conjunction with a toilet. In this application it is called a ballcock valve.

Ballcock valve. A ballcock valve is a float-operated valve especially designed for use in a toilet tank. When the water level drops, the valve is opened. When the level rises, the float rises, shutting off the water. Various types of ballcock valves, their design, operation, repair and replacement are discussed in Chapter 6, Toilet Repairs.

Check valve. A check valve is installed within a pipe run to limit the flow of liquid to one direction. It may be used, for example, in a waste pipe connected to a soil pipe to prevent soil from backing up into the waste pipe. Assume there is a wash tub in the basement of the house, and the soil pipe in this particular installation is well above the tub. A pump lifts the waste water from the tub overhead and into the soil pipe. A check valve in the waste line prevents soil from coming down into the tub. Situations such as this arise when a house is constructed on the down side of a hill, and the city sewer line is higher than the bottom of the cellar or basement floor.

Basically a check valve consists of a cap or plate supported by a hinge and held lightly in place by either a spring or the weight of the cap, which is of course the valve. When liquid flows in the desired direction, it pushes the valve open. When liquid attempts to flow in the undesired direction, it pushes the valve closed.

Check valves normally need no attention. When they leak, when they permit backflow, inspect for mineral deposits on the valve face and seat, for a weak or broken spring, deteriorated valve, worn valve face. Usually, it is best to replace the entire unit.

Automatic valves used in conjunction with heating and domestic hot-water systems and water-service lines are discussed in the chapters covering those subjects.

6

Toilet Repairs

The toilet is probably the most mysterious and troublesome portion of the entire home plumbing system. When it is working properly it gurgles vigorously and is then silent, but when it is out of order it may hiss all night, overflow, fail to flush and more. When you first remove the tank cover and study the levers and floats it is all very confusing. But when you take the time to analyze each portion of the mechanism you will find it is a simple, archaic contraption, but practical nonetheless.

The accompanying drawings show the basic design, the arrangement of parts and the sequence of operation of a flush-tank toilet with a plunger-type valve. This valve and this arrangement of parts will be found in most of the flush-tank toilets in use today. There will be some variations, of course, but essentially, this is what you will find.

When you turn the control handle on a toilet of this design, an arm raises a stopper to permit water to flow *out* of the tank and into the bowl. The descending water level in the tank lowers a float, which opens the plunger-type valve, called a ballcock, to permit water to flow *into* the tank. As the water is replaced, the float rises and closes the valve.

In this chapter we will describe the various malfunctions that can afflict a toilet, and explain how to deal with them short of calling a plumber.

TANK PROBLEMS

Water keeps running. If you are in doubt about this, place a piece of dry tissue on the back of the bowl and after a few minutes check to see if it's wet.

Remove the top of the water tank. If the water level is above the overflow pipe, the ballcock valve—the valve that admits water to the tank—has not closed. The valve washer and/or seat may be worn. Check this by lifting the float. If water flow stops, you can assume for the moment the valve is satisfactory. Now check the float by unscrewing it and shaking it. If you hear water inside, it has a leak and must be replaced.

How a Toilet Works

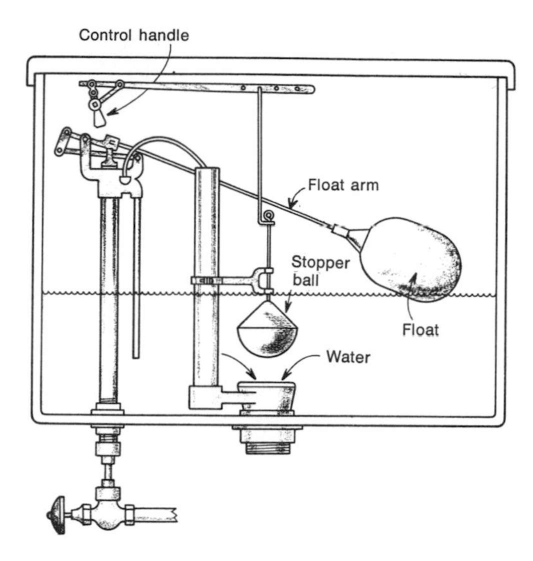

1. When the control handle is operated, the linkage raises the *stopper ball* from its seat, permitting water to pour out of the tank and into the toilet bowl. The *float* descends as the water level drops.

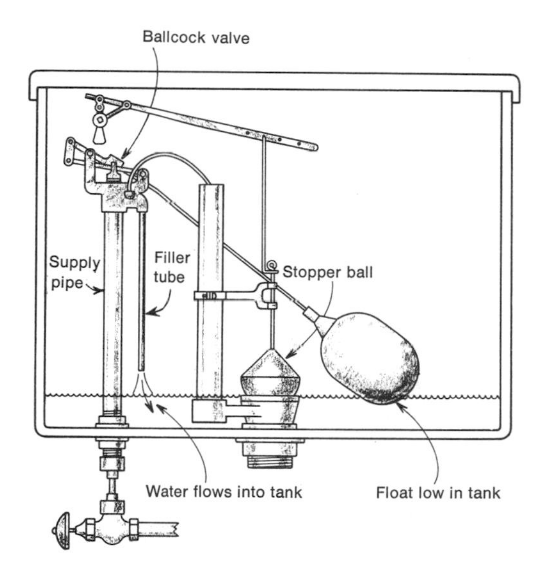

2. The descending float opens the *ballcock valve,* and water from the *supply pipe* is released through the *filler tube* and into the tank. The control handle has been released, allowing the *stopper ball* to settle into its seal and close the tank.

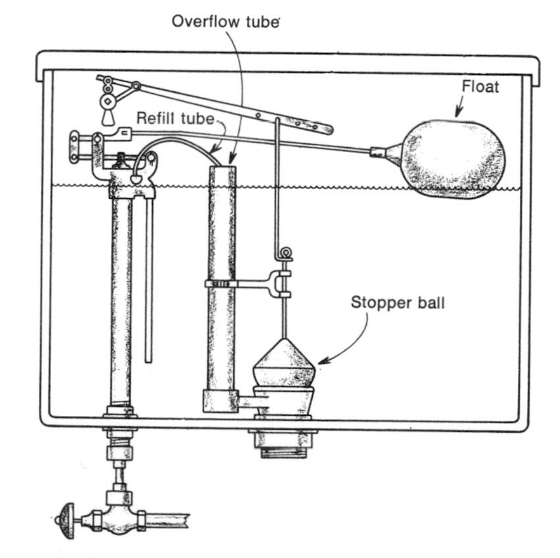

3. As the tank fills with water, the float ascends, closing the ballcock valve and shutting off the flow of water into the tank. If, for some reason, the ballcock valve does not close, the water will continue to rise until it can be drained off by the *overflow tube.* Meanwhile, the toilet bowl, which has been emptied, is partially filled again by the *refill tube,* which directs water into the overflow tube as the tank is filling.

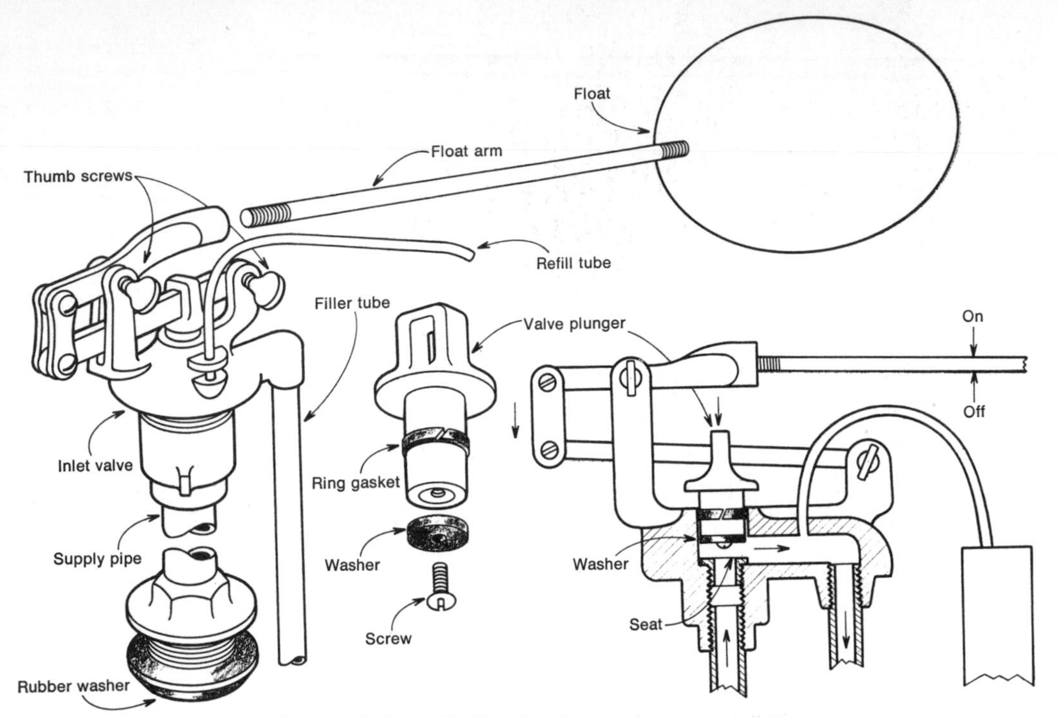

Exploded view of ballcock valve used on most toilets.

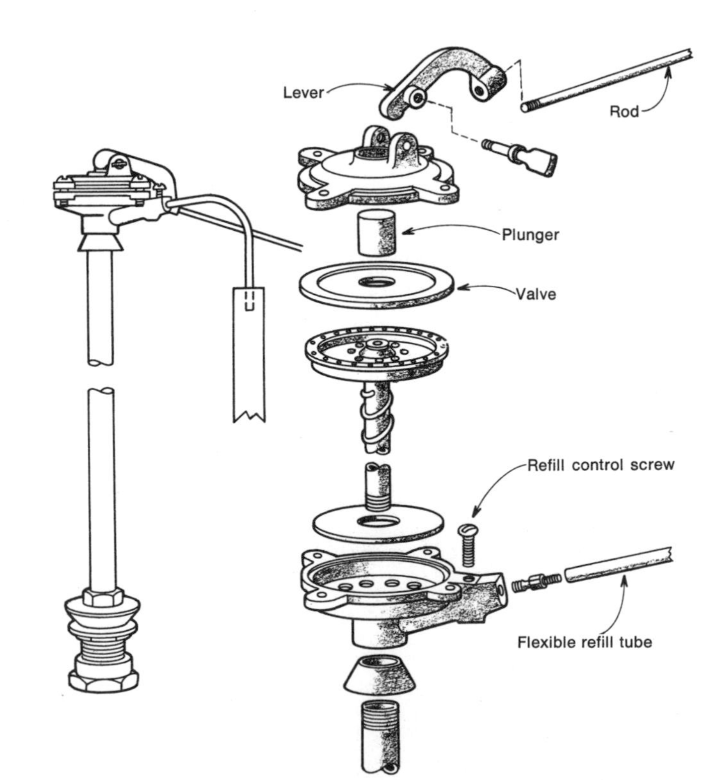

Major parts of a diaphragm-valve ballcock assembly. When rod is lifted upward by attached float (not shown), plunger pushes valve down against its seat, which is at the top of the single pipe that supports the mechanism and brings the water in. To adjust water level in toilets using this valve, bend the rod as you would with a ballcock valve.

If it hasn't a leak, its position in relation to the ballcock valve may be too high.

First check the rod that holds the ball. It may be loose and may have turned on its axis. Usually it is screwed into the ballcock valve mechanism, but in some valves it is held in place by a thumb screw. Tighten the screw. Next, lift the stopper ball by hand to let some water out of the tank. Push it down to stop the flow. Water will now enter the tank. The float will rise. If the inflow of water ceases about an inch below the top of the overflow pipe or at the level mark inside the tank, operate the handle a few times to make certain the parts do not touch one another.

If the inflowing water stops before reaching the proper level, carefully bend the float-support rod upwards. Judge the bend by the water height needed. If the water stops an inch short, bend the rod to raise the ball an equal distance.

If the water goes above the overflow pipe, bend the float rod downward.

If the level of the water rises above the float and the valve fails to stop the flow of water, the valve washer and/or seat are worn and must be replaced. Bending the float rod downward will not help, because once the float is completely submerged, it is exerting its maximum upward force (buoyancy). This is why lifting the float by hand is not a positive test for a ballcock valve. It is easy to apply so much upward pressure that the poorest valve will close tightly.

To test if the ballcock valve is closing and shutting off the flow of water, lift the float and listen. If a light upward pressure stops the water, just bend the float arm downward. If it requires considerable pressure to stop the water, the washer and/or valve seat in the ballcock valve need replacement.

To replace the ballcock washer, the flow of water to the valve must be shut off. Generally, there is a small shutoff valve to be found immediately below the toilet tank. If there is no valve there, the next nearest valve in the line must be closed.

Flush the toilet to be rid of the water in the tank. Then unscrew the thumb screws holding the lever mechanism atop the ballcock mechanism, and slide the lever out of the slot in the plunger. Remove the plunger and examine the washer on its bottom. If the washer is worn it must be replaced. The replacement must fit tightly enough to remain in place on its own. Now take a flashlight and look down into the tube whence the plunger came. If the small circle of metal you see at the bottom is rough or has tiny cuts running over its lip, the seat is no good. No washer will or can make a perfect seal with a rough, scored surface. In most instances the seat is not replaceable. The entire mechanism must be replaced. Usually, it is sold as a single assembly.

If the washer alone is at fault, replace and reassemble. The pantograph is a bit tricky; just remember that raising the rod connected to the float lowers the plunger.

Replacing a ballcock valve assembly. Start by closing the shutoff valve in the water feed pipe leading to the toilet tank. Remove the toilet-tank cover. Flush the toilet. Use a sponge to remove the last of the water remaining in the tank. Use a wrench to loosen and back off the nut holding the lower end of the water feed pipe to the shutoff valve. Use a wrench to loosen and back off the nut holding the top end of the water feed pipe to the bottom of the ballcock valve. Then loosen and back off the large nut holding the ballcock. (You may have to hold the ballcock valve with your hand to keep it from turning.) Upper nut removed, you can now lift the ballcock assembly up and out. Leave the feed pipe in place; do not let it turn. If it does, return it to its original position.

Next, transfer float and float rod to the new ballcock. Lower the ballcock into the tank, guiding the end through the rubber gasket and on to the top of the water feed pipe. Bring the top nut up against the end of the valve assembly and turn until it's finger tight. Next, turn the valve assembly to right or left until the float and float arm is positioned clear of the inner walls of the tank and clear of the stopper ball and its attached lever. Hold the ballcock steady with one hand and take up on the nut that tightens the ballcock to the bottom of the tank. Make this snug, but not too tight. Now replace the top and bottom nuts holding the water feed pipe in place.

Now open the shutoff valve. Let the tank fill and bend the float rod so that the water shuts off when it is about 1 inch below the top of the overflow pipe. That done, you can install the refill tube. This is usually furnished with the ballcock assembly. Screw it into place and then bend it so that it enters the overflow pipe.

Work the toilet a few times to make certain the ballcock mechanism doesn't hang up on the control lever or any of the other parts.

If the *water runs but the water level is not above the overflow pipe,* there is a leak between the bottom of the ball stopper and the valve seat. The ball and seat do not make a tight seal. Sometimes the cause is a partially disintegrated ball, sometimes it is a rough and corroded seat. Sometimes it is microbial growth. In the case of corrosion or growth, a little rubbing with steel wool or emery cloth will cure the trouble. If the ball is old, it must be replaced.

To replace the stopper ball, just unhook the upper lift wire and unscrew the lower wire from the top of the ball.

Handle must be held down. This is always due to a ruptured stopper ball. The ball is filled with water, or cannot hold air, and does not float on the surface of the water. Replace the ball.

Handle has to be jiggled to flush. The float arm or the float is interfering with the movement of the control or operating lever. Or the lever is so worn that it is not easily operated. Remove tank cover and inspect for interfering parts. Try handle and see if there is any side play. If so, the handle and attached arm should be replaced.

Handle has to be jiggled to stop water. This is caused by the stopper ball becoming stuck in the open position—above its seat. It may be that the stopper guide is corroded and prevents the ball from falling of its own accord. The lower lift wire may be corroded or bent. The guide may have moved so that the stopper ball is not centered over the top of the pipe leading to the bowl. It may be that someone bent the arm attached to the control lever, thus moving the stopper ball off center. Or, the upper lift wire is in the wrong hole. Or, the float and its supporting rod strike the stopper ball when they come down.

Wet tank exterior. This is usually caused by a worn ballcock valve plunger and/or its gasket. Water squirts upwards and out when the float falls and water is admitted. This will be immediately obvious when the tank cover is removed and the bowl is flushed. Replace gasket alone or entire assembly as necessary.

It should be noted that when the washer in the ballcock valve is replaced it will be necessary to rebend the rod holding the float. Also, always expect some "creep" in control mechanisms of this design. The final shutdown of inflowing water can take a minute or two even with all parts new and operating properly, and there will be a slow increase of water height during this time. This is normal.

Diaphragm valve. This valve is mounted on the top of its feed pipe and is closed by pressure produced by a floating ball attached to an arm. In this design the arm's pressure is reflected downward to a short rod called a plunger by the manufacturer, which presses upon a valve. The valve seats atop the silencer pipe. When the float is down, water comes up through the supply pipe, exits at the end of the pipe and flows downward through the holes over the diaphragm and into the tank.

Water level should not be permitted to go above the collar marked "Critical Level Band." If it does, there is a possibility of a cross connection, which means that, under certain circumstances, it is possible for water in the toilet tank to get into the drinking water. To vary the height of the water level within the tank, bend the float rod, or turn the adjustment screw, if there is one.

Assuming that the float is operating properly, failure to completely shut off water flow may be due to corrosion and foreign material on the underside of the valve or valve seat.

Then again, if the valve (diaphragm) itself is leaking, it will be impossible for it to do its job. If the plunger is sticking, there will be trouble. On some models the plunger is much thinner than the one shown. Occasionally, when the plunger has collected sufficient corrosion, it sticks in its guide and the entire assembly sings. This can sometimes be cured, without disassembling the mechanism, by using a pair of long-nosed pliers to work the rod around and up and down and by applying some powdered graphite (it comes in a little squeeze can) on the rod.

When installing a valve of this type, use a crescent wrench on the inner, bottom nut to hold the assembly firm while tightening the outer nut. Try to get no more than ½-inch of the refill tube inside the overflow tube. Again, this is done to eliminate a possible cross connection. The refill tube always discharges water while the ballcock valve is open. The tube's purpose is to fill the bowl with water after it has been flushed to make certain that the trap has not been emptied by the impetus of the moving glob of water or by a siphon action set up by two or more bowls being drained at once.

Replacement parts and kits for diaphragm valves are available from the manufacturer and at plumbing specialty shops. As there are many manufacturers in the field, many designs and sizes, it is best to bring along the old parts.

Toilet tank sweats. This is always caused by a temperature differential between the exterior of the tank and the air in the bathroom. The easy way to cure this is to lower the temperature of your home and let some dry air in. Usually, a sweating toilet tank indicates too much moisture inside the house. This will be accompanied by paint blistering from the exterior walls.

Placing a pretty cloth bag around the toilet tank will do very little to stop the sweating. The cloth provides some insulation, but with time the entire assembly drips.

One solution is to line the inside of the tank with a special plastic bag. The plastic acts as an insulator. The china doesn't get cold and water doesn't condense on the exterior.

The surefire cure is to use warm water in the toilet tank. A special "tempering" valve is sold for the purpose. It is connected to a hot-water line and mixes some hot water with the cold water entering the toilet bowl. The trouble is, that unless the valve and its hot-water line were installed when the house was first erected, it is necessary either to open a wall or run the pipe and valve on the surface.

TOILET STOPPAGE

As all the wastes normal to the toilet are water soluble, stoppage—the failure of the water to flush the soil down the drain—is usually due to an obstruction and not a defect in the system itself. The only exception to this statement is low water in the toilet tank. When this occurs, one generally finds it is necessary to flush the toilet two or more times before the soil goes down. This can be remedied by adjusting the float rod, as previously explained.

Toilet paper will dissolve after a time in water. Paper napkins and the like, though soft, are formulated not to dissolve and will not disintegrate when wet.

A word of caution. One's natural instinct is to try a device when it is reported inoperative. *Do not do this when someone says a toilet is clogged.* The result may be a bathroom full of water and muck. The safe way to check the toilet bowl is to remove the top of the toilet tank, put it safely to one side, then hand-operate the stopper ball and let a little water out of the tank. If it is not going down, push the stopper ball down and stop the flow of water into the bowl.

At this point you can safely assume that there is either a glob of paper stuffing the pipe or something plastic or metallic, like a child's toy, backed up by a wad of paper and excreta. A chemical attack is usually worthless. It may dissolve the paper, but will not dissolve metal or plastic.

If there is any water movement at all, continue to let a measured quantity of water into the bowl until the bowl is filled with reasonably clear water. This makes it less unpleasant to operate a force pump. In some instances, repeated, careful filling of the bowl and letting the water run down will break up a minor obstruction. Generally, it doesn't help.

Using a force pump. The heavy-duty force pump with a tubular extension is best. Let sufficient water into the bowl to cover the top of the plunger. Before you insert the

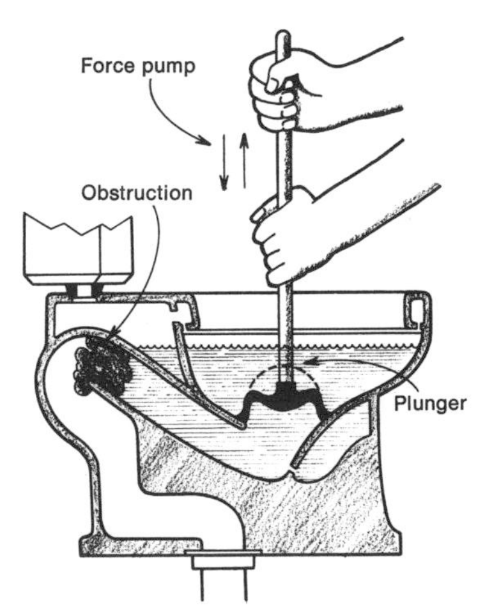

Using a force pump to remove an obstruction from a toilet bowl. Make certain there is sufficient water to cover lip of force pump, and pump steadily.

plunger, tip it up and pour clear water into its bottom. Then invert quickly into the bowl. Slide the plunger about a bit to fit the lip into the aperture. Then simply pump away, slowly and steadily. If the water level drops, you are winning the battle. If not, continue anyway; an obstruction sometimes gives way all at once. You can stop and rest if you wish. You will not lose ground. But do not give up until you have had at least a ten-minute go at it. The alternative to the force pump is the snake, and that is a bit mucky.

Should you pump the obstruction down, or should the water disappear, do not jump to the conclusion all is well and flush the toilet. All may not yet be well and you may still have a minor flood. Test your progress by admitting a small amount of water at a time.

If you find that pumping clears the trouble but that the drain clogs again after a day or so, the trouble may be a small obstruction across the pipe. In itself it does little to hinder the passage of soil, but with time, paper collects and forms a wad which effectively stops the flow of water. The force pump drives the paper free but the obstruction remains.

There are two ways a nonsoluble obstruction firmly caught in the drain pipe can be removed. One is by means of a snake; the other requires removing the toilet bowl to expose the end of the drainpipe. When this is done a snake may or may not be necessary. Sometimes the obstruction is visible and can be reached by hand alone.

Using a closet auger. All toilet bowls have traps within their porcelain base. The flow of water and soil out of a toilet bowl is up and then down. In some designs the flow is up the front of the bowl. In others the flow of soil is up and then down the rear of the bowl. Generally you can determine this direction by examination. In order to clear an obstruction in the toilet bowl's trap you must guide the snake up and over by hand. The trouble with this is that you have to put your hand in the water, and when you work the snake to and fro there is a possibility of chipping the lip of the trap. Once chipped, the spot will become dark and remain so.

The alternative approach is to use a closet auger, which is a short length of snake fastened to a handle and positioned within a bent length of tube. The tube enables you to direct the snake where you wish without placing your hands in the water and to get it past the lip without problems. The auger tube holding the snake is directed upwards into the trap. The auger handle, which is attached to the short snake, is turned while the snake end is urged forward. One can attempt to hook the obstruction on the end

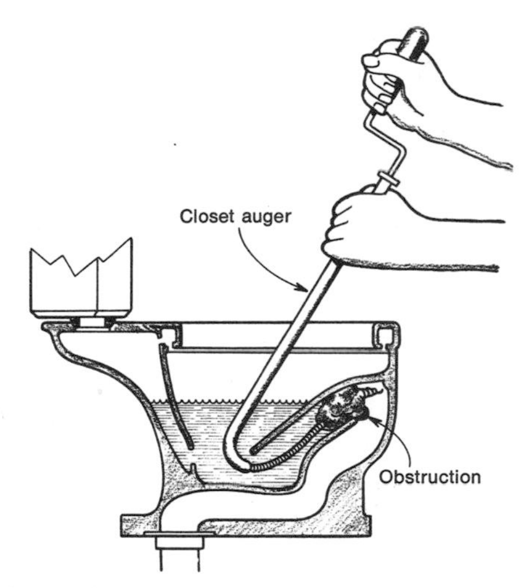

Using a closet auger. The bend on the end of the guide pipe prevents soiling your hands and helps guide the snake up and around the bend in the toilet drain. Handle on auger makes it easy to turn the snake as you guide it. Note that drain is at opposite side in this bowl.

of the snake and pull it backwards and out, and one can hope to push the obstruction clear and into the straight section of the drain, where, again, hopefully, it will continue on and out. As the trap portion of the toilet system is the most constricted and difficult to pass, this is where most foreign materials get stuck. Therefore, once clear of the trap, chances are the obstruction will pass out.

Using a regular snake. Should jiggling and cranking in both directions fail to hook or reach the obstruction, the closet auger must be replaced by a longer snake or the toilet bowl removed. When the toilet bowl and associate pipes are old, and old in a home means over twenty-five years of age, and the toilet bowl is already chipped, the better decision is to use the long snake. The reason is that removing the toilet bowl includes removing the toilet tank and disconnecting the cold-water pipeline. Often an old pipeline that was satisfactory until it was disassembled leaks upon reassembly. Also, the porcelain caps that cover the nuts holding the toilet bowl in place usually crack when they are removed.

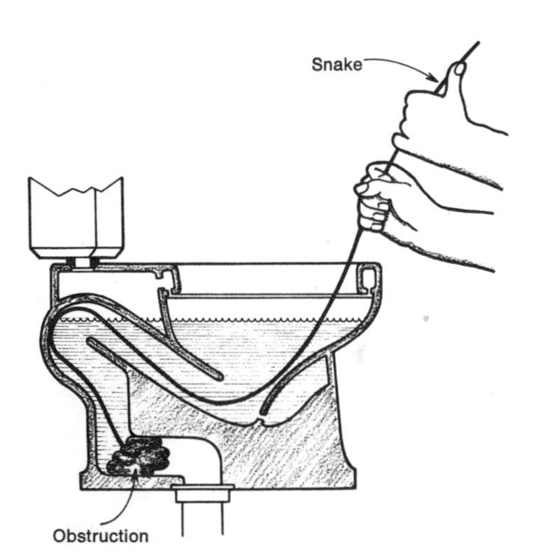

Using an ordinary snake to force the obstruction into the drainpipe. Getting snake past first turn must be done by hand. Note drain position in this bowl.

The long snake has to be hand guided up and over the lip of the trap, and once the snake has passed the trap it is necessary to continue to hand feed it into the bowl, guiding it past the lip. If this is not done, there is a chance the lip may be badly broken. Therefore, when you have to use a long snake in a toilet bowl it is always helpful to have an assistant on the coiled snake feeding and turning it while you guide it.

LEAKING TOILET BOWL

When water appears beneath the toilet bowl, and generally it will run down to the area below before it spreads around the bowl, and the drain is not plugged, the trouble is at the joint between the porcelain bowl and the top of the drainpipe. This is not a tight joint. It is not screwed fast or bolted. Instead the lower end of the bowl (horn) merely fits into a wax or putty sleeve that seals it to the drain. If you have easy access to the bottom of the bowl, there is a remote possibility the leak can be plugged without removing the bowl. But this is unlikely.

When all goes well you can remove a toilet bowl and replace it in about one hour. When you have difficulties, and some will be enumerated shortly, it can run into half a day or more. You will need a new wax ring when you replace the bowl so make certain you have one before you start. These are the steps.

Close the shutoff valve in the water feed leading to the toilet bowl. Flush the toilet. Use a plumbers friend to drive as much water down into the drain as you can. Remove the tank cover. Put it aside. Use a

Removing a Toilet

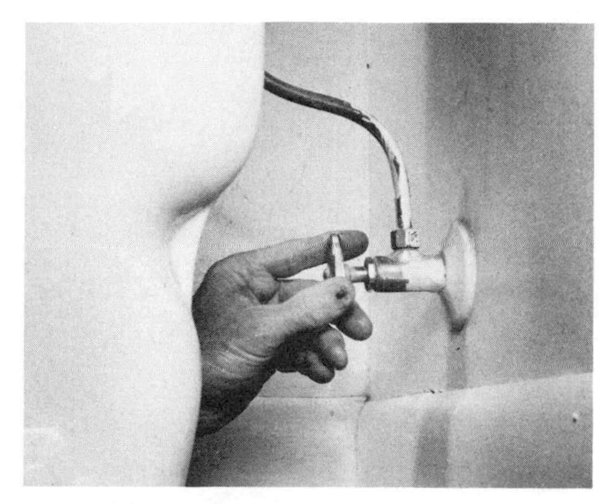

1. Close the valve in the pipeline leading to the tank.

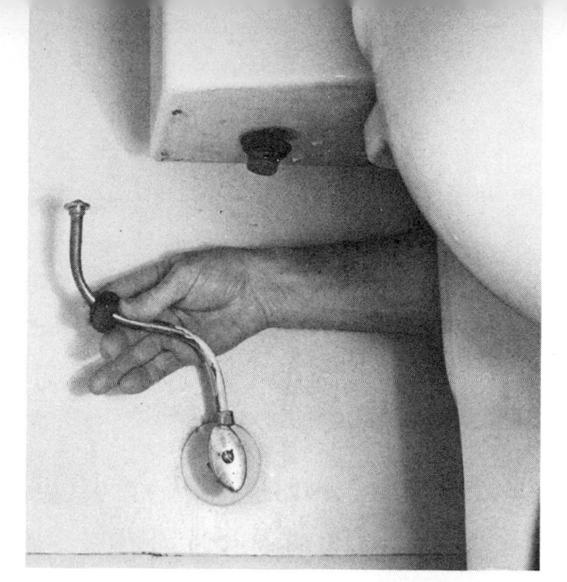

2. Flush the toilet and pump as much water down with a force pump as possible. Use a sponge to remove the water remaining in the flush tank. Then, as shown, disconnect the water feed line from the bottom of the ballcock-valve assembly.

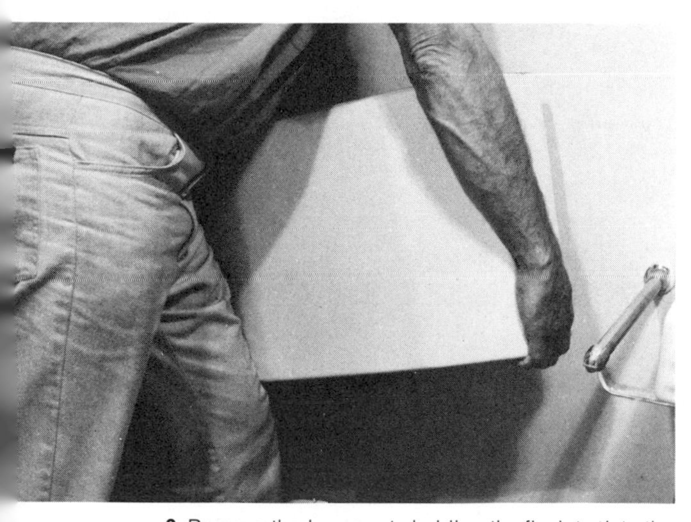

3. Remove the brass nuts holding the flush tank to the top of the toilet bowl. Lift tank and remove.

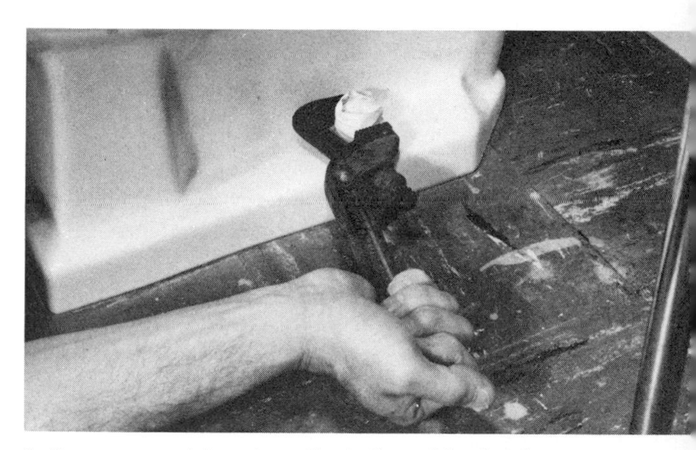

4. Remove porcelain nuts on the bottom of the bowl. Protect them with tape or a rag.

5. Next remove the brass nuts that are underneath.

6. Now lift the toilet bowl and remove it. *(continued)*

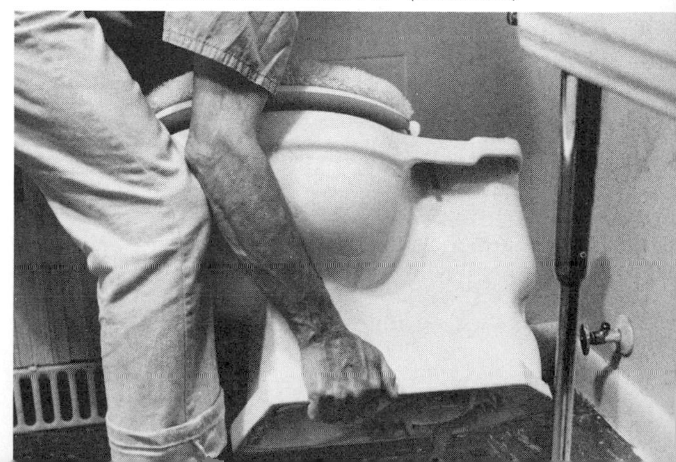

7. With the bowl out of the way you have access to the top of the soil pipe. Remove whatever obstruction may be present.

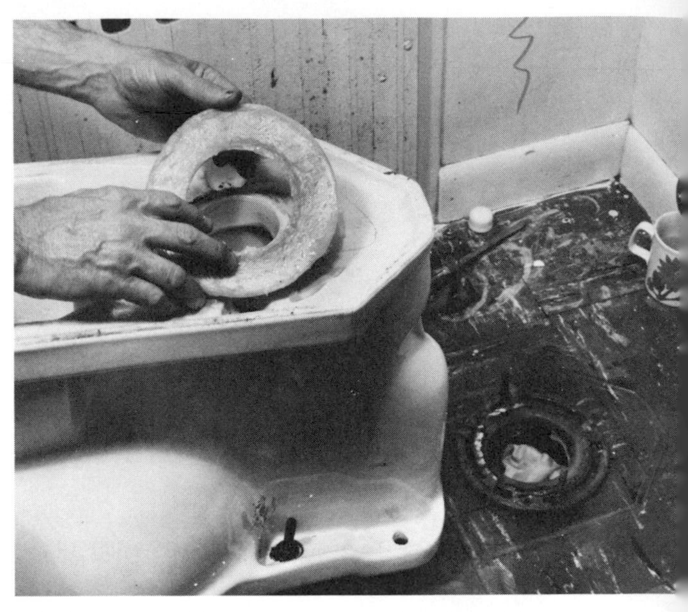

8. Before reassembling, remove the remains of the old wax ring around the horn on the bottom of the toilet bowl. Clean the horn and install a new wax ring.

sponge to remove all the water remaining in the flush tank. Loosen the nut on the feed pipe at the shutoff valve and back it completely off. Loosen and back off the nut holding the top end of the feed pipe to the bottom of the ballcock valve. Remove feed pipe.

Next use any wrench that will fit and loosen the two bolts holding the tank to the top of the toilet bowl. You may have to use a screwdriver on the bolts to keep them from turning. Now lift the tank up and clear of the bowl and put it to one side. Note the large rubber washer sitting on the top of the bowl. It has to go back the same way.

Now wrap some tape around the porcelain caps you find at the base of the toilet bowl. Use a stillson wrench to remove them and a crescent wrench to remove the brass nuts underneath. With these two nuts off, the toilet bowl may be lifted free and clear. When you do so, expect the water remaining in the trap to spill out on the floor. The bowl is not very heavy, about 35 pounds, but it is awkward and easily broken if dropped. It is porcelain.

The drainpipe under the bowl is now exposed. Often you can see the obstruction and remove it without further trouble. If not, see if the obstruction is still within the toilet bowl trap. If not there, run a snake down the drain until you clear it.

To replace the bowl you need to scrape off the old wax around the horn and replace it with a new wax ring. Incidentally, if you have been troubled with water appearing under the bowl, now is the time to check and see whether or not the drainpipe is actually soldered or screwed fast to the floor flange. The flange, sometimes called a toilet or floor flange, is that ring of metal that has the two bolts. If the flange is not joined properly, water will leak out and down when the toilet is flushed.

Wax ring in place, lift the toilet bowl high enough to clear the bolts and lower it carefully so that the horn on the bowl enters the hole in the floor flange and the two bolts come up through the holes in the bottom of the bowl. Seat the toilet with a little twist and line it up exactly as it was before. Look at the dirt marks on the floor. Now you can replace the brass nuts on the two flange bolts.

Next replace the tank and tighten the two nuts. Sometimes these bolts have rusted through and have to be replaced. Make certain the tank is centered on its large rubber spud washer. Don't overtighten the two holding bolts. You can crack the tank.

Finally, replace the feed pipe, taking care not to bend it nor turn it around. If you do it may leak at the lower joint.

Some tips on what not to do when removing and replacing a toilet bowl:

Don't bend the feed pipe if you can help it. It may leak at the joints afterwards.

Don't turn the toilet from its original position when replacing it; the feed pipe won't fit.

Don't overtighten the bolts holding the tank to the bowl, you can crack the bowl.

Don't let anything fall into the open drainpipe while it is uncovered.

7

Unclogging Drains

Drains become plugged for any number of reasons. Obviously, anything that will not dissolve and is too large to be washed down and out the pipe is going to plug up the line. Not so obvious is that the habit of cleaning one's hairbrush in the sink, or dumping cooking grease and small particles of food in the sink, will also clog the pipes in time. Hair gets caught along the way. It will not dissolve. Cooking grease solidifies when it becomes cold. Food particles often contain stems and shells that are not readily dissolved nor broken down by bacteria. Together, the hair, grease and food particles can plug a pipeline quite thoroughly.

Another point to bear in mind is that the sooner you get to work on a drain that is running sluggishly the easier the job will be. Don't wait until it is plugged solid, because then you'll have a job on your hands.

There are different ways to open a plugged line. Your choice will depend on your particular line, the substance that is plugging it and accessibility. A force pump will sometimes do the job when the obstruction isn't obstinate. Chemicals are satisfactory when there is some movement in the water. A snake is useful when you can readily put it to work. Sometimes the fastest method is to open the pipe at some convenient point. And sometimes it is necessary to use all three methods. For example, you may have to open a pipe to get a snake in, and then because the line is filled with grease, you may need to use a chemical to remove the grease.

Kitchen sinks and some washbasins can sometimes be opened with the aid of a force pump. Fill the sink with about 5 inches of water. Apply the face of the pump over the drain hole and work it up and down slowly and steadily.

Lavatories and washbasins that have overflow openings cannot be effectively pumped clear unless the openings are closed. This is difficult to do because the pump develops considerable pressure. However, try stuffing a wet rag into the opening. Or better yet, have a strong friend hold the rag firmly over the hole. It is worth a try.

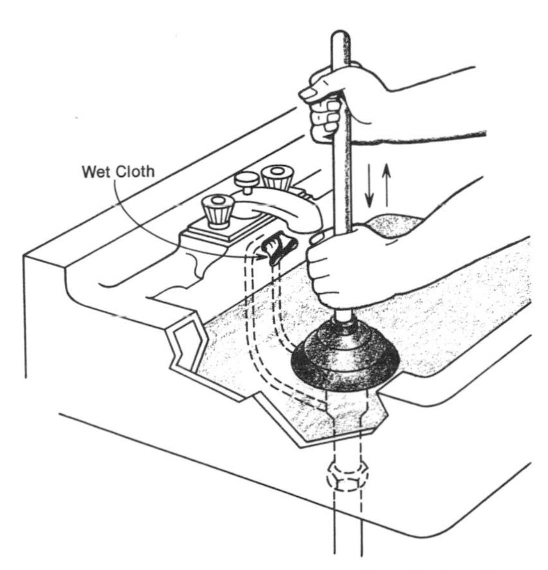

Opening a clogged lavatory with the aid of a force pump. Wet cloth is stuffed into overflow hole to keep the pump's pressure against the stoppage.

CHEMICALS

While chemicals seldom have much effect on a stopped-up toilet, they can be very effective in the small pipes that drain lavatories, sinks and tubs. Often you can avoid disassembling drainpipes and their traps by using chemicals.

However, the decision to do the job chemically should not be taken lightly. If it works, fine. If it doesn't, you have the problem of trying to work with a snake in a drain filled with strong chemicals, or opening a trap filled with the same.

If your problem is a sluggish drain you can safely use any of the chemicals made for the purpose. If the drain is moving very slowly you are taking a small chance that the drain may plug up entirely and no more waste water may flow out. In which case, you have a trap and drain filled with strong chemicals. If there is no drain movement at all, there is a good chance none of the chemicals will be effective.

There are two kinds of chemicals used for clearing clogged drains—acid and alkaline. Acid, generally concentrated sulphuric acid, is more effective, more costly and more dangerous. You must use great care with acid to prevent any of it from touching your skin. Acid is sold in a plastic bottle along with a long, screw-on spout designed to eliminate drips. Acid is usually not available in the corner hardware shop, but only in plumbing supply houses. At present, acid is sold under a variety of trade names including Clobber and Bust-Loose, at about $3.00 a quart.

Acid will dissolve grease and hair. Alkali will not dissolve hair, but is effective on grease. Remember, you cannot mix acid

The clog fighters. Clobber, containing concentrated, uninhibited sulphuric acid, is most powerful and dangerous. Always wear rubber gloves when you use it. 99 Brand is a strong caustic that many plumbers use. Drano is the least effective, but it will work when there is only a little soft grease in the line.

with alkali without getting a violent eruption. If there is some movement in the sink, you can use alkali, and if that doesn't work, you can continue to run water through (even though slowly) until the alkali is gone. Then you can try acid. If you pour acid on top of the alkali, the liquid will boil and erupt.

Several alkalis are currently used. Lye (caustic soda) is the least expensive drain opener you can purchase, about 30 cents a can. However, in many instances you will find it as much of a hindrance as a help. Like all alkalis, lye works by converting grease to soap. (People still make soap this way; directions are on the lye can.) Unfortunately, lye forms a hard soap, which can plug a drain as effectively as the original grease. Lye should never be used when there is no flow out of the sink, or when the flow is sluggish; it will plug the drain. You can use lye effectively when the drain is merely slow, *but definitely flowing.*

Drano is another alkaline drain opener containing sodium hydroxide (which makes a soft soap), sodium nitrate and a small percentage of aluminum flakes. Drano is much more effective than lye, partially because it forms soft soap and partially because the aluminum reacts with the caustic to form a gas that churns up the muck.

Another brand, called "99," is all sodium hydroxide in flake form. Personally, I have found it best among the alkaline drain openers.

Working with acid. Always wear long rubber gloves when working with acid. Always attach the spout, taking care to make it properly tight. Always clear the area so that you do not trip or foul yourself up when pouring. Always place a porcelain bowl under the trap before you add the acid. The acid sometimes eats its way sideways through the plastic washer under the trap nut. Drips and splashes on the floor will immediately bleach the floor and eat holes in cotton and some plastic rugs. Sulphuric acid generally does not attack wool as vigorously, but don't depend on it; the acid mixture may not be sulphuric. Drops on your skin will make you scream; it takes a few seconds before you feel the burning. *Don't panic.* You are not going to lose an arm. Put the acid down carefully and flush the spot with water, and keep flushing.

Remove as much water as possible from the drain and trap, because the less water there is the less the acid is diluted. When the trap is empty, you will need about a half quart of the acid to make certain you fill the trap and beyond. Give it at least thirty minutes to do its job. Don't touch the trap; it will be hot.

When you have finished pouring, stand back; there will be vapors and smells. Open all the windows and doors. Put the plastic bottle of acid away, but don't forget the drops that remain in the spout and on the screw top. Rinse the spout in another sink, while still wearing rubber gloves. Replace the cap on the bottle, then carefully rinse the top of the bottle to remove any acid that remains. Acid has an insidious way of entering a bit of cloth unseen and lingering. Weeks later holes appear, seemingly out of nowhere, especially in cotton.

Working with lye. Use a nonaluminum, nongalvanized pot for mixing lye. Never permit water to enter the can; you will have an explosion. Pour the granules in slowly, mixing vigorously with a wood or metal spoon. Do not let the granules accumulate; they will form a hard lump. You do not need gloves, but stay within reach of a sink.

Remove as much standing water as possible from the waste pipe and the drain. If the drain plug is within reach, remove it, drain the water, and replace the plug.

Use about half a can of lye at one time with as little water as possible. In this there are two limitations: you need enough water to fill the trap and beyond, and enough to dissolve the chemical. Generally a quart or so is enough.

Directions on the lye can call for cold water. I find that hot water works much better, though there are a lot more fumes. Lye granules and 99 flakes can also be mixed in the bowl or sink itself. Plug the drain hole and add the lye to the water, mixing as you pour.

However, never pour dry chemicals directly into the drain. They will coagulate and form a hard lump that will seal the drain. Always sprinkle them gradually into the water, mixing as you go.

The manufacturer of 99 suggests using very hot water, and tapping the side of the drain while the chemical is in place.

Drano is used a bit differently. Remove the water in the basin or sink, but don't empty the trap. Dump one heaping tablespoon down the drain into the water below. If that doesn't do it, use a second tablespoon. If two shots doesn't open the drain, give up. (Those are the instructions.) The manufacturer suggests cold water. Use hot, if there is room for a little water in the drain.

No matter what chemical you use, be certain to give it enough time. If you lose patience and turn on the water, you will wash the chemical down if the line is open. If the line is still plugged, you will dilute the chemical and reduce its effectiveness.

OPENING TRAPS

If chemicals do not improve drain flow, try using a snake. It is usually difficult to work a snake down a bathtub drain; the safer option is to try to open the trap itself. Every well-designed house provides access to the bottom of the tub and the trap. Look for a trapdoor in an abutting closet or in the ceiling of a closet on the floor below.

If you can gain access to the trap, simply back off the two large nuts at both ends of the trap and remove it. In some designs you can just loosen one nut, remove the other and swing the trap to one side. Possibly the trap will be clogged and cleaning it out will cure your troubles. If the trap is not clogged, run a snake up to the tub and then down toward the house drain. When you replace the trap, check the rubber compression rings. If they are old, now is the time to replace them.

If you can't get to the trap you have the difficult decision of chopping a hole in the wall or trying acid to clear the stoppage. If the acid works, fine. If not, fill the tub with water, add some caustic and wait until the entire mess seeps out into the drain before you cut a hole in the wall and disassemble the trap.

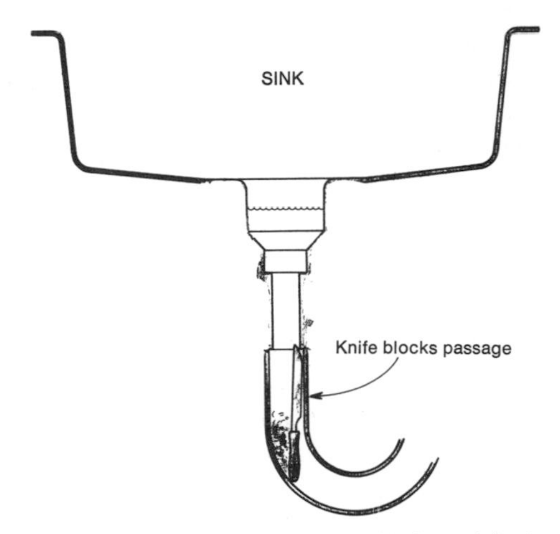

One of the major causes of partial kitchen-sink stoppages is a knife that lodges in the drain trap. Two knives can plug the drain completely.

You may be surprised the first time you encounter a knife at the bottom of the trap or connecting pipe beneath the kitchen sink. However, if you examine the drain opening you will notice that when the little cup is not in place, there is nothing but a metal cross protecting the opening. A table knife or its like can easily pass through. One can remain in place unnoticed for years. Usually, it is only when a second piece of flatware finds its way down into the drain that the problem becomes acute.

Therefore, if a thorough chemical cleaning does not open the kitchen drain completely, check the trap and associated pipes for metal. An open drain will, under normal conditions, easily carry off the full flow of both taps.

Traps beneath washbasins are usually fitted with a large, removable plug at the bottom of the loop. Some kitchen sinks are also fitted with plugs. Some traps in both types of installations have no plugs. If there is a plug, place a large basin underneath to catch the slop, and unscrew the plug. If anything remains in the trap you can remove it with a length of stiff wire, hooked at the end.

When there is no plug, remove the entire trap. In most instances the large sections of pipe are held together by slip joints and unions that are actually a form of compression joint. The seal between two lengths of pipe is effected by a large, ringlike washer, which is compressed when the nut is tightened.

While the washer retains its elasticity, which can be for many years, and while the brass pipe remains fairly firm, the joints can be made tight and firm by hand pressure. With the passage of time, the pipe and the nuts corrode; the washers decay. If you are going to work on an undersink drain system, do not start on the weekend. The pipe may crumble in your hands, and you won't be able to obtain a replacement.

Steps in opening a trap. Place a pot underneath the trap and use a wrench to remove the plug at the bottom. Use a wire hook to remove debris inside the trap.

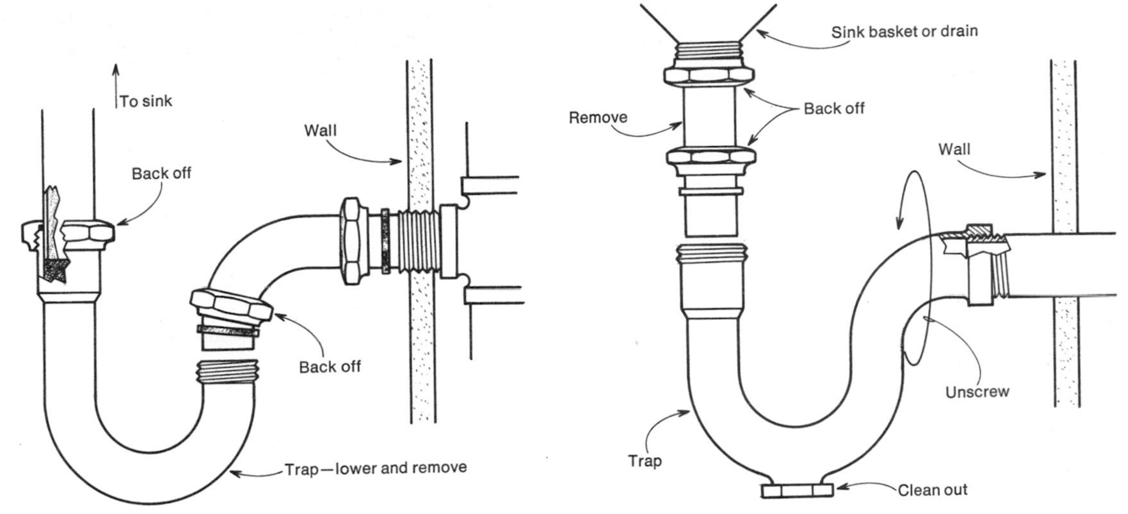

Removing two kinds of traps. Left: Back off the two large nuts on this trap and it can be lowered and removed. Right: First back off the two large nuts shown. Then remove the section of pipe between the sink and the top of the trap. Next unscrew the trap from the pipe coming out the wall.

STOPPAGE IN THE DRAIN SYSTEM

Although most stoppages occur in traps, sometimes they occur in the drainpipes leading to the main drain, also called the house drain or building drain. Sometimes waste water is blocked at the main trap at the end of the main drain, and sometimes in the sewer line, which is what the main drain is called after it passes through the house foundation on its way to the septic tank or city sewer main.

Stoppages occur most frequently in the fixture trap, in sharp bends and direction changes in the subdrain pipelines, in the main drain trap and in the sewer line.

Locating the stoppage. Before anything can be done, it is advisable to determine at which point the trouble lies. This is not always obvious, and the difficulty encountered varies with the arrangement of drainpipes in a particular house and with experience. Once you have had drain line trouble and cured it, recurrence usually follows the same pattern. The cure becomes routine.

If the stoppage takes place at the main trap, the trap in the main drain just before it passes through the foundation wall, water will back up in the drain system, but the water will not become visible until it leaves the system. Just where and when the waste water escapes depends on the height of the water level in the drain system and the type and position of the various fixtures (sinks, toilets, etc.).

When there is a fresh-air vent at the main trap, the first sign of trouble is waste water pouring out the vent. When there is no fresh-air vent—and they are by no means standard—the first sign of trouble may be the appearance of soiled water in the bathtub, or water leaking from beneath the toilet seat when the toilet is flushed, or very

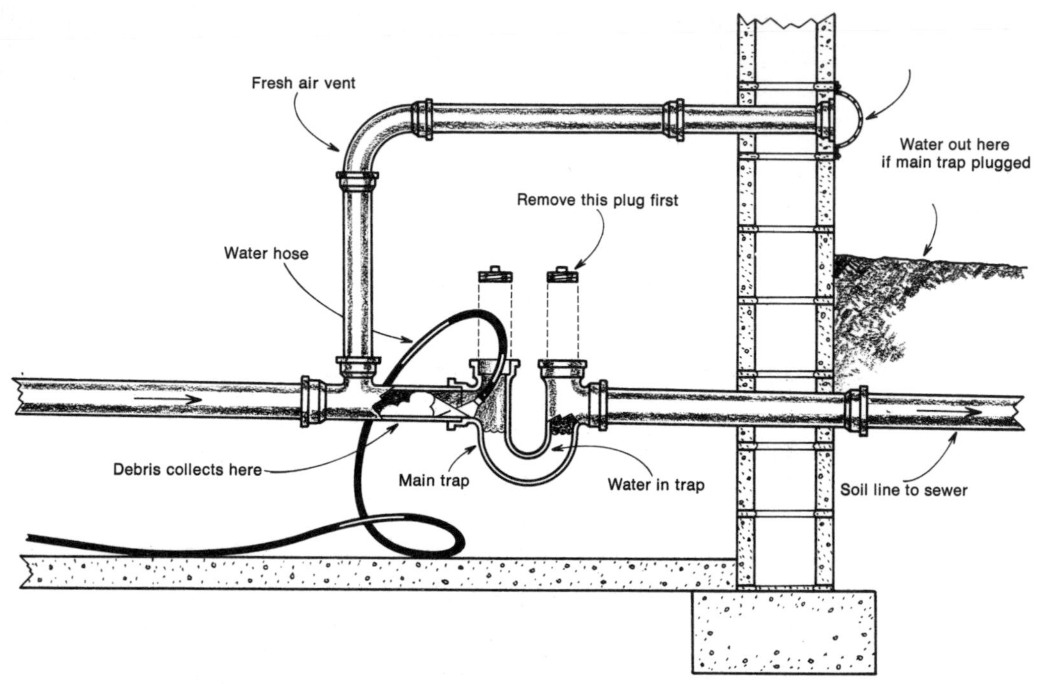

Side view of a typical main trap and fresh-air vent installation. When the main drain is plugged at any point beyond the Tee connecting the fresh-air vent to the drain, accumulating water will flow out of the fresh-air vent and onto the ground. As the blockage is often at the entrance to the main trap, to reduce the amount and chance of slop entering the cellar, first open the plug nearest the sewer line. Clean the trap very slowly, and then pierce the debris at the entrance, to allow the water to drain, and then remove the debris. Afterward, the drain system can be cleaned and tested for flow by inserting the end of a garden hose and running water through the system.

slow or no drainage in the toilet bowl. Sometimes water will drip from the cast-iron soil pipes in the basement.

Trouble at the main trap may not cause all of these symptoms to appear. It all depends on just how bad the stoppage is and the vertical disposition of the various plumbing fixtures. If there is merely a slowdown of waste-water flow, there may be little or no indication of trouble until the tub and the toilet are emptied at one time or until two toilets are flushed at one time. There may be sufficient flow to permit one fixture to drain properly at any given time.

When there is a stoppage in the submain drain line, overflow and sluggish flow may be limited to the one or more fixtures using that particular length of pipe.

Seepage from one of the cast-iron drainpipes is caused by a back-up of waste water in that pipe and one or more partially opened joints. Generally, the house is old and has settled a bit. The lead-caulked joints between sections of iron pipe open up just enough to let water out when the pipe is completely full and under some pressure. At other times there may be no sign of moisture whatsoever.

To summarize: If only one fixture does not drain or drains sluggishly, chances are that its trap is clogged. If the trap is open, the trouble is probably in the first sharp bend after the trap.

If two or more fixtures do not drain properly, the trouble is in the branch drainpipe that connects to the troubled fixtures. Again, the probable point of trouble is at a bend or a trap in the line, if there is one. Some subdrain lines or branch-circuit drain lines have traps.

If trouble shows at a toilet or at a toilet and the nearby tub, but does not show at the fresh-air vent, the trouble is probably in the line behind the fresh-air vent (on the side toward the fixtures).

If there is no main trap and no fresh-air vent, and water shows at the toilet and tub or drips from the main drainpipes, trouble could be anywhere in the main drain, but not very likely in the branch lines because that would require stoppage at two lines at the same time, which is unlikely.

Cleaning the drainpipe. In addition to the problem of finding the blockage and removing it, there is the problem of preventing the collected waste water and matter from flooding the house, or at least from splashing over the mechanic. With patience, this danger can be lessened if not removed entirely.

The trick is to do the work early in the morning after shutting off all the water the night before and flushing all the toilets. Most of the time there is some water movement down the drain. If no fresh water is permitted to enter the system for a dozen hours or more, the drain will usually be empty. It is necessary to flush the toilets because someone may forget and flush one of them while you are at work.

Start by opening the highest cleanout plug —highest meaning the plug that is the greatest distance above the level of the main drain. The higher you start the less chance of finding standing water. Remove the plug, and if no water comes out to meet you, look down the length of pipe with the aid of a flashlight.

If you can see water, the proper procedure is to run the snake down in an effort to find and remove the blockage. If the

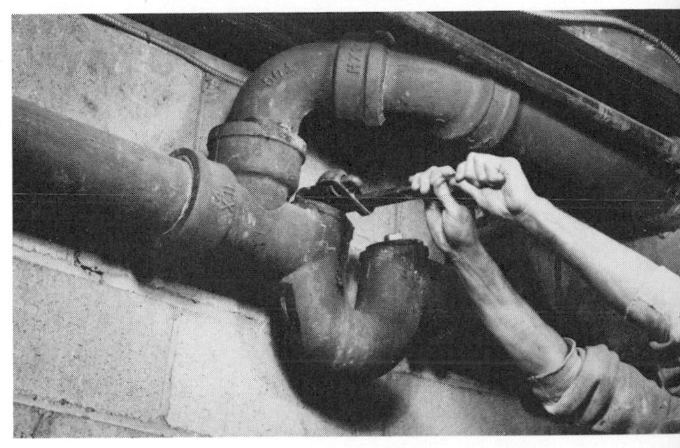

Removing the plugs on top of a running trap. The fresh-air connection is to the left of the wrench.

Use a stiff wire or thin rod to carefully probe the trap. If there is water behind the blockage, make a very small hole in it to let the water seep out. If you break the blockage abruptly you will be flooded.

blockage is not within reach, prepare yourself for a face full of dirty water when you open the next plug on the way down. If you do not see water, run the snake down anyway to make certain that no blockage exists along the line at this point and that the water did not seep through this section during the night, but runs freely.

If there are any more cleanouts on the way down to the main cleanout or main trap, it is advisable to open them first, just to make certain you do not release a small tidal wave when you open the main cleanout or trap.

To find the main trap in most homes, follow the soil stack down and around to the end of the house drain. The large, U-shaped trap is obvious enough. In other homes the trap may be hidden underneath a steel plate, the plate covered by an accumulation of old magazines. In still other homes a careless mason may have slopped concrete over the top of the trap, hiding it from view.

If there is a cleanout plug at the base of the soil stack, remove the plug and run your snake down the pipe until you strike a metallic obstruction. Mark the snake where it just enters the hole. Then remove the snake and measure from the mark to its end. That is how far away the trap is, if it is a trap you have struck.

Next, run the snake back down and bang its end against the trap. Have an assistant listen to the sound; he may be able to pinpoint the trap's location. If the snake's length inside the drainpipe is greater than the distance to the wall, the trap is outside the house—unlikely, but not unheard of.

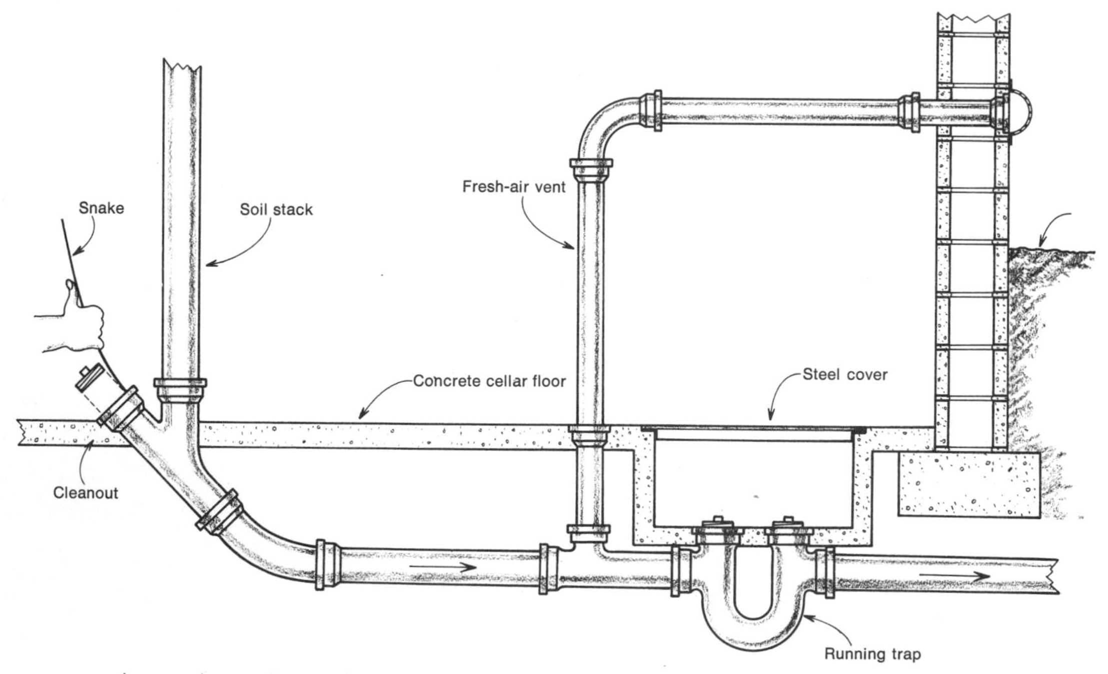

In some homes the running trap is in a well and covered with a steel plate. Use snake or rod to clear drain from base of soil stack to running trap. You can't run a snake completely through a trap of this kind.

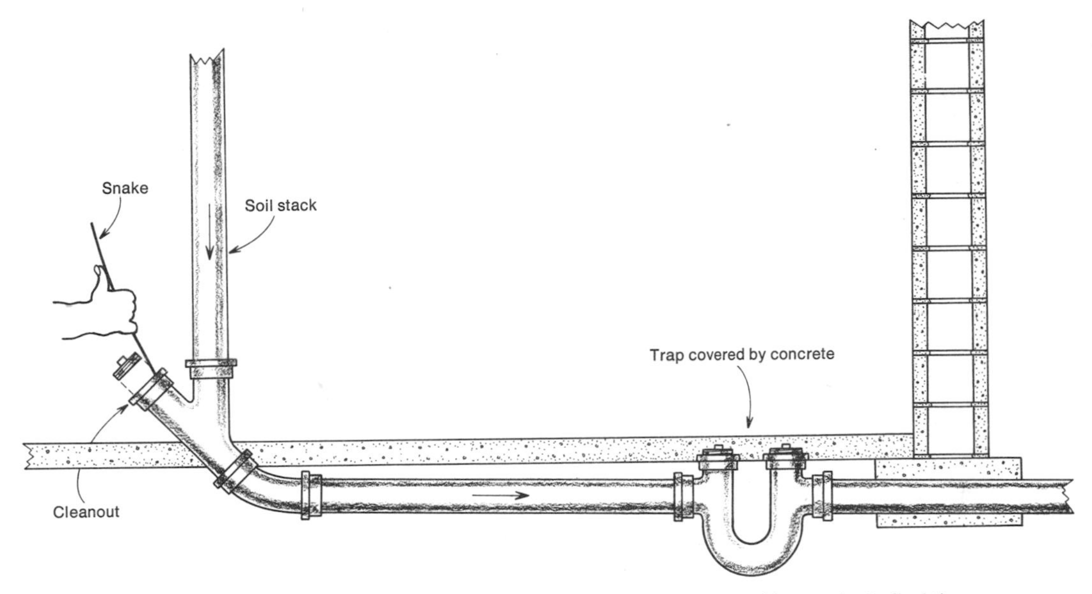

In some homes the top of the running trap was covered with concrete. Use snake to find the trap, then chop up the concrete over the plugs.

If, after a night of draining, there is no water in the main trap, in addition to the small "plug" of water at the bottom of the U-shaped curve, the stoppage may be at the lip of the trap or farther back up the main pipe. Or the trouble may be a sluggish, partially blocked sewer line.

Although there is no best sequence of tests, the following is least likely to give you a shower. If there is a U-shaped main trap with plugs at both ends, unplug the end toward the sewer line first. If no water comes out, use a flashlight to inspect the bottom of the trap, or as far into it as you can see. Take a stiff length of wire, not the snake, and gently work down and up through the bottom of the trap. If there is a glob of solid matter there, just make a small hole in it. You do not want to free it all at once, as you may release a flood.

If the trap is not stuffed up, remove the second plug. There will be no water here if the trap is clear. But there may be blockage at the entrance to the trap. Often the trouble is detergent that has resolidified into a massive lump. Use your flashlight for what it is worth, then try poking a wire up the drainpipe in a cautious exploratory move. If you feel a plug of material, make a small hole and let the water seep out. Wait until the flow settles down before enlarging the hole. When all the water is gone, remove the muck from the edge of the trap.

This done, if necessary, we can assume that so far as we know, all is well. But do not bet on it. Take a garden hose, connect it to a convenient outlet, remove its nozzle and direct a full stream of water into the trap. The trap and the sewer line following should easily accept the full flow of water. If any water backs up, the sewer line is partially stuffed. If you feed a septic tank, the tank may be full and in need of cleaning, or the drain field may be clogged, or a heavy rainfall may have saturated the field, keeping the tank full. Clearing a sewer line is discussed later in this chapter.

Assuming you have worked your snake through the upper reaches of the drain system and that the lower section, trap and associate pipeline appear to be free, and that you can direct a full flow of water down the sewer without it backing up, you are now ready for a cautious test. Remove the plug on the trap nearest the foundation wall. Then open a single faucet upstairs somewhere, and using the flashlight, watch the water trickle through the main trap. If this works satisfactorily, try flushing one toilet. If you see a rise in the trap as the water comes down, there is a good chance your sewer line is partially plugged, even though it will take the water from the hose without backing up.

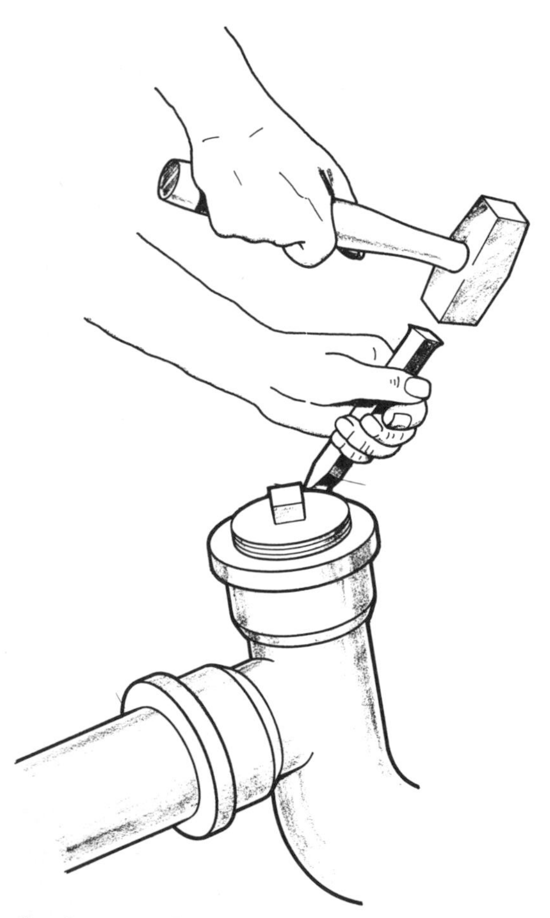

If a plug cannot be turned with a good-sized wrench, put some Liquid Wrench lubricant on the threads and then tap the edge of plug with a chisel and hammer.

If a plug is recalcitrant, do not slip a length of pipe over the handle of your wrench to give you added leverage. You may rip the boss right off. Instead, if the plug is of iron, take some rust solvent (Liquid Wrench or Quick-Turn) and apply. Then rap the plug a few times to help the solvent get into the threads. If that doesn't work, take a cold chisel and try to drive the plug around by striking its edge. Do not try to do the job with one blow, but keep rapping the plug lightly all around until it loosens.

Modern homes are erected according to code and the possibility of an inaccessible stoppage developing is close to impossible. But years ago, plumbers (and carpenters) followed their personal preference, and some of the older houses have drain systems with so many bends it is impossible for a snake to be forced through. In such cases, the better alternative to disassembling the system is to drill a hole at a suitable location for the admission of a snake. Start by center punching the spot and then using a small, sharp bit. Follow this by a larger bit. Select a size large enough to admit the snake and of a proper size to be later tapped and plugged.

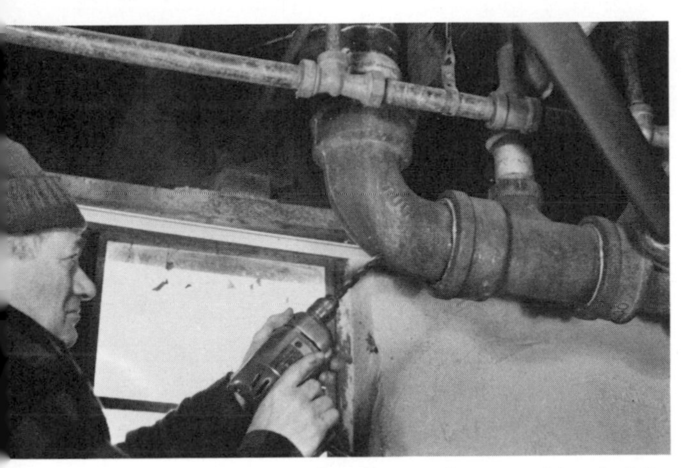

If there are no plugs that can be opened to admit the snake to reach a stoppage, drill a hole in the drainpipe and put the snake through the hole. Afterwards, tap the hole and close it with a bolt.

Opening the sewer line. The sewer pipe, which runs from the trap through the basement wall to the municipal sewer main, or to a private septic tank, can do its assigned task for countless years without a murmur and then balk. When it balks, the trouble may be an accumulation of debris, a bend in the pipe caused by the settling of supporting earth, or openings in the joints which have admitted tree roots.

If it is merely debris you can open a path for the water by using an ordinary snake, if it is long enough. And, possibly, a constant flow of water will enlarge the aperture. A lot depends on how much pitch there is and how many snake-throughs you do. If the debris is stubborn, you will not make much progress. The standard snake is much too flexible when pushed through a 4- or 5-inch pipe. You need a heavy-duty snake, sometimes called a "rod," because it is made of several rods fastened together. The rod and the motor to drive it can be rented by the day.

There are several advantages to using the motor-driven snake. The snake itself is thick and very stiff. You can attach a cutter to the end that is almost as large as the inside diameter of the sewer pipe, enabling you to cut away roots. And when you have a long run to the sewer main, the ½-inch diameter hand snake isn't much good. The motor-driven snake can be extended a couple of hundred feet and still do its job. Incidentally, you can "feel" the snake leave your sewer line and enter the main. There is no point in working beyond that.

When using a motor-driven snake, wear heavy, dry, waterproof work shoes so that you do not execute yourself should the motor not be grounded. Wear heavy leather gloves to protect your hands, and be certain to keep the switch under your foot so you can shut off the motor quickly. Do not let the motor run by itself while you handle the cable. If the end of the snake snags, the motor is quite capable of turning the reel over and breaking both your legs without even blowing a fuse. Remember, you have got to be able to shut off the motor the instant the snake jams.

Using a motor-driven snake. Note the switch under the plumber's right foot, his heavy shoes and leather gloves. Motor spins the snake; the cutter at the end hacks away at the obstruction as the plumber feeds the snake into drain. *Courtesy Conco, Spartan Tool Div.*

Clearing floor drains. Floor drains are installed in stairwells leading to a basement, in front of garages constructed underneath the building and in cellar floors. Outdoors they act to remove rain water. Indoors they serve to remove water spilled on the floor by accident or in the process of washing.

Outdoor drains usually go directly to a dry well, which is a hole in the ground filled with coarse stone. Should one of these fail to work properly, remove the cover and clean it out. If it still cannot absorb water fast enough either the ground is soaked, in which case the dry well won't function, or the dry well is clogged with debris. To make certain it isn't a clogged line, run a snake through.

Indoor floor drains usually have a trap either built in or immediately following. When these drains slow down, remove the cover and clean out the trap. If that doesn't speed the flow, run a snake down the following drain line. To test if the drain is clear, use a garden hose to shoot a stream of water into the trap.

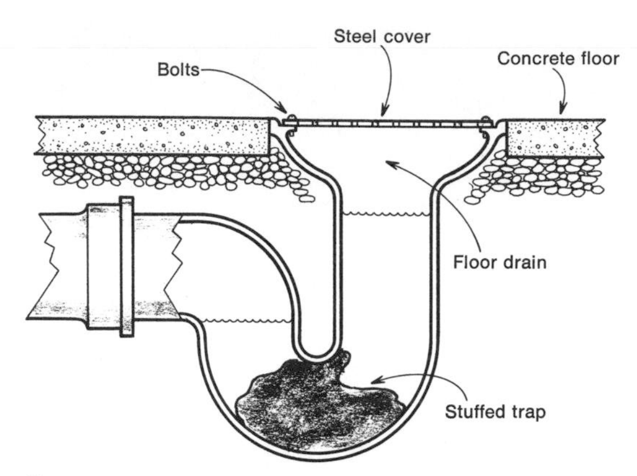

Cross section of a typical cast-iron floor drain. Remove the steel cover to get at the trap.

Repairing Leaks in Pipes

LEAKS IN THE DRAINAGE SYSTEM

As the pressure on the waste water in a drainpipe is no more than the weight of the water itself, leaks in the drainage system are rarely accompanied by large quantities of water, as are leaks in the hot- and cold-water pipes, which are under pressure. Most often a leak in a drainpipe is little more than seepage, which draws attention as a stain or a faint, background odor that grows more noticeable and offensive with time.

Leaks at cleanout plugs. One source of strong odor accompanied by minute leakage are the plugs on the cleanouts and the main drain trap. These are usually brass plugs in a cast-iron hole. At best the fit is not close. Often when the plugs have been removed and replaced a few times an airtight fit is no longer possible. Gas escapes and soon permeates the basement. One method of sealing the plugs without forever locking them in place is to use asphalt. Apply a little to the plug's threads; when it is screwed home the asphalt seals the joint yet remains soft enough to permit the plug to be removed. Asphalt can be purchased in small cans at hardware stores.

Leaks in lead pipes. These can be repaired quickly with a plumber's poultice, an old pipe-leak remedy. It can't be used for hot- and cold-water lines because it cannot withstand pressure; but it is close to ideal for repairing a leak in the lead bend under a toilet bowl.

The poultice consists of layers of cloth mixed with plaster. One simply applies some wet plaster to the pipe, making certain the plaster completely encircles the pipe and extends some 3 to 4 inches above and below the crack. The plaster is wrapped with a layer of clean cloth, much like a bandage. After a complete turn of the cloth, another thin layer of plaster is applied to the entire surface of the cloth. The cloth is given another turn and some more wet plaster is applied. Eventually, the poultice ends up as a sleeve made of a mixture of cloth reenforced with wet plaster about 1 inch thick. When it dries it resembles the kind of

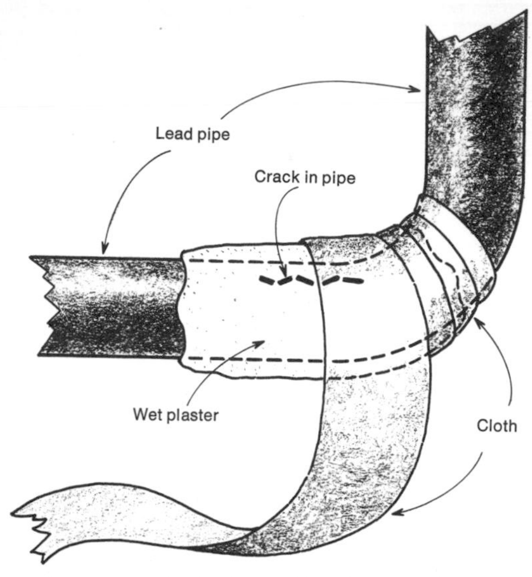

How to make a plumber's poultice. Cover a damaged drainpipe with alternate layers of wet plaster of paris and cloth. The plaster hardens into a cloth-reenforced bandage that will last for years.

"cast" used on a broken leg. Crude or not, the plumbers poultice will last for years and is about the only practical repair for a curved section of lead pipe.

When a caulked lead joint leaks, expand the lead with a caulking iron and hammer. Expand lead only enough to stop the leak; don't mash it.

Leaks in lead-caulked joints. When cast-iron pipes are moved in relation to one another, which is what happens when a house settles unevenly, the lead joints between pipe sections sometimes open a hair's breadth. Usually, these joints are easily closed by repeating the caulking process used to make them in the first place (see Chapter 4 for details).

Leakage from beneath toilet bowl. When water appears beneath the toilet bowl, and generally it will run down to the area below before it spreads around the bowl, and the drain is not plugged, the trouble is at the joint between the porcelain bowl and the top of the drainpipe. This is not a tight joint. It is not screwed fast or bolted. Instead the lower end of the bowl (horn) merely fits into a wax or putty sleeve that seals it to the drain. If you have easy access to the bottom of the bowl, there is a remote possibility the leak can be plugged without removing the bowl. But this is unlikely. (See Chapter 6 for explanation of how to remove and replace a toilet bowl.)

Leaks in brass drainpipes. When thin-wall brass pipes draining sinks, lavatories, tubs and showers leak, the pipes are no longer functional. Inspect them carefully to make certain the leak is in the pipe itself and not at the joints (joint leaks are discussed some pages on). Do not attempt to repair these leaks because they are invariably caused by corrosion, and if you touch the pipes they will collapse.

If the drainpipe is corroded, find the cause before you install a new pipe. An iron hairpin or nail may have been inside the pipe, or iron or galvanized wire may have been wrapped around it. As the inside of the pipe is almost always wet, the iron and copper or brass form a galvanic cell that corrodes the brass. When the outside of the pipe becomes wet from temperature changes, the same thing happens.

LEAKS IN PRESSURIZED PIPE

Pressure in a hot- or cold-water pipe can exceed 100 psi. Obviously, leaks in pressurized pipelines cannot be plugged with chewing gum and a little friction tape. Repairs have to be as strong as the associate pipes and fittings.

Small holes in copper pipe. Careless carpenters working in a finished home sometimes drive nails through in-wall copper pipes. These holes are readily soldered closed, but the water must first be removed from the pipe, and the pipe opened to the air to permit the generated steam to escape. Solder will hold pressure on hot- and cold-water pipes.

Large holes in copper pipe. Any hole larger than 1/16 inch cannot be easily soldered closed as it is. It must be patched. To do so, tin the area around the hole and tin the underside of a small, curved section of sheet copper or piece of copper pipe. Position the patch over the hole and solder away.

Leaks in thick-wall pipe. Leaks in galvanized, black, brass and cast-iron pipe can be plugged with factory-made pipe-leak clamps. They come in a number of sizes and will last for years. They consist of two curved metal sections that hold a sheet of rubber or plastic over the hole in the pipe.

Emergency clamps. There are probably as many ways of making an emergency pipe clamp as there are homeowners who find themselves in trouble on Sunday. Some of the suggestions in the accompanying illustrations may help you fix a leak with material you have on hand.

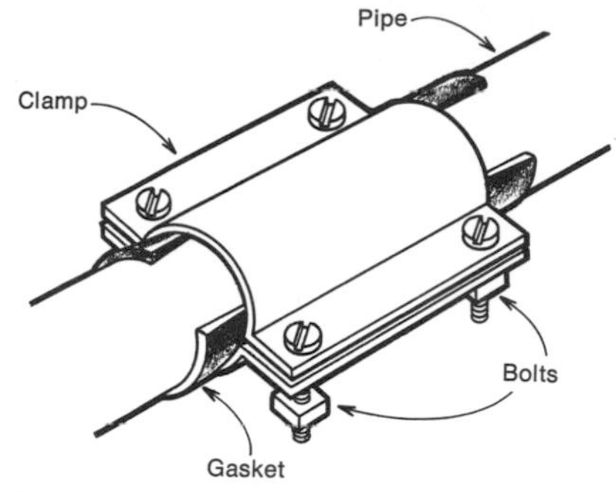

This type of clamp can be used to seal a leak in thick-walled pipe under pressure. Turn water off before applying the clamp.

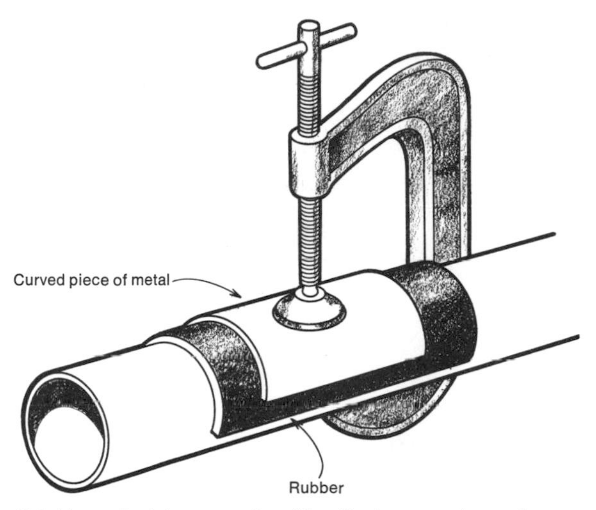

Patching a leak temporarily with a C-clamp, a piece of curved metal and a piece of rubber or gasket.

Another temporary patch, this one of wire, a wedge, a curved piece of metal and some rubber sheeting. The wedge is driven farther under the wire after the coils are fastened. Drive a small nail into the wedge at *X* to keep it from backing out.

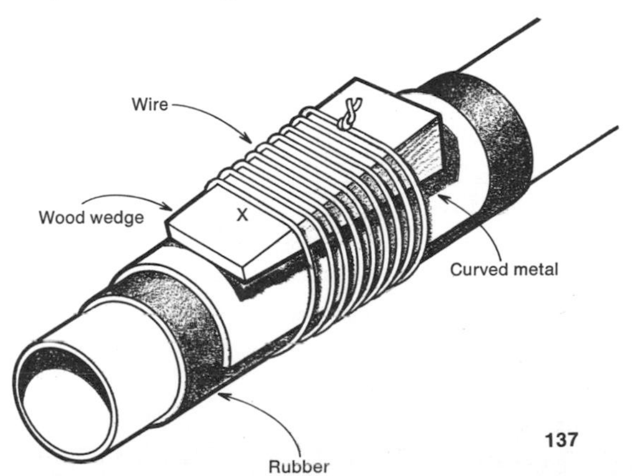

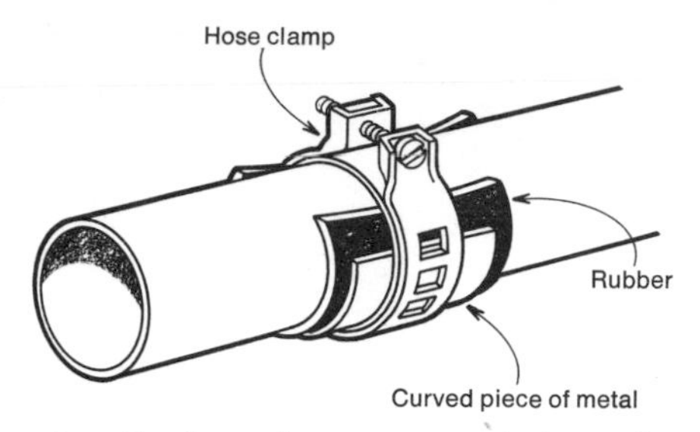

An automotive-type clamp, a curved piece of metal and some rubber or gasket sheeting can be used to make a permanent patch on a pressurized pipe.

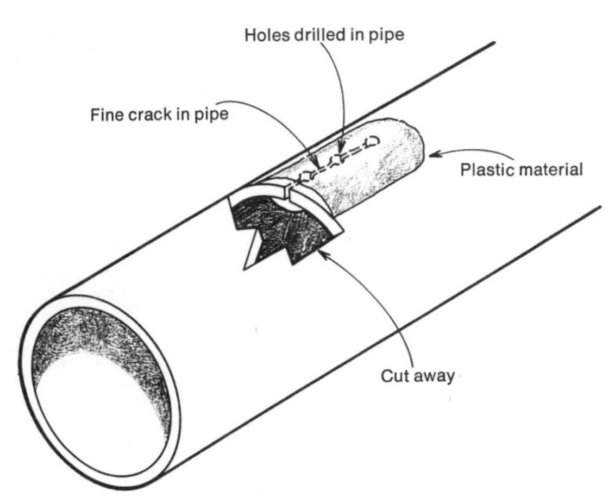

To seal a fine crack in a thick-walled pipe, drill holes at the ends of the crack and a few holes along its length. Roughen the surface with a file and apply a thick layer of pipe-sealing cement. Some of the cement will go through the holes and lock the patch to the pipe.

Cementing the leak closed. There are epoxys, liquid metals and other formulations on the market for sealing holes in plumbing systems. Some purport to work on wet pipe as well as dry. They are sold under various trade names but are used the same way.

The crack should be dry and clean—despite manufacturer's claims—just to be certain. The metal should be roughened with a clean file, and if the crack is very fine it is best to drill one, very small hole at the start and end of the crack, if that is possible, as well as holes along its length. These allow the patching material to get inside the pipe and "lock" the patch in place. The cement is then plastered over the crack.

REPAIRING JOINT LEAKS

Under-sink joint leaks. Leaks at the joints of thin-wall brass piping used to drain sinks, lavatories and the like are sometimes caused by loose compression nuts. Tighten them a little with your hand or a wrench. *Don't force them;* you can easily rip apart partially corroded thin-wall pipe.

Sometimes the leaks are due to corroded (and broken) compression nuts and sometimes they are due to dried out, flexible compression rings. In both cases it is simply a matter of taking the joint apart and replacing the nut and/or gasket.

Sometimes leaks are due to someone having knocked the pipes askew. The nuts have to be backed off and the pipes properly positioned. Sometimes when one or more pipes are bent, it is impossible to make a perfect joint. When this is true, replace the damaged section, or if there are many sections, solder several together. (See Chapter 4 for soldering instructions.)

Leaks at screw joints. Although screw joints are the easiest to "make up" (connect), they will also leak for a bewildering variety of reasons.

If the joint is made up dry, without pipe dope, it may leak because it isn't tight enough. Without the lubrication of pipe dope the joint heats sufficiently to prevent proper tightening. Without pipe dope, a properly tightened joint may leak because the threads aren't perfect.

Screw joints sometimes leak when they have been greatly overtightened, opened up and remade. Overtightening expands the female thread so much that normal tightening no longer holds.

Perfectly threaded screw joints, made with dope and properly tightened, sometimes unscrew when subjected to thermal cycling. They are next to a furnace and are alternately heated and cooled.

And some screw joints that were lightly tightened hold pressure for years until the pressure is increased or a faucet sets up a water hammer (see Chapter 19), and this causes it to leak.

Now that we have enumerated the reasons why screw joints leak, let's look at some ways of stopping the leaks.

Tightening a leaking joint. If the far end of one of the pipes connected to the leaking joint is attached to a union, loosen the union and tighten the offending joint. If there is no union in the line, tightening the joint at one pipe end backs off the joint at the other end. It would seem that you cannot win this way —closing one leak opens another—but this is not always true. Sometimes the nonleaking joint is so well made that you can back the pipe off a fraction of a turn and thereby tighten the leaking joints.

Gasket cementing a joint. If tightening a joint doesn't stop the leak and you do not have the time nor equipment to rethread the pipe, disassemble the joint, clean both the male and female threads and then reassemble using automotive gasket-forming cement on the threads.

Cementing a leaking joint. If the gasket-forming cement doesn't stop the leak, empty the pipe of water, clean and dry the outside surfaces of the joint and roughen them with a file. Then cover the entire joint with a generous coating of cement—liquid steel, or any of the cements made for this purpose. Let the cement dry hard before turning on the water.

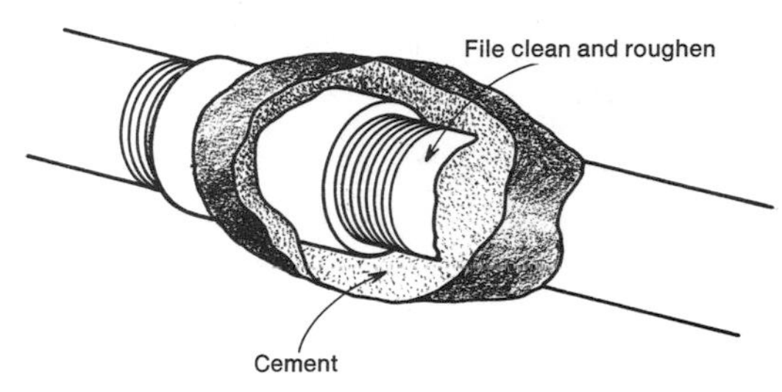

Using cement to seal a screw joint. The exterior of the joint is dried and roughened, then a generous layer of pipe cement is plastered all around.

Sure cure for screw-joint leaks. Take the joint apart and rethread the pipe end. To be certain, replace the coupling (if one is used), or any other fitting which has been leaking. Sometimes overtightening a length of pipe expands the female thread so much a tight joint cannot be made even with perfectly cut male threads.

Replacing a length of pipe. If there is a union in the line, the union is opened and the

Leak

When there is a small leak at a screw joint, try to tighten the pipe into the joint. In some instances the other end of the pipe can be backed out of its joint without leaking.

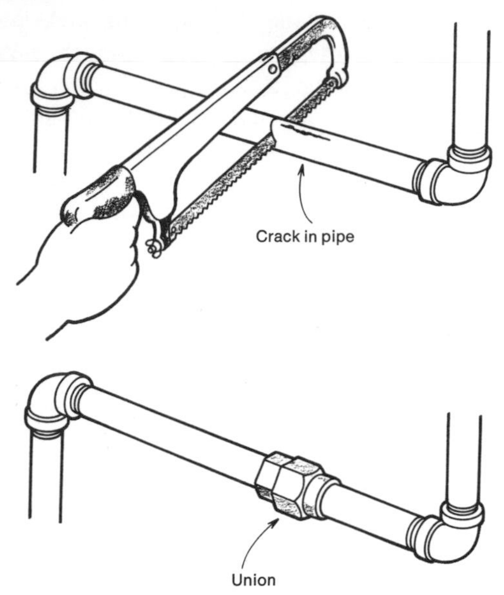

When you have to replace a length of pipe, cut it at an angle so you can swing the cut ends out of line and unscrew them. Then use a union and two pieces of pipe for the new section.

section to be replaced is simply unscrewed and another section screwed into its place. If there is no union and replacing a length of pipe requires disassembling a series of joints, it may be best to cut the offending pipe and just remove it.

To do so, cut the pipe at an angle so that you can bend one piece aside and unscrew it. The other piece is then removed. The old pipe is replaced by two lengths of pipe and a union.

Repairing a leaking solder joint. Should a soldered joint leak in copper pipe, drain the line and open communicating valves to let generated steam out. Then heat the joint and apply acid-core solder. (The acid is used to get the dirt or oxide out without disassembling the joint.) If this doesn't work, heat the joint again, disassemble it, clean the parts and try again with a noncorrosive flux and solid-wire solder.

Leaks at flare joints. If the flange nut is tight, the fault may be due to mechanical unevenness between the flare and its fitting, or dirt. If the joint has been leaking a long time, there also may be a groove present.

If there is sufficient tubing, it is best to cut off the old flare, remake it and reattach it to a new fitting, discarding the old. If there isn't sufficient tubing, you can make a perfectly tight joint by soldering the existing flare to the fitting. To do so the joint must be opened to the air—a valve at a distant end will do—and the pipe drained of water. When the joint has cooled, the solder must be filed back to permit the flange nut to be screwed back in place. Without the nut, solder joints of this design do not have much strength. Soldering the flange to its fitting prevents the joint from ever being disassembled without melting the solder.

The alternative to the use of solder, which always produces a watertight joint, is to try using automotive gasket compound on the inside of the flare where it contacts the fitting. Or wrap several layers of packing thread around the pipe before the flange nut is moved up and screwed into place.

Leaks at metal-ring compression joints. Again the first step is tightening the parts. If normal tightening does not stop the leak, try opening the joint and wrapping it with packing thread. If that doesn't do it, you can try making the joint even tighter. There is nothing to be lost unless you apply so much pressure you strip the threads off the fitting. Once a metal compression ring or gasket joint has been overtightened, it is almost impossible to open it up and reassemble it again without

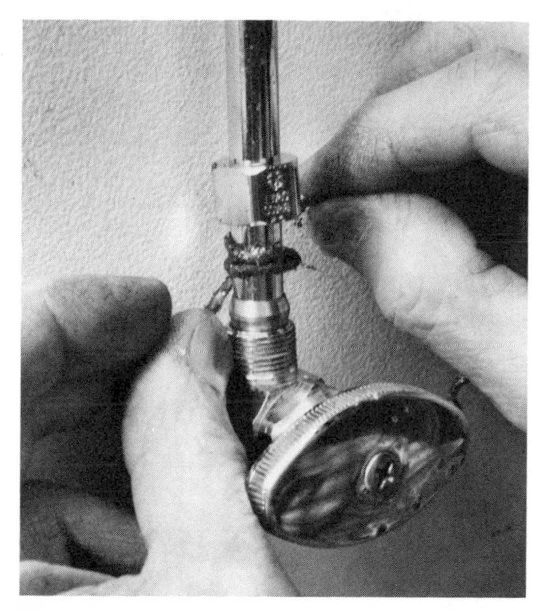

When a metal-ring compression joint leaks, back off the screw and wrap a few turns of packing thread around the pipe. Then reassemble the joint. This may stop the leak.

leaking. The high pressure forces the ring out of shape; at the same time it distorts the pipe within the ring. As a result a watertight seal is effected, but unless the joint is reassembled exactly as it was before it was taken apart, it will never hold water again.

Leaks at unions. If tightening the union does not stop the leak, it is at the metal-to-metal interface. The two metal surfaces are no longer perfectly flat and no amount of tightening will prevent water under pressure from escaping.

Open the joint and examine the metal faces. If there is a light layer of corrosion, rub both surfaces down with steel wool. If the water has cut a groove in one or both faces, there are several ways to repair the leaks.

The simplest is to find or make a circular gasket of rubber, automotive gasket material, lead sheet or automotive gasket compound and place it between the metal faces after they have been cleaned with steel wool. Whether or not the gasket will hold depends on the depth of the groove across the metal interface, and the pressure of the water.

Another method is to scour the groove clean and fill it with epoxy cement. When the cement is hard, carefully file it down even with metal. Unless the patch is made perfectly smooth, it is advisable to use a gasket with this repair.

Still another method consists of filling the groove with solder. This is a dependable method but the fitting must be removed as a lot of heat is required.

Two pipe wrenches are needed to tighten a union, on this run adjacent to a gate valve.

Designing a Cold-Water System

Each faucet in your home is connected to a pipe that in turn is connected to the water-service pipe leading to the city main. Water in the city main is under constant pressure. When you open a faucet, pressure in the main drives water out of your faucet and replenishes the water in the pipes. All this is of course obvious. It is the relationship between the pressure in the main, the resistance to water flow or passage presented by the pipe and the quantity of water that flows out of the faucet or faucets that requires a little thought. Assume the following:

All the faucets in the house are closed. Open one. Water flows out full and strong. Open a second faucet, the flow of water out of the first drops off a fraction. Open a third faucet, a fourth and so on. Each time you draw more water from the system, each time you increase the flow rate, you will find that the flow rate out of the first faucet decreases. In short, pressure decreases as the rate of flow or flow rate increases. The reason is that the pressure, the driving force in the water main, is fixed. The size of the pipe, the length of pipes, the number of bends, turns, valves and fittings are fixed; the more water that tries to get through at one time, the lower the pressure at all the openings in the pipe (faucets, toilets, washing machines, etc.).

In order to decide what size pipe will provide satisfactory flow from the various faucets and other points of water use, we need to know the pressure in the main, the desired flow rate and the maximum pressure loss that can be tolerated in the pipes and fittings. This may sound complicated but it's really not because we have tables that tell us how much pressure, at a comfortable minimum, we require at each fixture and how much flow in gallons per minute, gpm, it is nice to have at the fixture.

As we can easily find the pressure in the main, and as we know the pressure we require at the house ends of the pipes (faucets, etc.), and we know the desired flow, all we need do is jiggle about until we find the most economical pipe (smallest diameter) that will do the job. If we want to be certain, we simply install the next larger size of pipe. It costs more but it is a measure of insurance against

This water-service line of copper tubing is connected by flare fittings to a ground-key valve used as a curb stop. A protective metal tube will be placed over the valve before the earth is replaced.

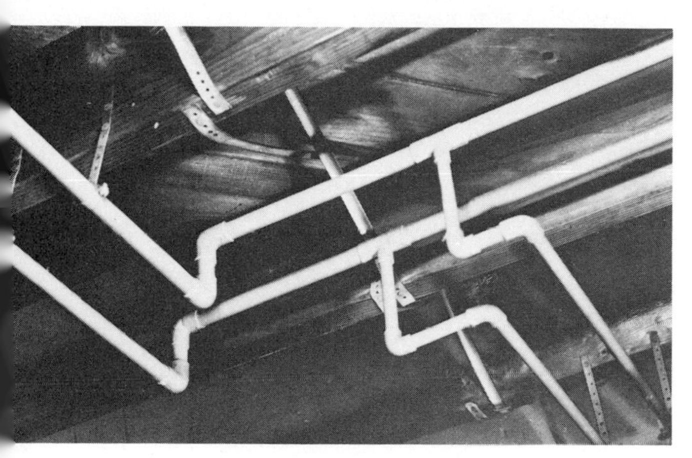

The use of rigid copper tubing and fittings in place of soft tubing without fittings reduces water pressure. In this instance the five fittings in each ½-inch line are equivalent to extending the line 5 feet.

lowered main pressure and a restricted service pipe in the future.

We proceed by finding the pressure in the main, estimating the probable demand for water and then go on to determine pipe size by using the tables, finding a duplicate situation or computation.

Determining water pressure at the main can usually be accomplished by a phone call or visit to the city's plumbing inspector or department. While you are there or on the phone find out just where your main is in relation to your home or projected home and just how many feet it is below the ground at the point you are or will connect to it.

DETERMINING PROBABLE DEMAND

When all the fixtures and appliances in a house are going full blast a condition of maximum water demand exists. When everything is shut down a condition of minimum water demand exists. Probable demand is some middle figure; the largest quantity of water that will ever be drawn from the cold- and hot-water system at any given time.

If the house is small, has few fixtures and is occupied by a large number of extremely clean, regular people, you just know all the faucets are going to be open all the time.

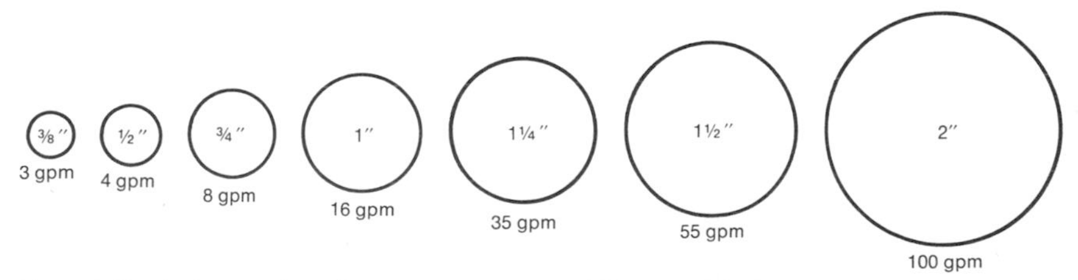

The relation between pipe diameter, pressure and flow graphically illustrated. Each pipe is exactly 100 feet long and the water pressure at the far end is 10 pounds per square inch. The flow in the 2-inch pipe is more than 30 times the flow in the ⅜-inch pipe.

If the house is very large, has lots of fixtures and a few, not particularly fastidious people, you know that it is going to be a special event when two or more faucets are open at one time.

Obviously, there is no exact method of predicting what people will do and therefore no exact formula for determining the quantity of water that will be drawn from any plumbing system at any given time. What we do is estimate with the help of two tables. One, Table 15, tells us approximately how much water flowing per minute in gallons (gpm) is considered satisfactory. For example, if there are 5 gpm coming out of your shower head you will have what most bathers consider to be a satisfactory shower bath. The gpm figures make no distinction between hot and cold water, but since there will be periods when nothing but cold water will be drawn and periods when nothing but hot water will be drawn, we must accept the full gpm figure when computing the cold-water system and when computing the hot-water system.

Table 15

Type of Fixture	Minimum Required Pressure	Desired Flow Rate
Lavatory faucet	8 psi	3 gpm
⅜ sink faucet	10	4.5
½ sink faucet	5	4.5
Bathtub faucet	5	6
½ laundry tub	5	5
Dish washer	8	5
Clothes washer	8	5
Shower	10	5
Toilet tank	10	3
Toilet flush valve	15	15–40
Sill cock plus 50′ Garden hose	30	5

Suggested pressures and flow rates for various fixtures. Figures are based in part on the National Plumbing Code. The ⅜ sink faucets applies to faucets that are supplied by a length of ⅜-inch pipe running from the shutoff valve to the faucet connection. If your pressure is higher you will naturally have more water flowing out per minute than if the pressure is lower. In no way are these figures critical. The 30 psi is required at the garden hose sill cock only if you want 5 gpm.

Table 16

Total Number of Fixtures	Probable Simultaneous Use
1 to 25	50 to 100%
5 to 25	35 to 50%
25 to 50	25 to 35%
50 and more	10 to 25%

Estimated simultaneous use of all fixtures in a building.

Table 16 provides a guideline for estimating the probable demand when the total demand is known. Again it is little more than a guess, and you may want to modify the number you use.

To find maximum water demand, make a list of all the fixtures in your home. Look up the desired rate of flow for each fixture in Table 15. Add the gpm to get total demand in gpm. A word of caution. If you live in an area of frequent rain and rarely use your garden hose, there is little point to including this figure in your computations. On the other hand, if you expect to use your hose frequently and do not want to be annoyed by the drop in water pressure and flow that may accompany its use, include the hose in your computation.

Then, to find probable demand, refer to Table 16. This gives probable simultaneous use of the various fixtures in percentages, based on the number of fixtures in the house. For example, if you are going to install plumbing in a moderately sized house having a total of 20 fixtures you would use 40 percent. Thus if your total desired flow was, for example, 80 gpm, your probable demand would be 40 percent of 80 gpm or 32 gpm. This means that at the busiest time of day, water would be drawn at a rate of 32 gpm. On some days there may be short periods of greater momentary flow, and on some days less. But most of the time, the flow rate would not exceed this figure.

As explained previously, the pressure on any single faucet in a house depends on the pressure at the city main, the size of the pipe connecting the various fixtures to the main and the number of fixtures in use (open faucets). The more fixtures in use at any given moment, the larger the pipe has to be. As we are strongly conditioned against spending money unnecessarily, we strive to determine just how many faucets will be open at any time. There is no point in selecting pipe large enough to supply 25 fixtures at one time when common sense tells us that no more than 15 will be used at one time. On the other hand we do not want to install pipe that cannot feed more than 10 fixtures at any time when it is likely that 15 will be working.

DETERMINING PIPE SIZE

Table 17 gives the size of galvanized pipe most often used for small homes. High pressure would be around 85 psi, medium pressure about 50 psi and low would be about 35 psi. If you run copper tubing you can use the next lower pipe size. However, it is inadvisable to use anything less than ¾-inch tubing or pipe for the water-service line.

The pipe sizes given for the fixtures assume that the pipe will run from the fixture supply line to the fixture's shutoff valve. Although there are no sizes given for the main supply line, it is assumed that the service-pipe diameter will be continued past the water meter and well into the house where the branch lines—which are given as fixture supply lines—are attached.

Table 18 suggests suitable diameters of copper tubing for connecting fixtures to their shutoff valves.

If your neighbor is tied to the same water main you are tied to or plan to connect to, pay him a visit and see what he has in the way of fixtures and what pipe size he is using and what the results are. This is probably the fastest, most dependable way of determining pipe size. It is the method used by most plumbers—they use the size pipe they used before and which gave them satisfactory results.

Table 17

Application	Pressure at the Water Main High 85 psi	 Medium 50 psi	 Low 35 psi
Water service	¾	¾–1	1–1¼ inch
Bath or shower	½	½	½ inch
Lavatory	½	½	½ inch
Flush tank, toilet	½	½	½ inch
Flush valve, toilet	1	1	1¼ inch
Water heater	½	½	¾ inch
Kitchen sink	½	½–¾	¾ inch
Laundry tub	½	½–¾	¾ inch
Clothes washer	½	½–¾	¾ inch
Dish washer	½	½–¾	¾ inch
Sill cock	½	½	¾ inch

Galvanized pipe sizes commonly used in a small home. Pipe diameters suggested are on the skimpy side. If you use copper tubing you may go to the next smaller size suggested. However, *do not use ⅜-inch tubing* for any application, except the last 2 feet or less when supplying a lavatory, toilet or similar fixture.

Table 18

Fixture	Copper Tube Size, inches
Drinking Fountain	⅜
Lavatory	⅜
Water Closet (tank type)	⅜
Bathtub	½
Dishwasher	½
Kitchen Sink	½
Laundry Tray	½
Service Sink	½
Shower Head	½
Sill Cock, Hose Bibb, Wall Hydrant	½
Washing Machine	½
Flush valve, toilet	1

Courtesy Copper Development Association.

Suggested minimum sizes of copper tube to be used to connect the fixture to its shutoff valve. Tube length should be under 2 feet.

If you are still a bit uncertain, discuss your pipes and fixtures with your local plumbing inspector or department. In most towns they will help you. As an alternative or as a check, you can also compute your required pipe size.

Start by drawing a simple sketch of the elevation of the building, showing in rough proportion the position of the water main, curb stop, main valve, water meter and the cold-water pipes. Add a list of all the fixtures as they are to be connected on their respective floors. Now add all the elevations (heights) and the distance the water-service line has to run from the city water main to the water meter.

Using Tables 15 and 16 add all the gpm listed for all the fixtures you are going to have on each floor. Then use the appropriate percentage figure and find the probable flow rate on each floor and the total probable flow rate that has to come through the meter to satisfy the fixtures that are working.

For example, you are going to provide cold- and hot-water lines for a two-story house with a basement. The top floor has a toilet, tub and lavatory. The first floor has the same plus a kitchen sink and dishwasher. The basement has a tub, washing machine and sill cock for a garden hose. This comes to a total of ten fixtures, if we ignore the garden hose.

Top floor: toilet, bath, lavatory	3 fixtures
First floor: Toilet, bath, lavatory, kitchen sink, dishwasher	5 fixtures
Basement: Laundry tub, clothes washer	2 fixtures
	10 fixtures

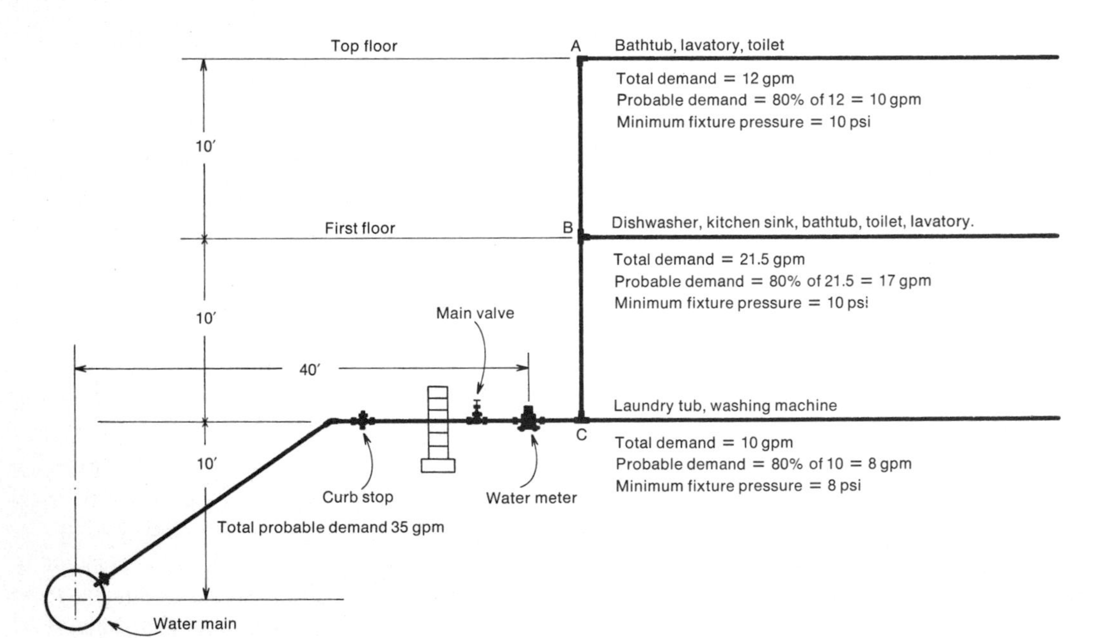

A simple sketch illustrating the position of the various fixtures in our example. This shows total and probable demand, and minimum pressure required by fixtures on each level.

According to Table 16, probable demand in a house with under 25 fixtures is between 50 and 100 percent. Let us play it cautiously and use 80 percent.

According to Table 15, a toilet, bath and lavatory require a total flow rate of 12 gpm. Using our 80 percent, this works out to a probable demand of .80 × 12, or 9.6 gpm, which we will round off to 10 gpm for the top floor.

The first floor draws a total of 21.5 gpm when every fixture is in use. At 80 percent probable demand, it works out to about 17 gpm.

The laundry tub and clothes washer draw a total of 10 gpm. Again, according to our estimate, probable demand will be 80 percent, which gives us a figure of 8 gpm. Thus the *probable* demand, that is to say the *probable maximum flow* of water on each floor and through the water meter (total) will be as follows:

Top floor	12 gpm maximum	10 gpm probable
First floor	21.5 gpm maximum	17 gpm probable
Basement	10 gpm maximum	8 gpm probable
	43.5 gpm maximum	35 gpm probable demand

Next we write the probable demand figures down on our sketch at the appropriate floors and at the water-service line. At this point we know how much water we need in gpm at each floor and from the service pipe.

DETERMINING NECESSARY PRESSURE

We know what fixtures are to be installed and connected on each level in our building. By referring to Table 15 we find the pressure requirement of each fixture. Obviously, if there is sufficient pressure for the fixture requiring the highest pressure, there will be sufficient pressure for the other fixtures on the same level. (At this point we are not thinking about gpm or friction, just pressure needed at each fixture.) In our example it works out this way:

Top floor	toilet tank	10 psi (highest on level)
First floor	toilet tank	10 psi (highest on level)
Basement	clothes washer	8 psi (highest considered)

The top- and first-floor figures are straightforward: the toilet flush tank requires the highest pressure of all the fixtures on each floor. The basement figure requires a bit of explanation.

According to Table 15, the sill cock, which is to be installed and connected to the water line at the basement, requires 30 psi. We have ignored this requirement and substituted that of the clothes washer (8 psi) in place of the sill cock because we are only going to use the garden hose once in a while. We really don't care too much whether or not we discharge 5 gpm or less at the end of the hose that will be connected to the sill cock. As there will be times when little else in the house will be used, there is a good chance that there may be close to 30 psi available for the garden hose much of the time anyway.

Effect of "head" on water pressure. At this point we have a rough sketch of the projected plumbing system showing the various elevations. We know the desired flow rate at each floor and the desired minimum pressures required at each floor. Now we will make a rough estimate of the minimum pressure necessary in the water main to do the job. By doing so we will quickly know whether or not the existing pressure is practical for what we have in mind.

First we add all the vertical steps between the main and the top floor. In our example the main is 10 feet below the water meter. We use a distance of 10 feet between floors. (Actually it is closer to 9 feet.) Thus the water has to rise 10 feet to get to the basement, 10 more to get to the first floor and 10 more to get to the top floor, a total of 30 feet. The distance the water may travel sideways in ascending to the top floor is unimportant in the following calculation.

Next we refer to Table 19, Water Pressure Conversion. This table conveniently converts "head" in feet to psi (pounds per square inch) and back again. Head is the weight of a column of water one inch square. Head is the dead weight of water that has to be lifted by the water pressure. In our example we face a 30-foot head to reach the top floor. Thus we need at least 12.99 psi just to go from the water main to the top floor. As we also need 10 psi to power our toilet tank we need 12.99 plus 10 psi or 23 psi (rounded) in the city main *plus* the pressure that is definitely going to be lost in friction when the water flows. As you can see, by making these simple calculations we can get a quick appreciation of how much pressure at a minimum is needed in the main.

Table 19

Water Pressure Conversion

Head in Feet	Pounds per Sq. In.	Pounds per Sq. In.	Head in Feet
1	0.43	1	2.31
2	0.87	2	4.62
3	1.30	3	6.93
4	1.73	4	9.24
5	2.17	5	11.55
6	2.60	6	13.86
7	3.03	7	16.16
8	3.46	8	18.47
9	3.90	9	20.79
10	4.33	10	23.09
20	8.66	20	46.19
30	12.99	30	69.29
40	17.32	40	92.38
50	21.65	50	115.48
60	25.98	60	138.57
70	30.31	70	161.67
80	34.64	80	184.76
90	38.97	90	207.86
100	43.30	100	230.95

Courtesy Josam Mfg. Co.

FINDING PIPE SIZE FOR TOP FLOOR

We do this in several steps. First we measure the length of pipe connecting the top of the riser or vertical feed pipe to the shutoff valve on the farthest fixture. We measure every inch of pipe and count all the fittings along the way. If your house is up, you can measure along the course your pipe will take. If not, scale the dimensions from the plans.

In our example the total distance is 37 feet. We ignore the branch piping. As all the water does not travel through the entire 37 feet of pipe, we reduce this figure by roughly a third, estimating that most of the water will travel through 24 feet of pipe. If all the fixtures were grouped around the end of a long run of pipe, we would use the entire run of pipe in our calculations. But as our fixtures in our example are spread out along the run of pipe we have estimated that 24 feet at full flow is a reasonable representation of the conditions.

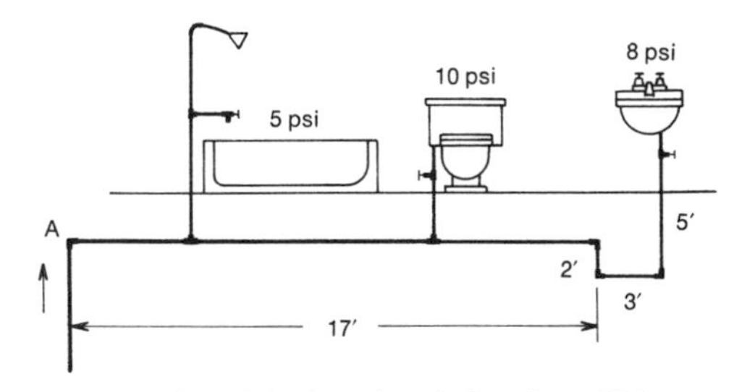

Representation of the lengths of pipe that will be used to connect the fixtures on the top floor. The fixtures are shown occupying their approximate final positions.

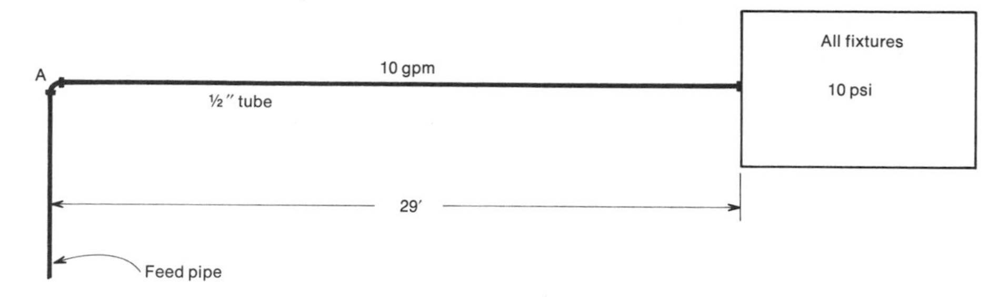

Simplified representation of top-floor piping for the purpose of calculating cold-water pipe size. Note that all the fixtures are considered as being positioned at the end of the pipe, and that the pipe is only 29 feet long. The total probable flow, 10 gpm, is shown as flowing through the entire length of pipe. The fittings are not indicated.

Table 20

Fitting Size, Inches	Standard Ells 90°	Standard Ells 45°	90° Tee Side Branch	90° Tee Straight Run	Coupling	Gate Valve	Globe Valve
⅜	0.5	0.3	0.75	0.15	0.15	0.1	4
½	1	0.6	1.5	0.3	0.3	0.2	7.5
¾	1.25	0.75	2	0.4	0.4	0.25	10
1	1.5	1.0	2.5	0.45	0.45	0.3	12.5
1¼	2	1.2	3	0.6	0.6	0.4	18
1½	2.5	1.5	3.5	0.8	0.8	0.5	23
2	3.5	2	5	1	1	0.7	28
2½	4	2.5	6	1.3	1.3	0.8	33
3	5	3	7.5	1.5	1.5	1	40
3½	6	3.5	9	1.8	1.8	1.2	50
4	7	4	10.5	2	2	1.4	63

Equivalent Length of Tube in Feet

Friction loss in copper valves and fittings given in equivalent lengths of copper tubing. Allowances are for streamlined soldered fittings and recessed threaded fittings. For threaded fittings, double the allowances shown in the table. Courtesy Copper Development Assoc.

Table 21

Fitting Size in Inches	Standard Ells 90°	Standard Ells 45°	90° Tees Side Branch	90° Tees Straight Run	Couplings	Gate Valves	Globe Valves
½	1	.6	1.50	.30	.30	.2	8
¾	2	1.2	3	.6	.6	.4	15
1	2.50	1.5	4	.8	.8	.5	20
1¼	3	2	5	.9	.9	.6	25
1½	4	2.4	6	1.2	1.2	.8	36
2	5	3	7	1.6	1.6	1	46
2½	7	4	10	2	2	1.4	56
3	8	5	12	2.6	2.6	1.6	66
3½	10	6	15	3	3	2	80

Equivalent Length of Tube in Feet

Friction loss in galvanized-iron valves and fittings given in equivalent lengths of galvanized-iron pipe of the same size.

Our example has six fittings. As all the water will not pass through these fittings, we have opted to use only five fittings in our calculations and consider each fitting a 90-degree elbow.

Trial run. Let us assume we will use ½-inch copper tubing for the top floor and compute the pressure that would be lost in the tubing.

We know we are going to send 10 gpm through the tubing. We have estimated that the effective pipe length is 24 feet and that there are effectively 5 elbows in the line. Now we will compute pressure drop under these conditions.

Referring to Table 20 (Table 21 if you plan to use galvanized-iron pipe), we find that a ½-inch, 90-degree copper-tube elbow offers as much resistance to water flow as 1 foot of equal diameter tubing. Since we have 5 fittings, we add 5 feet to our 24 feet to come up with a total effective pipe length of 29 feet.

Next we refer to Table 22 and find that 10 gpm through 100 feet of ½-inch copper tubing suffers a pressure loss of 46.6 psi. As our pipe is only 29 feet long, our pressure loss will be $^{29}/_{100}$ths or 29 percent of 46.6, which works out to 14 psi (rounded.)

Repeating for clarification:

Total actual pipe length from feed to last shutoff valve	37 feet
Estimated effective pipe length	24 feet
Total number of fittings	6
Estimated effective fittings	5
Resistance of each fitting (per Table 20)	1 foot
Five fittings @ 1 foot	5 feet
Effective pipe length	24 feet
Total effective pipe length	29 feet
Pressure loss in 29 feet of ½-inch copper tubing (29 percent of 46.6)	14 psi

Thus, if we use ½-inch copper tubing for the top floor we will need:

Fixture pressure	10 psi
Top-floor pressure loss	14 psi
Total	24 psi

The 24 psi is what we will need at the point where the top-floor pipe connects to the riser or vertical feed pipe coming up from below (*A* on the accompanying drawing). Since the top floor is 30 feet above the level of the city's water main, and as we will have approximately 60 feet of pipe between the top floor and the water main, the 24 psi requirement is fairly high and only practical where the pressure in the main is high.

Let us try ¾-inch tubing. Without bothering to account for the slight difference between ½-inch and ¾-inch fittings, let's see

Table 22

Flow, gpm	Pressure Loss per 100 Feet of Tube in psi Standard Tube Size—inches						
	⅜	½	¾	1	1¼	1½	2
1	2.5	0.8	0.2				
2	8.5	2.8	0.5	0.2			
3	17.3	5.7	1.0	0.3	0.1		
4	28.6	9.4	1.8	0.5	0.2		
5	42.2	13.8	2.6	0.7	0.3	0.1	
10		46.6	8.6	2.5	0.9	0.4	0.1
15			17.6	5.0	1.9	0.9	0.2
20			29.1	8.4	3.2	1.4	0.4
25				12.3	4.7	2.1	0.6
30				17.0	6.5	2.9	0.8
35					8.5	3.8	1.0
40					11.0	4.9	1.3
45					13.6	6.1	1.6
50						7.3	2.0
60						10.2	2.7
70						13.5	3.6
80							4.6
90							5.7
100							7.5

Courtesy Copper Development Association.

Pressure lost due to friction in copper tubing. Note that the loss of pressure increases at a tremendous rate as the quantity of water flowing through the pipe increases. For example, when the flow through 100 feet of ½-inch tubing increases from 5 to 10 gpm, pressure loss jumps from 13.8 psi to 46.6 psi.

what Table 22 says for 10 gallons through 100 feet of ¾-tube. The pressure loss given is only 8.6 psi. Twenty-nine feet, or 29 percent of 8.6 comes to 2.4 psi; quite a difference compared to 14 psi drop for ½-inch tube. However, as ¾ costs much more than ½-inch copper, let us consider a compromise.

If we feed the tub with ¾-inch tube and continue on from there with ½-inch we will cut our pipe costs. Since we do not need accuracy, we will assume that the tub is in the middle and that we are going to run 15 feet of ¾-inch tube and 15 feet of ½-inch tube. Again, we are speaking of the effective length, which is based on an estimate. Pressure loss in the ¾-inch tube will be 1.2 psi (15 percent of 8.6). Pressure loss in 15 feet of ½-inch tube will be 1.07 (15 percent of 13.8).

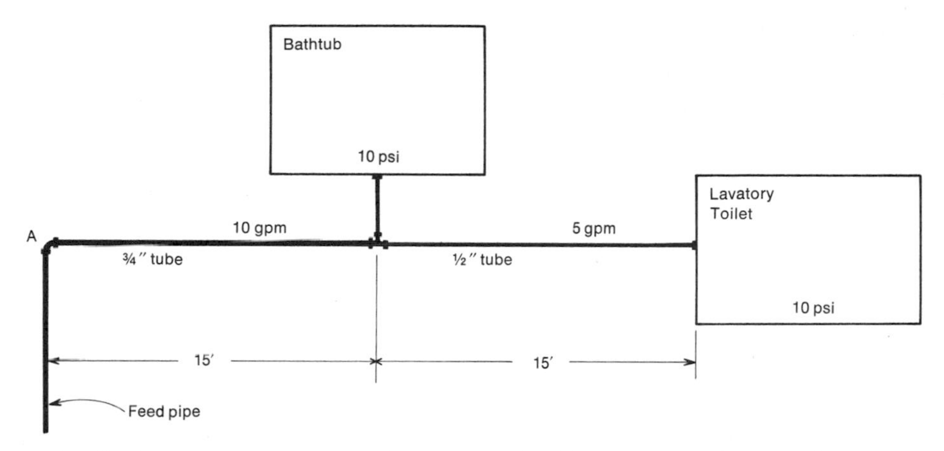

Simplified representation of the top-floor piping using $\frac{3}{4}$-inch copper tubing for half the pipe run and $\frac{1}{2}$-inch tubing for the balance of the run. For simplification the tub is shown and computed as being exactly in the center of the pipe run and the two remaining fixtures at exactly the end of the run. No fittings are shown.

Thus, total pressure lost will be 2.27 psi (1.2 + 1.07).

The pressure drop of 13.8 per 100 feet was used in the above calculation because we are only running 5 gpm through the section of pipe from the tub to the last fixture. Table 22 gives 13.8 psi as the loss for 5 gpm through 100 feet of ½-inch copper tubing.

As can be seen, the pressure required at the top floor (*A* on drawing) is now considerably lower.

Fixture pressure	10 psi
Top floor pressure loss (rounded)	2 psi
	12 psi

By simply including 15 feet of ¾-inch tubing in place of ½-inch tubing of equal length we have reduced our pressure requirement at the top floor from 24 psi to 12 psi.

Since we are planning to use ¾-inch tubing at the start of our top floor piping we cannot use smaller pipe to feed it. Therefore the feed or riser coming up from the first floor to the second, must at a minimum be ¾-inch tubing.

Table 22 states that 10 gpm through 100 feet of ¾-inch tube drops the water pressure by 8.6 psi. As our pipe is only 10 feet long, the pressure drop will be 10 percent of 8.6 or .86, which we round to 1 psi.

Since our water must rise 10 feet to go from the first to the second floor (From *B* to *A* in our drawing), we must include the pressure lost to a 10-foot head, 4.33 psi (Table 19) in our calculations.

Pressure lost to friction in riser	1 psi
Pressure lost to 10-foot head (rounded)	4 psi
Total pressure lost in riser	5 psi

To determine pressure needed at first floor Tee (*B* on drawing), we add:

Total pressure lost in riser	5 psi
Pressure required at top floor	12 psi
Total pressure required at first floor Tee	17 psi

FINDING FIRST FLOOR PIPE SIZE

We will follow the same procedure we used for the top floor and split the pipeline in half, using ¾- and ½-inch tubing.

There are 6 fixtures with a probable total flow of 17 gpm. The actual length of the pipe is 42 feet. As all the water does not travel the full distance we shall estimate the effective pipe length to be 30 feet. There are 6 assorted fittings and we shall estimate that in effect there are 5, 90-degree elbows.

For simplicity, we will split the system into 15 effective feet of ¾-inch tubing and 15 effective feet of ½-inch tubing and assume that there are 3 fittings in each section.

To simplify further, we shall assume that 17 gpm flows through the first 15 feet of tubing and that 7 gpm flows through the second section.

Total effective pipe length in first section:

15 feet of ¾-inch tubing	15 feet
3 fittings @ 1.25 feet	3.75 feet
Total effective tube length	18.75 feet

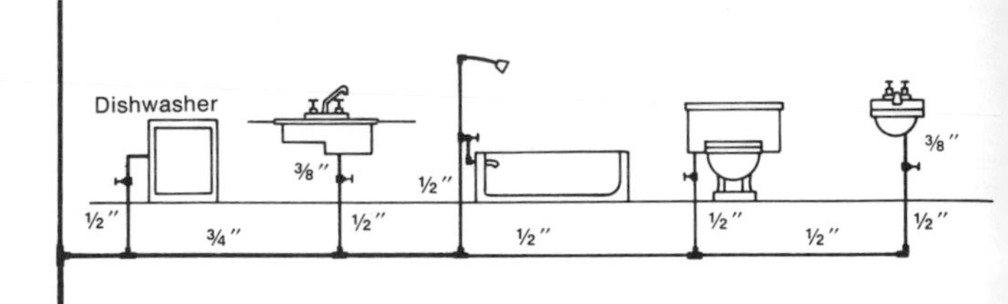

Approximate representation of the lengths of pipe that will be used to connect the first-floor fixtures. The fixtures are shown occupying their approximate final positions.

To find pressure lost in this tubing we refer to Table 22 and estimate that the pressure loss for 17 gpm through 100 feet of ¾-inch tube is 22 psi. Multiplying by 19 percent (18.75 rounded) we get 4.1 or 4 psi as the pressure lost in ¾-inch section of the first-floor cold-water piping.

To find the pressure lost in the ½-inch tubing section of the first-floor piping, we first find the total effective pipe length, and then, as before, add the effect of the fittings.

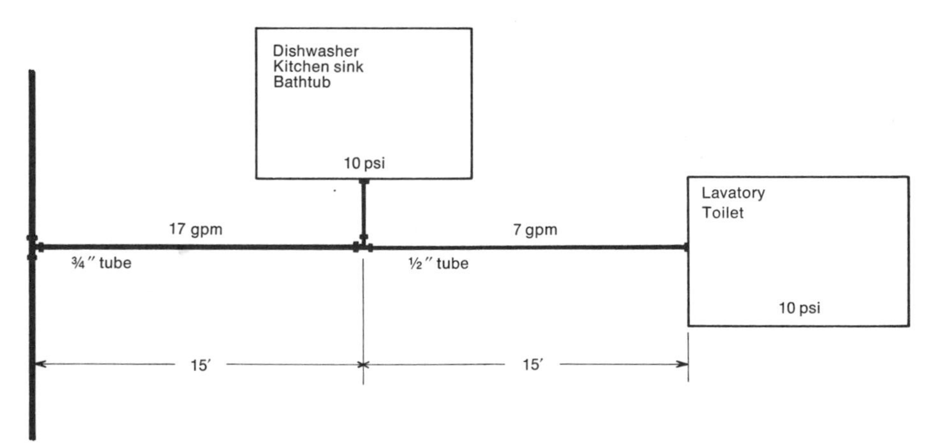

Again we have simplified the piping for the purpose of computing the first floor. We have laid out the pipe in two equal sections of ¾-inch tubing and ½-inch tubing. The dishwasher, kitchen sink and bathtub are computed as if they are all connected to the center of the run. The toilet and lavatory are computed as being at the end of the run. Actually the flow to the last two fixtures will be a bit less, but this division makes for easy computation.

Total effective pipe length	15 feet
3 fittings @ 1 foot each (Table 20)	3 feet
	18 feet

To find pressure lost in this tubing we refer to Table 22 and estimate the pressure lost when 7 gpm flows through 100 feet of ½-inch tubing to be 18 psi. Multiplying by 18 percent we get 3 psi (rounded).

Thus the pressure lost in the first-floor piping between the Tee that connects it to its feed pipe and the last shutoff valve in the pipeline will be:

Loss in the ¾-inch tube	4 psi
Loss in the ½-inch tube	3 psi
Total first floor pressure loss	7 psi

As the highest required pressure on the first floor is the toilet at 10 psi, our first-floor pressure requirement at the feed pipe Tee (*B* in the drawing) is

Fixture pressure	10 psi
Pressure loss	7 psi
Required first-floor pressure	17 psi

Relating top-floor to first-floor pressure. In our example we require 17 psi at the first floor Tee (*B* in drawing) to supply the top-floor water requirements. By chance we also require 17 psi at the same point for the first-floor water supply. As both pressures are to be equal there is no problem.

If there weren't more than 10 or 15 psi difference at this point, there still would be no problem so long as the higher pressure requirement was met. The result would simply be a little more pressure than planned for the other floor.

Finding first-floor feed-pipe size. We know the top floor will draw 10 gpm and that the first floor will draw 17 gpm, thus the total flow through the feed pipe going from the basement to the first floor will be:

Top floor	10 gpm
First floor	17 gpm
Total first-floor feed-pipe flow	27 gpm

The pipe length is 10 feet and we use the entire 10 feet in our calculations because the entire quantity of water flows through the entire length.

Referring to Table 22, 27 gpm through 100 feet of ¾-inch tube works out to an estimated 45 psi. For 10 feet we take 10 percent of 45 to get 4.5 psi, which is the friction loss. Now we need the head loss, which again for a 10 foot rise is 4.33 psi. This remains the same regardless of total water flow. Thus, first-floor feeder pressure lost is:

Pressure lost to friction	4.5 psi
Pressure lost to 10-feet head	4.33 psi
Pressure lost in first-floor feed pipe (rounded)	9 psi

Thus if we use a ¾-inch tube for this run our pressure loss will be 9 psi from the meter (*C* in drawing) to the first-floor Tee (*B* in drawing).

Finding pressure required at water meter. In our imaginary house the pipe feeding the first and second floors is connected immediately after the water meter (*C* in drawing). Therefore pressure required after the meter will be the pressures we just calculated.

Pressure required at first floor Tee	17 psi
Pressure lost in first-floor feed pipe	9 psi
Pressure required at meter	26 psi

FINDING BASEMENT PIPE SIZE

Ignoring the sill cock, we have two fixtures in the basement drawing a probable flow of 8 gpm. The maximum necessary pressure is 8 psi. Let's try ½-inch copper tubing.

The actual pipe length from the meter to the farthest fixture shutoff valve is 25 feet. There are 6 fittings through which the water has to pass. Let us assume an effective pipe length of 20 feet and 5 fittings.

Estimated effective pipe length	20 feet
Five ½-inch fittings @ 1 foot each	5 feet
Total effective pipe length	25 feet

Table 22, pressure loss per 100 feet of ½-inch tubing is an estimated 33 psi. Twenty-five percent of 33 = 8.25.

Pressure needed for basement fixtures	8 psi
Pressure lost in ½-inch tube and fittings	8.25
Total pressure needed for basement piping	16.25 psi

As previously calculated, we need 26 psi at the water meter to supply the first and top floors. As we will connect our basement pipe to the same point, the basement pipeline will be supplied with the same pressure; therefore, the use of ½-inch copper tubing for the basement will be fine. There will simply be more pressure than we need and this will do no harm.

FINDING THE SIZE OF THE WATER-SERVICE PIPE

We know that all our fixtures together will draw 35 gpm. We know that we need 26 psi at the house side of our water meter. We know that the pipe run from the city main to the water meter is 40 feet. We know that the water will have to pass through the corporation stop, the curb stop, the main valve and the water meter. In addition, the water will have to overcome a 10-foot head to reach the basement.

Trial run. Let us try 1-inch copper tubing for the service line and see how it works out. For simplicity we will assume that the corporation stop and curb stop offer the same resistance to water flow as the main valve, which is a gate valve.

3, 1-inch gate valves (Table 20) @.3 feet	0.9 feet
40 feet of pipe (water flows the entire length)	40 feet
Total effective service pipe length	41 feet (rounded)

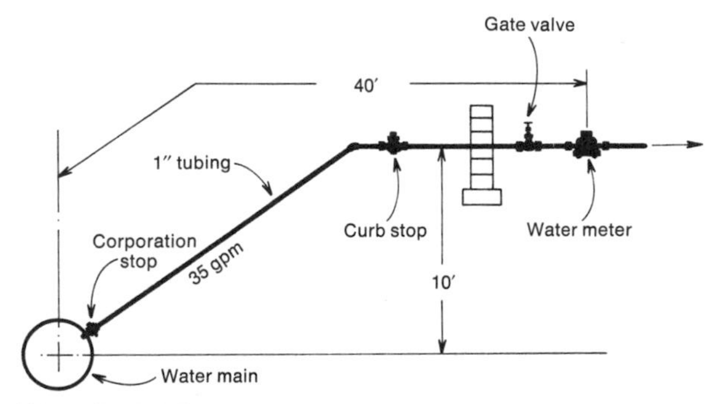

Water flowing from the water main loses pressure as it passes through the corporation stop, water pipe, curb stop, gate valve and water meter. In addition it loses pressure due to its rise in elevation. In this example, the rise is 10 feet, causing a pressure drop of 4.3 psi.

Table 23

Tube Size, Inches	Approx. Flow Rate, gpm
¾	22
1	35
1½	63
2	100
3	190

Courtesy, Copper Development Assoc.

Pressure loss in water meters will equal 10 psi on the average under the conditions listed above.

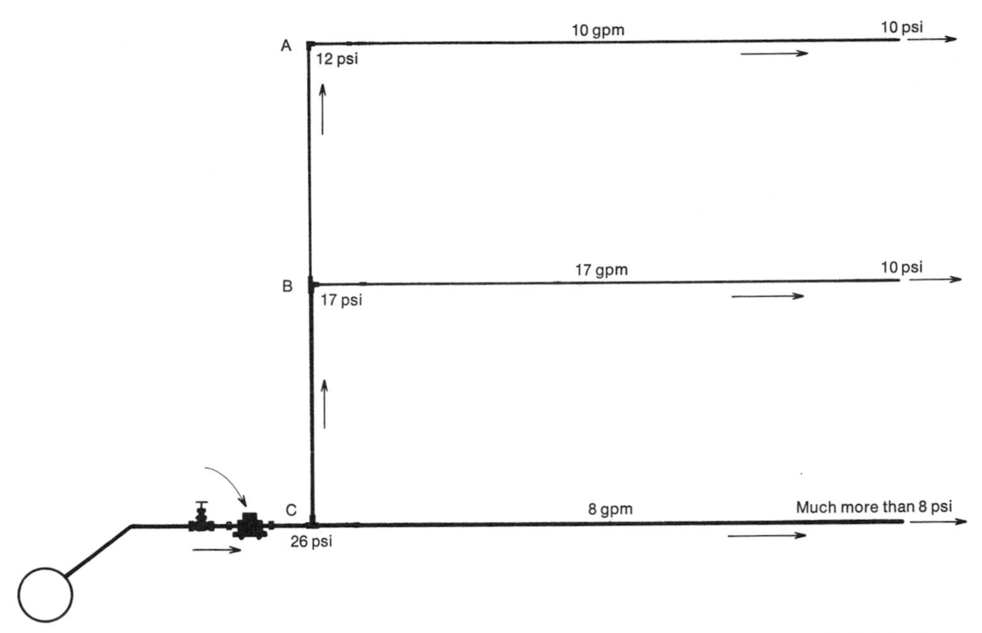

The major pipes composing the cold-water system, showing how the pressure will divide if we use the pipe sizes calculated and if there is 51 psi in the city main.

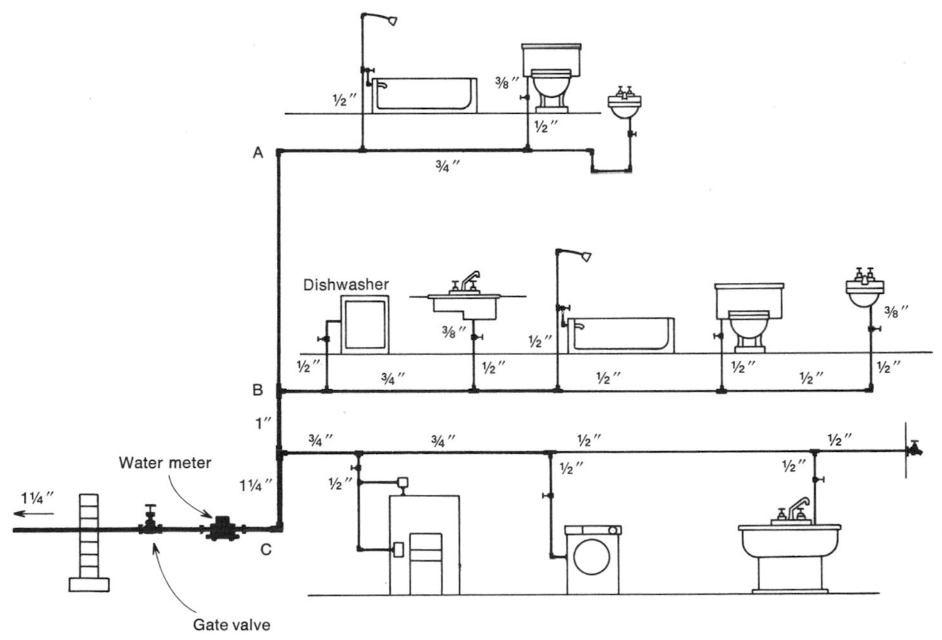

The pipe sizes that will be used in our example. Smaller pipe will reduce flow and larger pipe will increase costs.

Pressure lost in 41 feet of 1-inch main carrying 35 gpm = 41 percent of 25 psi (estimated from Table 22)	10.2 psi
Pressure lost to 10-foot head	4.33 psi
Pressure lost in water meter (Table 23)	10. psi
Total pressure lost from main through meter	24.53 psi

Main pressure needed with 1-inch copper service. Our previous calculations indicated that we needed 26 psi at the house side of the water meter to supply our needs. Since we can expect to lose 25 psi (24.53 rounded) in our 1-inch copper service line, we will require:

Pressure needed at meter	26 psi
Pressure lost in service	25 psi
Pressure needed at city main	51 psi

If we have 51 psi or thereabouts in the city main we can use the pipe sizes we have calculated. If the pressure is lower we can make do with sometimes unsatisfactory water flow or go to a larger service pipe or a larger basement-to-first-floor feed pipe.

CURING HIGH PRESSURE AND PRESSURE FLUCTUATION

Should your pressure be above 80 psi or so, or should your pressure fluctuate sufficiently to cause a nuisance, it is advisable to install a pressure-regulating valve in the service line. A pressure-regulating valve automatically holds water pressure to any desired value (within its range) below that of the incoming water despite pressure variations.

Generally 40 psi is about right for a one-story building and 50 psi or so is satisfactory for a two-story building.

Beside spraying a room with water when one is not careful, higher than necessary water pressure shortens the life of faucets and valves, increases the possibility of leaks, produces water hammer if none is present, and increases water hammer if it is present. There are no advantages to water pressure that is higher than needed. Water hammer and the installation and servicing of pressure-regulating valves are discussed in Chapter 18, Water Hammer.

Setting pressure regulators. A pressure regulator acts to reduce incoming pressure to a lower, preset value. It cannot increase pressure and it cannot hold pressure steady if its capacity is exceeded—you try to draw more water than it can pass. And it cannot hold pressure if you try to draw more water than the water-service pipe can supply at the rate you are drawing it.

To actually ascertain the pressure of the water leaving the regulating valve it is necessary to attach a water-pressure gauge to the pipe-line. This is connected by a Tee fitting.

While it is nice to know the pressure, it is not actually necessary. As long as there is an adequate flow of water to the highest fixture and as long as there isn't excessive pressure producing hammer and sudden showers when a faucet in the basement is opened, the pressure may be deemed satisfactory.

Proper regulator setting can be found by experiment, assuming the regulator is of sufficient size. Generally, if its pipe size matches that of the water-size line, its capacity is reasonably correct.

To adjust a pressure regulator without a pressure gauge, about one third of all the faucets in the house are opened. The washing machine(s) should be included in the count, and the faucets opened should be equally distributed between all levels. The purpose is to attempt to duplicate normal water usage during a busy period in the house. This done, the regulator is adjusted until the flow of water from the highest faucet is satisfactory. Following, all the faucets are closed with the exception of a single faucet at the lowest point in the plumbing system. If there is now too much pressure on this faucet, pressure is reduced to a compromise between too little at the top floor and too much (with nothing or little else running) at the lowest level.

LOW PRESSURE

Low pressure in the system is evidenced by reduced flow when several faucets are opened at one time. Low prèssure may be due to lowered pressure in the main, increased water usage, constriction in the water-service line, undersized pipes.

Low main pressure is easily checked. Usually a phone call is all that is necessary. Increased household usage or a change in house habits leading to multiple fixture use at one time is as easily checked. If pressure in the main is up to its usual mark, and you haven't taken in boarders, the problem is unfortunately to be found in the water-service line.

There is no way of opening a constricted water-service line, the trouble can only be cured by unearthing the pipe and replacing it. However, as replacing a service line is a lot of work, it is wise to ascertain whether or not pressure relief can be obtained by replacing valves or pipe within the building.

To save money plumbers will sometimes install a globe valve ahead of the water meter in place of a more expensive gate valve. At a low flow rate the difference is not noticeable; when the flow rate is increased pressure drop can be considerable. For example, a globe valve will reduce water pressure by about 10 psi with a flow rate of 40 gallons per minute. At the same flow rate a gate valve of equal size would reduce the pressure only .7 psi.

Pressure is also lost in the water meter. Increasing meter size from 1 inch to 1¼ inches will reduce the loss to half at this point.

For example, a disc-type, 1-inch water meter causes a pressure loss of about 15 psi at a flow rate of 40 gpm. The same type of meter, designed for an 1¼-inch pipe causes a pressure loss of only about 7 psi. A 1¼-inch meter can be used on a 1-inch pipeline by using reducers in the pipeline at the meter.

Another possible caùse of pressure loss within the house itself is the fittings. Each time water takes a turn, pressure is lost. By replacing needless fittings with a single length of copper tubing, more pressure can be saved.

In some instances, where the loss of pressure due to a constricted or undersized service pipe is not too great, improvements in the plumbing system within the building can sometimes bring the pressure up to requirements.

When it is impractical to replace a constricted or undersized water-service pipe, or the pressure in the main is low and going to remain so, the answer, other than just waiting for the water to dribble into the tub, is to install a pump and tank. During periods of no or low water consumption, street water pressure fills the tank and pressurizes the cold-water system. When the demand for water brings the pressure below a preset figure, the pump goes into action. It injects water drawn from the storage tank into the cold-water system, thus increasing the rate of flow out of the faucets and by relieving the demand on the city supply line, enables it to supply the flow desired.

Designing a Hot-Water System

This chapter will help you cope with most of the problems that may arise with your domestic hot-water system; help you to plan additions and improvements in your present system; and help you to plan and design a completely new domestic hot-water system for a new house.

First it may be advisable to define domestic hot water. Domestic hot water is the water used for washing, cleaning and cooking. Domestic hot water is so called to distinguish it from the water that may be used in a hot-water space-heating system or a steam-heating system. Domestic hot water is domestic cold water that has been heated. Domestic hot water gets its pressure from the municipal main or a local well pump.

The domestic hot-water circuit begins with a pipe at the water-service main that leads cold water to the hot-water heating device. From there the now hot water is piped to wherever it is required.

Designing a domestic hot-water system requires consideration of the following:

Pipe type
Pipe size
Probable hot water demand
Heater type and size

You can use any metal pipe for running hot-water lines. The use of plastic pipe for hot water depends on whether or not your local code will permit it, and the water temperature. CPVC plastic pipe, which is presently the most heat-resistant plastic used for piping, cannot be used with water over 180 degrees F. Other types have even lower heat resistance. You can, however, run copper to the heater and use plastic farther along where the water is cooler.

The easiest way to design a plumbing system is to use the same pipe for both cold and hot water. However, if you decide to run galvanized for one and copper for the other, take care they do not touch. In the presence of moisture this will cause rapid corrosion.

The same methods and tables used to compute cold-water pipe size may be used to compute hot-water pipe size. The same prob-

An electrically powered, storage-type hot-water heater connected by plastic pipe. Note unions in line which permit the heater to be easily removed and replaced. *Courtesy B.F. Goodrich Chemical Co.*

able demand figures, less the water not drawn by toilets and sill cocks, because they do not draw hot water, may be used. Thus in most instances it is unnecessary to go through the entire series of calculations (if you are calculating and not simply imitating what your neighbor uses). In most instances the pipe sizes used for the domestic hot-water system are identical to those used for the cold, with possibly a reduced diameter section here and there.

COMPUTING HOT-WATER DEMAND

In our example we had a probable cold-water flow rate of 35 gpm. Our probable hot-water flow rate will be the same less the two toilets, which do not draw hot water. As each toilet draws 3 gpm, our hot-water flow rate is 35 gpm less 6 gpm, or 29 gpm.

Hot water at a constant flow of 29 gpm is a frightening amount of hot water. The equipment necessary to provide this flow would be very large and expensive. So instead of buying and installing and powering a unit of this size we try to estimate the probable demand for hot water in gallons and work to satisfy this figure.

Repeating for clarity: We use the probable hot-water flow figure (gpm) to make sure our pipe size is correct to give us the flow we want. Then we try to estimate *how much* hot water will be drawn from the system. Knowing how much hot water we will need helps us select a hot-water heating unit that is a happy compromise between all the hot water we would like to have and the hot water we can afford.

Table 24 lists estimated quantities of hot water drawn by various fixtures and household appliances. These are average figures and will permit you to estimate your hot-water needs in gallons per hour or day.

Table 24

Fixture or Appliance	Gallons
Bathtub	30
Shower bath, 5 min.	15
Lavatory	2
Kitchen sink, 1 meal	3
Dishwasher, 1 load	4
Clothes washer, 1 load	40

Approximate quantity of hot water used by various household fixtures and appliances.

Plotting the hot-water usage curve. Table 24 gives us the one-time use per fixture in gallons. If we take the trouble we can easily find the total number of times the tub will be used each day, the total number of times the dishwasher will be used and so on. Adding all these figures tells us how much hot water our family will probably use per day. This is not enough to enable us to select the best hot-water heater for our needs. We must also find out just when and how much hot water will

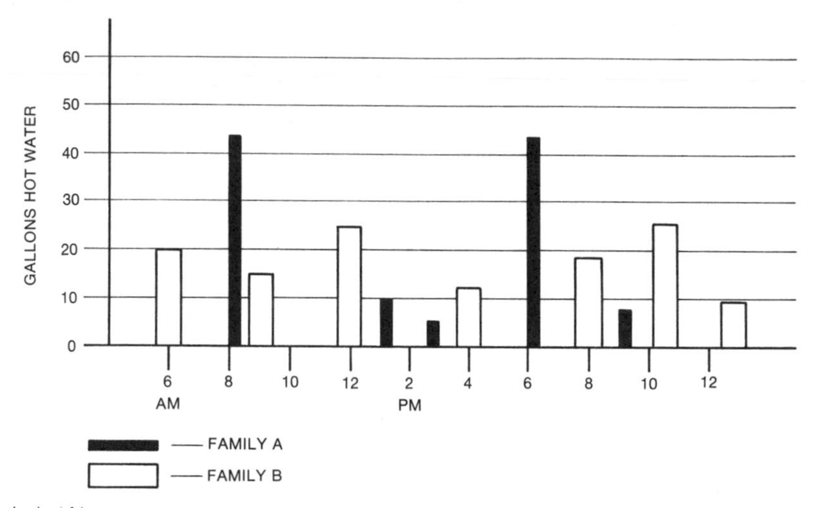

Periods of hot-water usage in gallons by two families over an average day. Note that family *A* draws most of its hot water at two periods of the day, while family *B* draws its hot water during many periods over the same day.

be required over the day. The reason is the nature of heaters, as shall be explained.

The accompanying graph plots the hot-water usage of two families over an average day. Note that family A uses about 90 gallons of hot water each day, and that family B uses about 110 gallons over the same period. At first glance it would appear that family B required a larger capacity hot-water heater than family A. Actually it is the reverse. Family A draws all its hot water in two gulps of 45 gallons each. Their hot-water heater must be capable of delivering 45 gallons of hot water over a short period, of, say 15 minutes. While family B draws a greater total quantity of hot water over the day, their demand is spread out and is, according to the chart, never more than 25 gallons at any interval. Family A probably requires a 70-gallon, tank-type heater, whereas family B's needs can probably be adequately satisfied with a 35-gallon, tank-type heater.

This is the reason you must know how much hot water will be drawn at any given time as well as the total number of gallons of hot water you may need over the day.

INSTANTANEOUS WATER HEATERS

Water heaters fall into two general groups: instantaneous, sometimes called tankless; and storage, sometimes called tank type.

Instantaneous hot-water heaters, as the name suggests, produce "instant" hot water. When the device is operating, cold water enters one end and is instantly converted into hot, and the process goes on continuously. No matter how long you permit the hot-water faucet to remain open, hot water continues to pour out.

It would appear that the instantaneous heater is the best type to have; you never run out of hot water. This is true enough if you don't draw too much. Instantaneous heaters are rated in gallons per minute for a specified temperature rise. For example, if the unit will produce 2 gpm and raise the water's temperature 100 degrees and the water enters

An immersed, tankless hot-water heating coil partially removed from its niche in a hot-water space-heating furnace. *Courtesy Weil-McLain Corp.*

at 80 degrees, you can draw 2 gpm at 180 degrees out. If you draw more water the temperature will go down. If the incoming water is colder than 80 degrees, the water coming out will be cooler. If you draw more than the rated 2 gpm the water will be cooler.

While there are electrically powered instantaneous water heaters manufactured and used, the types almost always used in the home are "powered" by the house heating furnace. These are available in two designs.

One type is called direct, or immersed. It consists of a copper coil immersed within the hot water inside the heating boiler, which may be a steam boiler or a hot-water space-heating boiler. The copper coil is in direct contact with the boiler water. The rate of heat transfer is high.

The other type is called indirect or take-off. It consists of a copper coil wrapped around

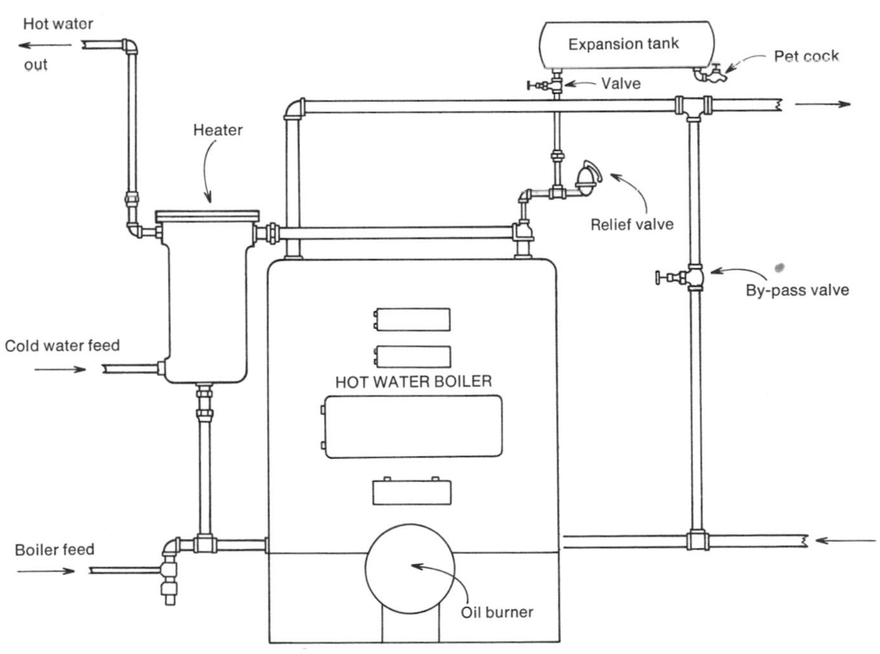

Typical indirect, tankless hot-water heater. Note the position of the bypass valve and its associated piping.

a heating pipe leading from the heating boiler to the radiators and in some designs back to the boiler itself. This type is called indirect because the copper coil carrying the domestic hot water is not in direct contact with the water in the boiler.

The indirect type is far less efficient than the direct, immersed coil type, and while both types require the furnace to be working in order to heat the water, the indirect type "throws" much more heat into the cellar. This can be a nuisance in the summertime. The only good thing that can be said for the indirect type is that it is more easily repaired and replaced.

Both types are called tankless because normally they are not used in conjunction with a tank. However, the coils themselves and the associated piping do store a quantity of hot water so that the tankless heater always has a small amount of hot water ready to be drawn off.

TANK-TYPE HEATERS

A tank hot-water heater incorporates a hot-water storage tank and a heating means. In the days gone by the heating means was a coil of copper pipe over an open gas flame or a little pot-bellied, coal-burning stove. The tank was of bare iron and separated from the gas or coal fire. Today, coal is no longer used. The tanks are insulated and the gas flame or electrical heating coil are built into the tank.

A new, electrically heated tank is close to 100 percent efficient in transferring electrical power to hot water. A new gas-fired tank is rated at 70 percent efficiency. The efficiency of an oil-burning furnace is around 75 percent. However, most of the heat goes to heat your house, so that the efficiency of an oil-burning furnace in relation to the domestic hot water it produces is very low.

ESTIMATING COMPARATIVE HOT-WATER HEATING COSTS

Table 25 lists the Btu's produced by gas, electricity and heating oil. If you know the cost of gas and the cost of electricity, you can readily make a comparison between these two fuels. Just be certain to figure the efficiency quotients of the two different fuels.

The comparison between the cost of heating domestic hot water by tank heater and by a "furnace" heater is difficult to make. As Table 26 indicates, oil produces a lot of heat, but little of it ends up heating the domestic hot water. No hard figures are available, but the consensus is that it is far cheaper to use the tank than the furnace for water alone. Thus the tank heater is cheaper in the summer, in most installations and localities, and the furnace is more economical in the winter when you are heating hot water in addition to heating your home.

SELECTING A HOT-WATER HEATING UNIT

Tank vs. furnace. If you live in the South where you rarely need to heat your home, obviously you will select a tank-type water heater. If you live up North where you will

Table 25

Electricity	**3413 Btu/kilowatt hour**
Natural and mixed gases	800 to 1200 Btu/cubic foot.
Sewage gases	600 to 800 Btu/cubic foot.
Lp gases	2500 to 3300 Btu/cubic foot.
Number 2 heating oil	140,000 Btu/gallon.
Number 4 heating oil	148,000 Btu/gallon.
Number 5 heating oil	152,000 Btu/gallon.

Heat produced by electricity and various gases and oil. A Btu (British thermal unit) equals sufficient heat to raise 1 pound of water 1 degree.

Table 26

HIGHBOY MODELS

	TANK CAP. GALS		**30**	**30**	**40**	**40**	**52**	**52**	**66**	**66**	**82**	**82**	**120**	**120**
	Elements		1	2	1	2	1	2	1	2	1	2	1	2
	Hgt. of Heater		44⅞	44⅞	58½	58½	58⅜	58⅜	58½	58½	59½	59½	61¾	61¾
	Jacket Dia.		17¾	17¾	17¾	17¾	19¾	19¾	22¼	22¼	24¼	24¼	28¼	28¼
Heating Unit Wattages	**N.E.M.A. Standards 230 V.A.C.**	Upper		1000		1250		1500		2000		2500		3000
		Lower	1500	600	2000	750	2500	1000	3000	1250	4000	1500	6000	2000
	Max. U.L. Approved 230 V	Upper		6000		6000		6000		6000		6000		6000
		Lower	6000	6000	6000	6000	6000	6000	6000	6000	6000	6000	6000	6000

LOWBOY MODELS

	TANK CAP. GALS		**20**	**20**	**30**	**30**	**40**	**40**	**52**	**52**
	Elements		1	2	1	2	1	2	1	2
	Hgt. of Heater		31½	31½	29½	29½	31½	31½	39¾	39¾
	Jacket Dia.		17¾	17¾	22¼	22¼	24¼	24¼	24¼	24¼
Heating Unit Wattages	**N.E.M.A. Standards 230 V.A.C.**	Upper		1000		1000		1250		1500
		Lower	1000	600	1500	600	2000	750	2500	1000
	Max. U.L. Approved 230 V	Upper		6000		6000		6000		6000
		Lower	6000	6000	6000	6000	6000	6000	6000	6000

Courtesy Rheem Mfg. Co.

Dimensions and performance of typical tall and squat electric hot-water heaters.

be heating your home much of the time, a furnace-type hot-water heater is your best bet. If you live somewhere between these two extremes, you may find it economical to install a tank heater in addition to installing a furnace with a hot-water coil, in one form or another. Doing so enables you to switch from tank to furnace with the seasons.

Gas vs. electricity. The choice of gas or electricity depends on which Btu's are less expensive—those produced by gas or those produced by electricity. However, economics are not the only consideration. There is also the matter of the location of the hot-water heater.

Electrically heated hot-water tanks can be located anywhere you can run an electrical line. They may be in direct contact with your home's walls. Gas-fired heaters must be ventilated. This means there must be provisions for the admittance of fresh air and a code-approved flue to bring the exhaust fumes safely outside the building. In addition the heater must be clear of all combustible materials by a code-specified distance, and the walls, floor and ceiling may have to be of fireproof material. You can get all this information from your local building department. Thus, while you can install your electrical unit anywhere, you are somewhat limited as to where you can place your gas-fired heater.

Table 27

	Tank Cap. Gal.	Heat Input in Thousand Btu's. Nat.	Mxd.	Mfd.	L.P.	Rec. 60° Rise G.P.H.	Rec. 100° Rise G.P.H.	Approx. Ship. Wt. (lb.)
HIGHBOY	30	45	45	45	—	63.0	37.8	116
	30	—	—	—	38	53.1	31.9	
	40	45	45	45	—	63.0	37.8	150
	40	—	—	—	38	53.1	31.9	
	50	50	50	50	—	70.0	42.0	225
	50	—	—	—	45	63.0	37.8	
LOW-BOY	30	36	36	36	36	50.4	30.2	128
	40	38	38	38	—	53.1	31.8	190
	40	—	—	—	36	50.4	30.2	

Courtesy Rheem Mfg. Co.

Performance data of gas-fired, tank-type heaters.

Typical gas-fired storage-type hot-water heater. Note metal flue pipe and the fact that heater is positioned near a nonflammable concrete wall.

Tank capacity. If you have plotted your family's use of hot water you know how many gallons they need at various times of the day. If most of the water is used once or twice a day with several hours between, select a unit that contains 140 percent more hot water than you need at those periods. The reason for the increase is that no tank furnishes more than 70 percent of its holding capacity since cold water enters as hot water is drawn out. Therefore, when you have quickly drawn off some 70 percent of the tank's contents the water runs cold. Thus, if you need 70 gallons in a short period, you must install a 100-gallon tank. In addition, I strongly suggest you buy a tank that is oversized by 25 percent to take care of heating losses due to aging.

Now let us assume another household condition. Instead of requiring 70 gallons of hot water once in the early morning and once late at night, your family needs 50 gallons of hot water at 10 AM and another 50 gallons at 11 AM. During the rest of the 24 hours very little hot water is used.

Table 28

Number Bathrooms	Number Bedrooms	Laundry	Size (in gallons)
1	1 or 2	No	30
		Yes	40
2	3 or 4	No	40
		Yes	50*
2	4 or 5	No	50
		Yes	75*
3	4 or 5	No	75
		Yes	100*

* In many instances the use of two water heaters at point of use will be more practical. If two water heaters are used, the combined capacity should equal or exceed the size shown above.

Courtesy American Gas Association.

Suggested gas-fired, tank-type water heaters for various family needs.

Well, a heater with a tank of 70 gallons will easily meet your needs at 10 AM. But, *will there be another full tank* by 11?

Recovery time. Domestic hot-water tanks and associate heating equipment are rated by two figures—tank capacity and recovery time. Capacity is obvious. It is measured in total gallons and should be given on the tank's nameplate. What isn't obvious and rarely mentioned is recovery time; the time it takes for the tank to come up to temperature when it is filled with cold water. When it is mentioned it is usually given as "fast" or "rapid," which means nothing. What you need and should secure before purchasing any domestic hot-water heater is the actual time it will take that tank to heat a given quantity of water a given number of degrees.

In our example, you required 50 gallons of hot water at 10 and 11 AM. A 70-gallon tank will do fine for the first load because it has had all night to heat up, but in order to deliver another 50 gallons of hot water an hour later it must be able to "recover" fully in one hour. So when you select your tank you must also consider its recovery time as well as total capacity. If the recovery rate you require is faster than that which is available, you will have to go to a larger tank to satisfy your needs.

At the same time, it is uneconomical to purchase a tank with a far higher recovery rate than you need. The faster the tank recovers the more energy it requires. In the case of an electrical tank this means larger diameter wires. In the case of gas, larger diameter pipe, and in both cases a more expensive unit.

Instantaneous heater capacity. Direct and indirect tankless hot-water heaters are rated in gallons per minute. This figure is dependent upon boiler water temperature, which in the case of a hot-water boiler is usually 200 degrees F, so far as the specification is concerned. If you set the thermostat on your boiler to a lower figure, the hot-water coil will produce less hot water per minute.

As you rarely draw water directly from a tankless heater, but mix it with some cold water, you can figure that the manufacturer's figure can be increased by perhaps 25 percent. If the water in the boiler is 200 degrees F, the water in the coil will be approximately 10 to 25 degrees lower in temperature. As anything over 130 degrees is close to scalding, you wouldn't adjust your mixing valve to permit water much hotter than 150 or 160 degrees to enter your domestic hot-water system. The higher temperature is acceptable as the water will cool somewhat before reaching a faucet.

If you refer to Table 15 in Chapter 9 you can compute just how much hot water you will need on a gallon-per-minute basis. For example, a tub requires 6 gallons of water per minute. If you mix the hot water coming out of the coil with 25 percent cold water, your hot-water requirement is down 25 percent or 1.5 gpm. If you further mix cold water with the hot at the tub, your instantaneous heater capacity is further reduced, perhaps another 1.5 gpm. Thus an instantaneous heater with a capacity of approximately 3 gpm would take care of the bathtub.

What happens if your instantaneous heater's capacity is exceeded? The hot water divides among the open faucets and runs lukewarm. As it is financially impractical to install an instantaneous heater that can supply all the hot-water faucets at one time, this is always a problem with instantaneous heaters. On the other hand, a tank can supply all the faucets with hot water at one time, but it soon empties.

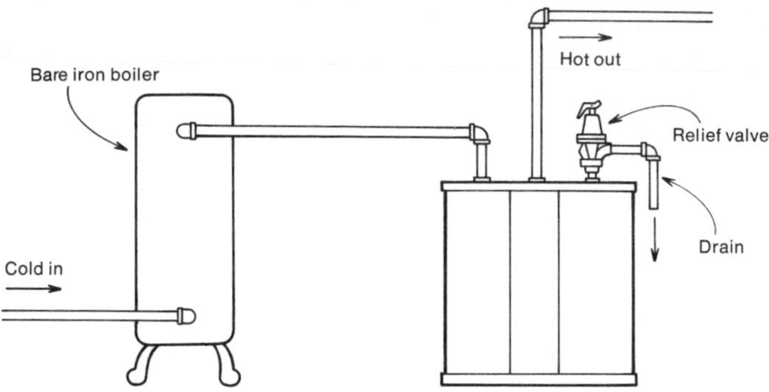

How a bare metal boiler is used to preheat cold water before it enters a water heater. In this illustration the heater is a storage-type heater, but the same arrangement may be used with any type of domestic hot-water heater.

CONNECTING HOT-WATER HEATERS

Hot-water heaters of all kinds are connected to the hot-water system by the simple expedient of connecting a pipe to the cold-water supply and a second pipe to the hot-water system. The cold-water connection is made wherever convenient. The same is done with the hot-water connection.

The only precaution you must keep in mind is that the heater should not "hang" from the pipes—that is, there should be no stress on the pipes leading into the tank or furnace coil.

You may connect as many tanks and coils as you wish between the cold-water supply and the inlet to the tankless or tank-type heater. Do not use plastic pipe as it is self-insulating and do not use pipe insulation or insulated water tanks for the same reason. There is no limit as to the size of your preheating assembly, just do not make the pipe so long or of so small a diameter that there is no pressure left to supply the faucets.

Two tank heaters can be connected; the best way is to connect them in series. Connect the hot-water outlet of one to the cold-water

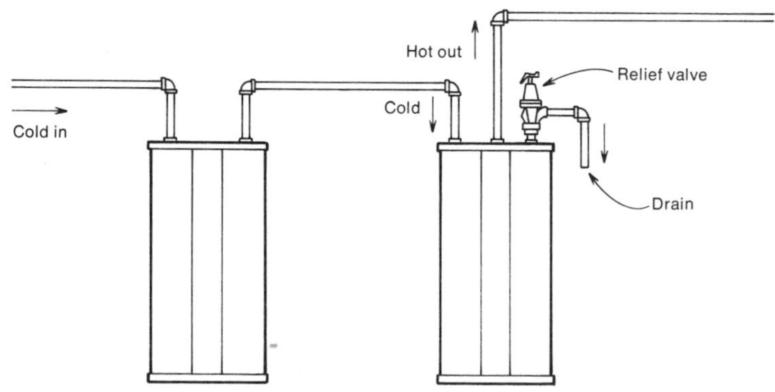

How two storage heaters may be connected in series to provide a greater capacity of hot water than possible with only one. Heater at left does most of the work, so it must be connected so it can readily be interchanged with unit on right.

inlet of the succeeding unit and then on to the hot-water pipes. With this arrangement the full capacity of the tank nearest the hot-water line is available plus 70 percent of the following tank.

The tank close to the cold-water feed pipe will do most of the work because its temperature will drop first. If you install this arrangement, use unions so that you can switch the tanks around every year or so. This arrangement can be used with tanks of any size; they needn't be alike in size or even heating means.

If you connect your tanks in parallel, the two tanks should be close to the same size and have approximately the same rate of recovery. If one tank is much smaller than the other it will run out of hot water first. If additional hot water is drawn, cold water coming in through the smaller tank will chill the hot water still coming out of the larger tank. The result will be that you will not have a total hot-water capacity much greater than that of the small tank alone.

The accompanying drawing shows the connections for switching from a tank to a tankless heater. Use gate valves if you do not want to lose pressure; globe valves if you do. Remember that you must be certain to turn off the heat under the tank or coil when it is not used. If you shut off the valves with the heat on, there will be an explosion.

HOW TO GET MORE HOT WATER FOR YOUR MONEY

Several methods are used to increase or stretch the quantity of hot water available. One, which was mentioned briefly before, is mixing. This method is almost always used

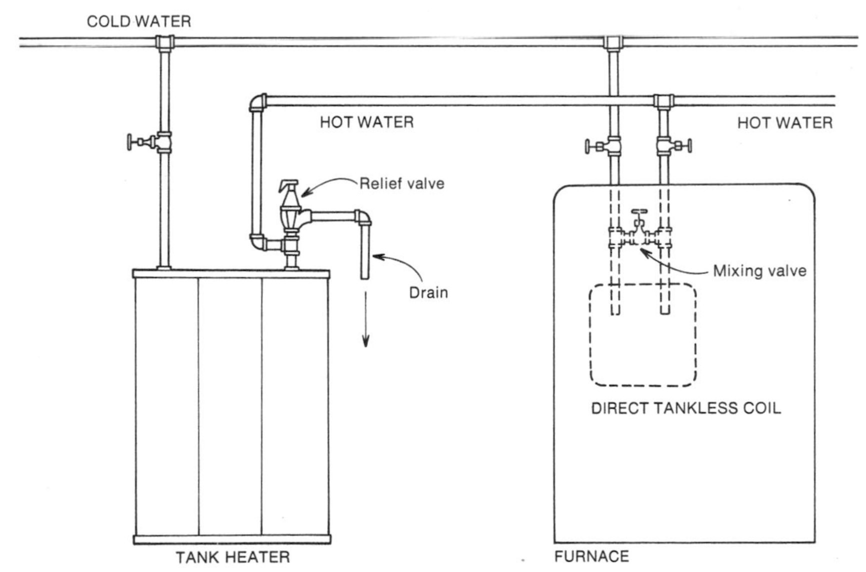

How a storage heater may be connected in parallel with an instantaneous, or tankless, heater for switching with changing seasons. Valves leading to the tankless coil must never be closed unless the furnace is shut down and cold. Neither can the valve leading to the tank heater be shut until it is shut down and cold. Operating either heater with the valve or valves closed can cause a dangerous explosion.

with tankless heaters. A special "mixing" valve is installed connecting the cold- and hot-water pipes leading in and out of the water heater. This valve differs from the ordinary in that to some degree it compensates for the changes in the temperature of the hot water coming out of the heater.

Mixing valves are generally not used with standard tank heaters as their output temperatures are too low. However, there are tank heaters made to operate at 180 degrees F, and mixing can be used advantageously at this temperature.

As a rough rule of thumb, the use of an 180-degree tank in place of a 140-degree tank can result in an approximate 40 percent flow increase based on 40 degrees water coming in and water at about 140 degrees going out.

Note carefully, however, that the 40 percent increase in hot water is not free. You pay for this in increased energy input and slightly increased energy waste during standby periods. The hotter tanks lose more heat by radiation. The only savings is in unit size. If you are pressed for space, this may be the way to go.

A second method of increasing available hot-water supply is to preheat the water fed to the heater. This may be done by connecting a bare (uninsulated) boiler in the pipeline ahead of the heater. "Waste" heat in the cellar warms the supply water a little. In warm climates it is customary to install "hot-water" coils on the roof of the building. Water so warmed needs very little additional heat to become "hot" water.

The third method is slightly dishonest. You don't provide or make more hot water, you stint on its use. This is done by deliberately inserting a section of small-diameter pipe in the hot-water feed line, or a globe valve. Both act to reduce pressure in the hot-water line. The result is that less is used. If this seems unfair, remember that at 10 psi a shower will release 5 gpm. When you think about your children taking an hour-long shower, it doesn't seem dishonest at all. Incidentally, plumbers very often use this method.

If your hot-water heater is a long distance from the fixtures it serves, it is good practice to insulate the pipelines. There is pipe insulation made for small-diameter water pipe.

LIMING

The most common problem in the hot-water system is loss of heating capacity. The water gets as hot as it ever did but there just isn't as much water available. This is most often caused by a condition called liming.

Liming is the formation of a hard, insulating layer of "lime" inside the heating tank or coil. Liming is heat dependent. Minerals dissolved in the water deposit out on the hottest portion of the metal the water contacts. The rate of deposition depends on temperature and water mineral content.

You can draw a little more hot water out of limed equipment by raising the temperature to which the tank or furnace thermostat is set. You can adjust the mixing valve, if there is one, to pass more hot water and less cold.

At best these efforts are short lived and to some extent self-defeating. The rate of liming increases with temperature, so when you up the temperature of your hot water you also increase the rate of liming. Also, as the temperature increases you reach a point where the hot water emerging from the faucet is dangerous.

Generally no effort is made to remove the lime from the inside walls of a storage-type hot-water heater. It is replaced when the layer of lime is so thick the tank does not perform properly. Direct and indirect, instantaneous hot-water coils can sometimes be freed of lime. The method used is as follows.

In the case of both a steam furnace and a hot-water space-heating furnace, the furnace is shut down and permitted to cool to the point where you can touch the boiler itself.

The valve in the feed pipe bringing cold water to the hot water coil is closed. If there is a second valve to the other side (hot side) of the coil, that too is closed.

If there is no second valve adjoining the hot-water coil, open a hot-water faucet on the first floor and *then* open a second hot-water faucet in the basement. Wait until all the water drains from the hot-water pipes.

Next, disconnect the ends of the coil from its associated piping. Unscrew the unions or unsolder the joints as the case may be. Blow through one end of the coil to remove most of the water. Take a cork and seal the end of the lower pipe.

Take a short length of copper, brass or plastic pipe and connect it to the higher, still-open end of the hot-water coil. Use the union, if there is one, or sweat the length of pipe in place. Do not use iron, galvanized or chromium-plated pipe. Bend the pipe you have added so that its open end is 6 inches or higher than the coil end. Now place a plastic funnel in the open end of the pipe.

Next put on a pair of rubber gloves and pour a mixture of hydrochloric acid and water—one part acid to three parts water—very slowly into the coil. If there's a lot of foaming and gasing, stop. Wait until the action slackens and then pour in a little more of the acid and water. Foaming indicates the presence of lime. When the coil is full and there is no more foaming the acid has dissolved as much lime as it is going to. Let the mixture remain for a half hour or so, then drain and rinse with cold water for ten or fifteen minutes.

You can if you wish check on whether or not there is any more lime inside the coil by adding a fresh mixture of acid and water. If foam appears, there is more lime. If no foam or gas appears there is little or no lime present.

There is no doubt that you can remove all traces of lime from inside the coil. However, as the temperature of the coil is not even, the lime's thickness will not be even. Therefore, there is a strong possibility that the acid will contact and dissolve the copper. The possibility of the acid eating a hole through the copper tubing increases with repeated applications and extended exposure to the acid.

In the case of a steam boiler, you may be able to tolerate a little leakage. All that will happen is that water will be introduced into the boiler by other than normal means. As a small quantity of water is always added to a steam boiler, this usually does no harm.

In the case of a hot-water space-heating boiler, the normal pressure of the domestic hot-water line is usually higher than the normal pressure in the boiler. Water from the leaky coil will enter the boiler and cause the pressure-release valve to open and water to drip out. You cannot operate your boiler this way for any length of time because the release valve will deteriorate and the leak will increase.

But hold on a moment. If your expansion tank is filled with water the same result will occur. So make certain your tank is empty before condemning the coil. See Chapter 15 on Hot-Water Space Heating Systems.

Removing all the lime from inside a hot-water heating coil does not necessarily complete the job. There is always the possibility that the outside of the coil, immersed in hot water, is also limed up or badly corroded. There are two ways to find out. Remove the coil and inspect it, or try it. Generally, it is easier to try it. To do so, the attached pipe is removed and the coil reconnected as it was before.

If the results are not satisfactory, or if the inside of the coil is so heavily limed as to make it next to impossible to fill with acid, it is necessary to remove the coil from its niche within the boiler. When you have it out, you can either treat the coil further or replace it.

To remove a direct, tankless heating coil, the valve in the cold-water feed line is closed. If there is a second valve, *it* is closed. If not, the entire hot-water line is drained. Following, the furnace is shut off and permitted to cool sufficiently to permit you to touch the iron boiler itself. If you are working on a steam boiler, shut the boiler feed valve and drain a few gallons of water from the boiler itself (see Chapter 16).

If you are working on a hot-water space heating boiler, do the same, but as you will have to drain the entire radiator system to get the water in the boiler below the hot-water coil, connect a garden hose to the boiler drain connection and lead the water outside. In addition, open one or more of the radiator bleed valves to speed draining.

At this point your furnace is cool. Its boiler is empty of water to a point below where you judge your coil to be. The hot-water heating coil is disconnected.

If your coil is mounted on a plate held in place by a number of bolts, unscrew the bolts with a wrench. If the bolts cannot be turned, apply liquid wrench or any other rust solvent, tap the side of the bolts lightly with your wrench and try again. Do not "strong arm" the bolts. If you break one off you will have to drill it out and retap it. Otherwise there is a good possibility you will have a leak under the missing bolt when you reassemble the hot-water heater.

If your coil is mounted on a huge iron plug, loosen and remove the bolts holding the plug by its "ears." Now tap the side of the plug lightly with a hammer. If it doesn't come out, apply a generous dollop of liquid wrench to the joint between the plug and the wall and tap some more. *Do not apply leverage beneath the plug's ears.* The plug is of cast iron and you will break it.

With time, lots of tapping and plenty of rust solvent the plug will come out.

The copper coil is brazed to the plug or the pressed metal cover. You can purchase a replacement coil mounted on a plug from your local plumbing supply house. But if you take the trouble you can find a shop specializing in brazing coils of this type. The savings will be spectacular. Be certain, however, when securing a replacement coil that it is of ridged, finned copper. A smooth coil does not have the necessary surface. Be certain it is as long or longer than the old coil and that it is no larger in overall dimensions or it may not fit into its space.

Reinstalling the hot-water coil. If your coil is mounted on a cast-iron, tapered plug, take some steel wool and polish the edge of the plug and the edge of the furnace wall where they mate. Rub until the metal is clean, bright and rust-free. Check the temperature of the plug and of the furnace wall with your hand and wait until they are both nearly the same. Insert the coil part way and turn the plug until its pipes line up with the pipes to which they must connect. Then push the plug all the way in and use a wooden mallet to gently tap the plug home. Following, replace the holding bolts, taking great care not to overtighten them and break the cast-iron ears.

If your coil is mounted on a flat metal plate that is sealed to the furnace wall with a gasket, remove the old gasket and replace it with a new one after making certain the metal underneath is clean. If you can't secure the proper gasket, you can use automotive gasket-forming cement.

Deliming indirect coils. The same methods used with the immersed, direct coil may be used with the indirect, exposed hot-water heating coil. Since these coils are exposed you can get to them fairly easily. In some designs the coil can be separated from the pipe. With these you just shut off the hot water before going to work. With other designs the heating coil is integral with the pipe leading to the furnace. If it is a steam furnace, just shutting it down and letting it cool is all you need do before disassembly. If it is a hot-water heating system, you have to drain the boiler, as previously explained.

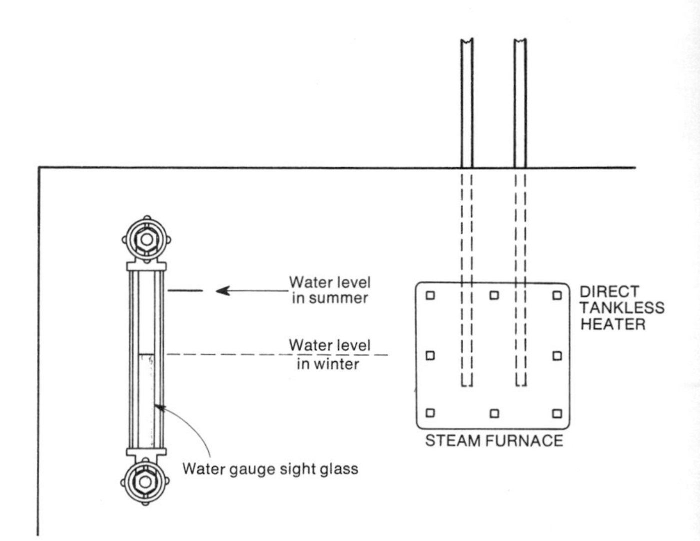

Proper water level in a steam boiler during winter and summer. Water is raised to summer height by manually opening the cold-water feed valve.

OTHER CAUSES OF REDUCED HOT WATER

Low boiler water. Normal winter operation usually calls for the water level in a steam boiler to appear at midpoint on the sight glass. If the water level drops below this point the immersed hot-water coil is above the water and heat transfer is severely reduced. Persistent low boiler water is a fault of the automatic feed mechanism and should be corrected. See Chapter 16, Steam Heat.

In the summer you will get more hot water if you raise the level of the water in the boiler. You can do this by manually opening the water-feed valve.

Closed bypass valve. Modern hot-water space-heating systems have a pipe that connects the radiator feed pipe to the radiator return by means of a pipe and valve. You can easily recognize these pipes as they are relatively large in diameter in comparison to the domestic water pipes.

This pipe and its valve is called a bypass, and its purpose is to permit a portion of the water driven through the heating system by the circulating pump to go around the furnace. Should this valve be fully closed when your thermostat calls for heat and the circulator starts, all the cold water in the radiators will pass through the furnace, the immersed coil will be chilled and the hot water will suddenly run cold.

To prevent this from happening a portion of the radiator flow is shunted around the boiler through the bypass. The proper position of the bypass valve is something you have to find by experiment. Generally, about half open is satisfactory for most homes.

Incorrectly adjusted mixing valve. When the mixing valve we discussed a few pages back is improperly adjusted or defective it will permit too much cold water to mix with the hot water. The result is that you never have really hot water no matter how well your hot-water heater is working.

Adjusting the mixing valve. Valve shown has three pipe connections. Two pipes at top are hot-water inlet and outlet; pipe at bottom is cold-water connection.

Try turning the valve handle all the way in one direction and checking the results. Then try it all the way in the other direction. If there is no change in the temperature of the water coming out of the nearest hot-water faucet, the valve is defective and should be replaced. If there is a change, try various positions until you find what is best for you.

POPING VALVES

This is the term applied to the practice of opening a safety valve to test it. The correct procedure is to increase pressure and/or temperature until the valve opens. This is difficult and dangerous in the home. Instead, blow the valve mechanism clean every once in a while. Make certain nothing is piled on it. Test it once every six months or more by very gently lifting the lever until water flows. Let the water flow a few minutes to make certain the valve faces are clean. Release gently. There is no need for daily or even monthly poping. And remember, if you lift the little arm too high you will permanently damage the valve.

SAFETY

All the tank-type hot-water heaters must be protected by a combination temperature and pressure-relief valve. The temperature sensing element must be in contact with the top 6 inches of water and set to release at no higher than 210°F. The pressure-relief element must be set to open at not less than 25 psi above working pressure, and always below the tank's working pressure, whatever that may be. The valve must be sufficiently accurate to work at no more than 10 percent of the pressure to which it is set.

When the combination pressure/temperature sensing valve opens, it releases water from the tank. The drainpipe leading from the safety valve is never directly connected to the house drain system. Instead it must empty into an open tub, or as is common, onto the cellar floor. Having the safety valve drain into the open air eliminates the possibility of a plugged drain preventing the safety valve from doing its job.

11

Designing a Drainage System

When you add a couple of sections of pipe to your domestic hot- or cold-water line, or even install a completely new system, it is pretty hard to go far wrong. At worst, you may not have as much pressure at the faucets as you wanted or there will be less hot water than you anticipated. When you err in the planning and designing of a drainpipe or a system, lots can go wrong.

Unless the drain, be it single pipe or complete system, is properly designed there is a good chance the new drain will admit sewer gas, fail to drain properly, clog up intermittantly and even admit soil backing up from a nearby toilet.

Drains are fairly simple, but they must be designed to follow the principles developed over the centuries by countless generations of plumbers and as specified by local or national code.

As with hot- and cold-water design, you have several approaches to drain design. You can see what your neighbor has done. You can discuss your system with your local building department, or you can plan your own drain system following the few general rules and using the tables provided in this chapter. If this is your first plumbing job you would be well advised to do all three; see what is being done, plan your system and then go over it with your plumbing department. But under no circumstance should you merely toss the drainpipes in under the assumption that, everyone knows water runs downhill.

SEWER GAS

Sewer gas is a strange, poisonous, explosive and smelly substance. Its smell is its least dangerous quality. Sewer gas is mainly the result of anaerobic bacteria digesting excreta, waste food and other matter that enters a sewer main and septic tank.

Sewer gas consists of methane (natural gas) hydrogen sulfide, gasoline vapor, carbon monoxide, acetylene, hydrogen, ammonia, carbon dioxide, nitrogen and oxygen. Usually the oxygen content is very small. Men entering a closed sewer without first thoroughly ventilating it have died from asphyxiation. Methane, gasoline vapor, hydrogen and acetylene are highly explosive in the presence of oxygen so that partially ventilating a sewer can make its content liable to explosion. Hydrogen sulfide and carbon monoxide are extremely poisonous. Hydrogen sulfide and other gases found in sewer gas are highly corrosive.

The strangeness of sewer gas lies in its varied effect upon humans. Some men have worked for years in moderately ventilated city sewers, breathing in some sewer gas with each working day, their health apparently unaffected. Other men have died from gas poisoning shortly after entering a manhole.

Traps. To prevent sewer gas from entering a building, a number of traps are installed in the drainage system. Each trap consists of a prefabricated bent section of pipe. The bend is positioned so that a portion of the water flowing through the trap remains in the bend. The plug of water normally suffices to seal the drain and prevent the low pressure sewer gas from passing. The trap in the drainpipe leading from each fixture is called a drain trap. The large trap in the house drain just before it passes through the foundation wall is called the main trap, house trap or running trap.

VENTING

The drainage system is vented—connected to the open air— by one or more vent pipes for several reasons. Venting prevents the water seal in a trap from being blown or sucked out by siphoning, aspiration (jet pumping) and/or wind pressure. Venting cannot, of course, prevent loss of seal by evaporation. But this is rarely a problem in an occupied building.

Venting also reduces corrosion by releasing the aforementioned hydrogen sulfide gas and by discouraging the growth of anaerobic bacteria which produce sewer gas. Venting releases to the open air whatever gases and smells may form within the building's drainage system so that in the event of trap-seal failure at a fixture, less or no smell and gas will enter the building.

Inlet
Outlet
Depth of seal
Weir level
Cleanout plug

Basic parts of a trap. Note that weir level is as high as water will of itself remain in the trap.

Typical plastic pipe stub-out for draining a kitchen sink. When the wall is in place a plastic or thin-wall brass trap will be connected between the stub-out and the sink. Venting is accomplished by the reduced plastic stack which continues up through the roof.

Table 29

	Pipe Diam.	Maximum Fixture Units Allowed: Fixtures on Same Level ⅛-in. Pitch	¼-in. Pitch	½-in. Pitch	Fixtures on 2 or More Levels ⅛-in. Pitch	¼-in. Pitch	½-in. Pitch
No toilets	1¼	1	1	1	1	1–2	1–2
	1½	2	2	3	2	2½–5	3½–7
	2	5	6	8	7	9–21	12–26
	2½	12	15	18	17	21	27
	3	24	27	36	33–36	42–45	50–72
No more than 2 toilets	3	15	18	21	24–27	27–36	36–48
Any number of toilets within fixture unit capacity	4	82	95	112	114–180	150–215	210–250
	5	180	234	280	270–400	370–480	540–560

Maximum capacity of various diameter drainpipes at various pitches as measured in fixture units. Table may be used for branches, house drains and sewers.

And venting also aids the flow of large volumes of liquid through the drainpipes by preventing air pressure from building up ahead of the water as it moves downward.

BASIC DRAIN COMPONENTS

It can be seen that there must be a drainpipe leading from each fixture to the main sewer or septic tank in order to remove unwanted waste and soil. It can be seen that a trap is necessary in the drainpipe to prevent sewer gas from entering the building. And as explained, it is necessary to vent the drainage system to prevent removal of the water seal in the trap and to air the drain pipes. Therefore, a modern drainage system consists of drainpipes, traps and vents properly interconnected.

SIZING DRAINPIPES

Choice of pipe size for fixture drain, branch drain, house drain and sewer, which is simply a continuation of the house drain may be made with the aid of Table 29. The table is based on fixture units, which are given in Table 30, and the pitch of the pipe as well as its size. Note also that there is a very definite distinction between waste discharge and soil (toilet) discharge.

Table 30

Fixture	Fixture Unit Values*
Bathtub (with or without shower)	2
Dishwasher (home)	2
Drinking Fountain	1
Kitchen Sink	2
Lavatory	1
Laundry	2
Service Sink	2
Shower (separate, each head)	2
Washing Machine	2
Water Closet (tank)	3
Water Closet (flush valve)	6
Bathroom Group (tank)	6
Bathroom Group (flush valve)	8

* Values are for total discharge, hot and cold. For either hot or cold discharge by itself, use 75% of the total fixture unit value.

Courtesy Copper Development Association.

Estimated fixture unit values. Each unit is equal to a flow of 7½ gallons of water *out* of each fixture.

Table 31

Diameter of Pipe	Velocities 1/16-Inch Fall per Foot	1/8-Inch Fall per Foot	1/4-Inch Fall per Foot	1/2-Inch Fall per Foot
Inches	fps	fps	fps	fps
1¼	0.80	1.14	1.61	2.28
1½	0.88	1.24	1.76	2.45
2	1.02	1.44	*2.03*	2.88
2½	1.14	1.61	2.28	3.23
3	1.24	1.76	2.49	3.53
4	1.44	*2.03*	2.88	4.07
5	1.61	2.28	3.23	4.56

Approximate velocity of soil and waste flow in pipes of various diameters at different slopes. The National Plumbing Code requires a minimum flow of 2 feet per second.

At this point two ideas may come to mind: Why not use larger-diameter drainpipe and run it at a greater pitch, thus eliminating all possibilities of drain and sewer clogging? Unfortunately it doesn't work out that way. As shall be explained, you can increase the possibility of drains clogging up by using oversize pipe and by running them at too steep a pitch.

Pipe diameter and flow. Though it is true that the larger the pipe diameter, the more liquid it can handle, it has been found that optimum drainage condition occurs when a pipe is about one-third full. Maximum scouring action is produced by the flow and the solids ride high and clear. When the pipe diameter is increased for an equal gpm flow, the water moves more slowly, the solids tend to cling to the bottom of the pipe and it soon clogs. It is a mistake to install larger than required drain pipe.

Pitch. All drainage pipe lines are pitched downward towards the sewer pipe. All properly installed drainpipe has a pitch of no less than 1/8 of an inch and no more than 1/2 inch to the foot. When this pitch is impossible

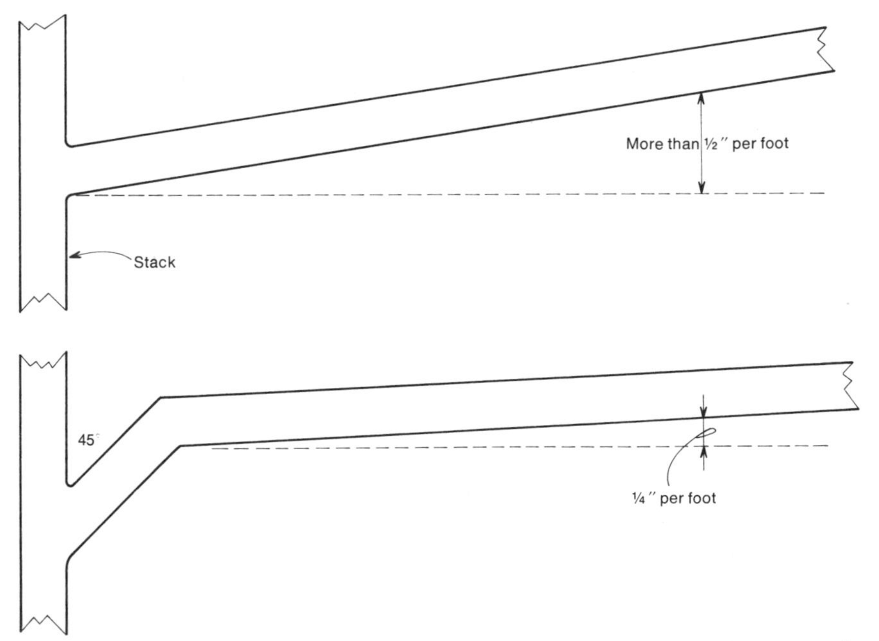

If you have to pitch any drainpipe more than ½ inch to the foot to reach a stack, run a section of the pipe at 45 or more degrees and the balance at ½ or ¼ inch to the foot.

or impractical, the pipe is pitched at 45 degrees (1 foot drop for every foot of pipe length) or more.

It has been found by means of a series of noxious tests that soil flows best in a pipe pitched at ¼ inch to the foot. A pitch greater than ½ inch to the foot causes the liquids to run off and leave the solids behind. In time the drain will plug up. Pipes pitched at less than ⅛ inch to the foot do not provide sufficient water velocity and the solids tend to settle and clog and there is insufficient scouring action. Given a choice of the two pitch limits, it is wiser to opt for the ½-inch pitch, it is easier to install. The ⅛-inch pitch require extreme care to prevent "valleys" in the pipe run. Pipes pitched at 45 degrees or more have no problem. The solids slide right down.

Lower end of a plastic soil stack. Note that the trap is vented to take a washing-machine outlet hose.

STACK PIPES

A stack is a pipe stood on end. If it is a drainpipe carrying off nothing but water it is called a waste stack. If it carries water and excreta it is called a soil stack. If it merely vents the drain system it is called a vent stack.

Every building that has a drain is required by the National Plumbing Code to have a stack, which under certain conditions and local codes permitting may be reduced to half its diameter above the topmost fixture.

The stack serves several purposes. It is a convenient, central drain for the fixtures on various floors and it serves as a vent for the fixtures on the topmost floor.

Most local codes limit waste stacks to a minimum of 2 inches and soil stacks to a minimum of 3 inches. However, it is recommended that no soil stack be installed less than 4 inches in diameter of cast iron or its equivalent in copper or plastic.

When all the fixtures are on one floor and drain into the stack more or less at one level, Table 32 may be used. The table applies to both waste and soil stacks. When there are two floors or more, a branch interval exists between branch drains and Table 33 may be used. This gives branch interval capacities. To use the table select the pipe size recommended for the floor with the highest number of fixture units. The total number of fixture units discharged is ignored. The problem of discharging water into a vertical pipe is the pile up of water as it enters. The capacity of the vertical pipe itself is very high.

Table 32

Stack Diam. Inches	Total Length Feet (Max.)	Maximum Fixture Units With Only Y-fittings	With Sanitary Tees
1¼	50	1	1
1½	65	12	8
2	85	36	16
3	212	72	48
4	300	384	256
5	390	1020	680

Stack capacities in total fixtures units discharging into the stack at any given point.

Bends and offsets. If the stack makes a 45-degree or lesser turn that section is considered part of the stack. If the stack makes two turns in opposite directions the section above the "return bend" is reduced in drainage capacity by half. If the stack runs horizontally for a distance the horizontal section must be pitched and sized exactly as a horizontal drain must be.

Fittings. Only standard sanitary drain fittings can be used. Long sweeps (gently turning elbows) are recommended. Double Tees and hubs pointing downward should not be used.

Cleanouts. Everytime a stack or a drain makes more than a 45-degree turn, a cleanout fitting should be installed. If there is 45 feet or more of drainpipe following a cleanout, a second cleanout should be installed.

Stack termination. The upper end of all stacks—waste, soil and vent—should be at least 1 foot above the surface of the roof to prevent the wind from blowing objects into the open end of the pipe. When the stack terminates near a window, ventilator or door, the pipe end should be brought at least 3 feet above the opening to prevent sewer vapor from entering.

This offset soil stack starts as a cast-iron pipe and is then reduced to copper. Note lead toilet bend (upper left) and the wiped lead joint connecting it to the copper pipe. Cast-iron pipe is 4 inches in diameter; the copper pipe 3 inches.

At a minimum the stack should continue on through the roof without any reduction in diameter. In cold climates and when the pipe end is less than 3 inches in diameter, it is advisable to use an "increaser" at the top end of the pipe as it passes through the roof. A

Table 33

	Stack Diam. Inches	Maximum Fixture Units in One Interval—8 Ft. Length: With Only Y fittings	With Sanitary Tees
No soil discharge	1¼	1	1
	1½	4	2
	2	15	9
	3	45	24
Waste and soil	4	240	144
	5	540	324
	6	1120	670

Stack capacity per branch interval. Read the table in fixture units per floor. For example, using 4-inch pipe, and Wye fittings, 240 fixture units of discharge can be accommodated on each of several floors.

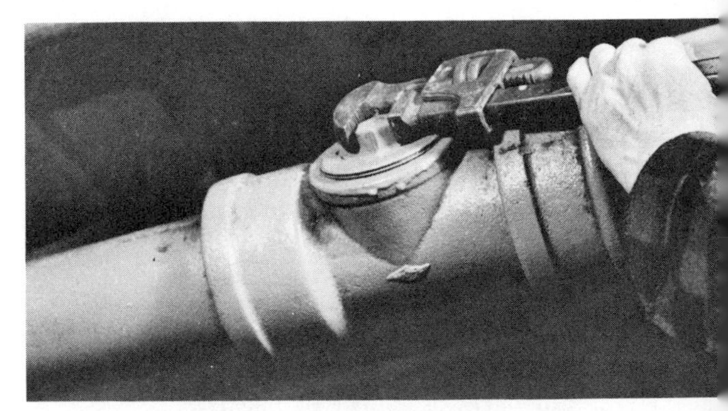

Cleanout installed in a building drain with plug positioned at an angle to the wall so it can be easily removed and the cast-iron pipe cleaned.

stack pipe will never freeze as long as the building is occupied and in use. However, the warm moist air can ice up the opening if it is small. With a 3-inch opening or larger, there is little danger of this happening. The end of the stack pipe is never closed. There is no harm in snow or rain entering the stack.

Some localities still permit storm (rain) water to be combined with soil and waste water and drained off into the municipal sewer. When a building has a flat roof, this is a great economy. The roof is pitched into the top of a drain stack and the rain water flows quietly out to sea. However, it is an extreme case of penny-wise dollar-foolish thinking. Should there be a rain storm when a house sewer is clogged, there will be a horrible flood. A combination of soil and rain water will pour out of every toilet, every tub, every sink and lavatory.

SIZING DRAINPIPES

Each fixture is drained by a length of pipe that connects to a trap incorporated (as in the case of a toilet) into the fixture or follows the fixture and leads to either the stack or a branch drainpipe that connects to the stack. To size the fixture drainpipe, one simply looks up the recommended pipe size in Table 34.

Pipe sizes shown have been selected to provide good velocity with the pipe about one-third full. This leaves room for larger than usual discharges. Neither smaller nor larger than recommended pipe sizes should be used. The smaller tend to clog and the larger size pipe slows the flow, reducing scouring action and increasing the tendency of the drain to clog.

When two or more fixtures discharge into any pipe other than the stack, that pipe is called a branch and must naturally be larger than the pipes that feed it. The junction between a fixture discharge pipe and the branch is always a troublesome spot. Reduced speed of flow tends to lower water temperature, and in the case of a kitchen sink discharge, the juncture between the small and large pipe becomes a collection point for grease. In effect a grease trap. For this reason it is always good practice to install a cleanout plug that is readily accessible at this fitting.

Table 34

		Pipe Size in Inches
Bath tub, bidet		1½
Shower stall		2
Lavatory		1¼
Toilet (direct to stack)	(copper)	3
Washing machine		1½
Dish washer		1¼
Kitchen sink		1½
Two sinks, one trap		2
Floor drain		2

Recommended pipe sizes for draining various fixtures.

LENGTH AND PITCH

Having studied the previous paragraphs and tables we know what size the drainpipe should be. What we don't know, and need to know for reasons that shall be explained, is just how far we can run horizontal drainpipe before we either vent it or connect it to the stack, and how much total drop we can allow. Total

Table 35

Waste Pipe Diameter	Maximum Units Per Branch	Soil Pipe Diameter	Maximum Units Per Branch
1¼	1	3	20
1½	3	4	160
2	6	5	360
3	32		

Capacities of horizontal drainpipe of various diameters at a slope of approximately ¼ inch to the foot, as measured in fixture units. Note that pipes are rated differently for waste and soil.

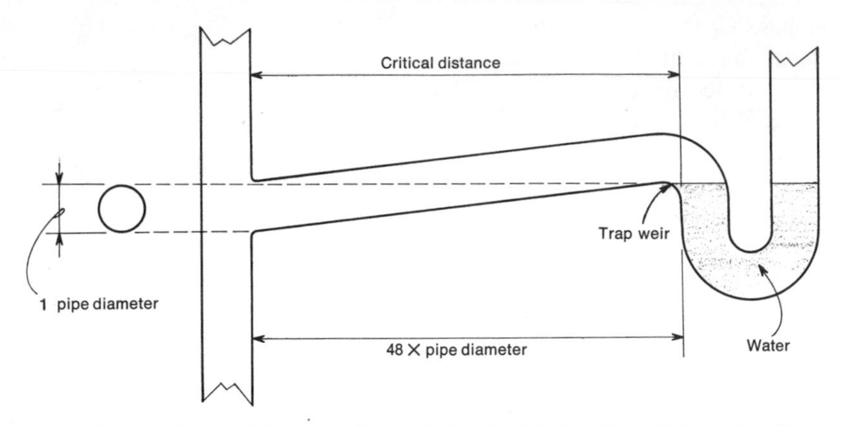

When the distance from the fixture trap's weir to the stack is less than 48 times the diameter of the drainpipe, and the drainpipe drops no more than once its own diameter between the weir and the stack, that fixture is within its critical distance and may be vented by the stack alone if there are no fixtures higher up the stack.

drop meaning how far below the weir in the trap we can bring the drainpipe before we either connect it to the stack or vent it.

The reasoning for all this is quite straightforward. There must be sufficient pitch to enable the waste and soil to scour the drainpipe and trap. But if the horizontal drain empties too far below the trap, or is permitted to descend too far before being vented, the trap will be emptied by siphoning. On the other hand, if the fixture trap is positioned too close to the stack, the trap will not work properly. A gob of water coming down the stack may empty the trap by aspiration (jet pumping), or some of the muck may be thrown up into the trap, clogging it in time.

The general rule for determining the limits of slope and length for a horizontal drainpipe is predicated on pipe diameter. The maximum drop from the weir in the trap to the stack or a vent should be no more than one pipe diameter. If you are running 2-inch pipe, a 2-

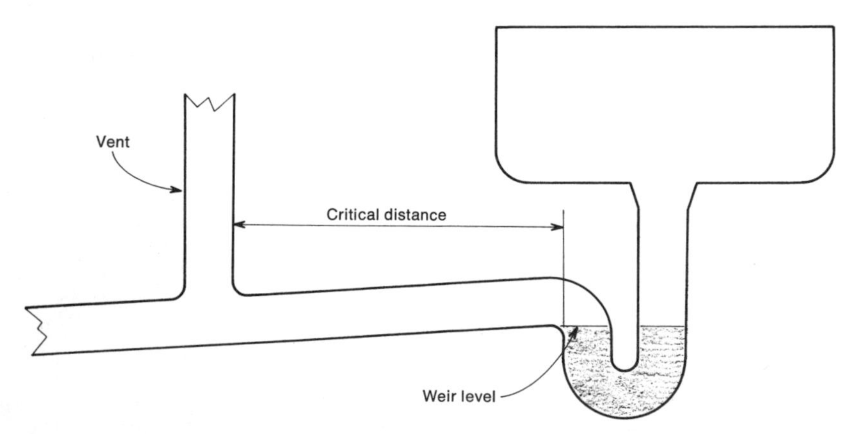

When a fixture's drainpipe must extend beyond its critical distance, the fixture must be vented at a point no farther from the fixture than its critical distance.

inch drop for the entire length of pipe up to the stack or vent is permissible. The minimum distance between the trap weir to the stack or vent fitting should be no less than two pipe diameters. Again, if you are running 2-inch pipe, you need a separation of at least 4 inches between the weir and the vent or stack fitting.

The maximum permissible length of pipe between the trap weir and the stack is often called the critical distance. It is limited to no more than 48 pipe diameters. For example, if you are running 2-inch drain pipe the critical distance is 48 × 2, or 96 inches. If you vent the pipe at this point or closer to the trap you can continue on as far as you like.

As for pitch, the larger the pipe diameter the smaller the pitch can be. But again, ¼-inch per foot is optimum with a maximum of ½-inch per foot. If you can't make it with a ½-inch pitch go to 45 degrees or more, but don't run the pipe at a pitch of ¾ or 1 inch to the foot.

Proper fixture position can reduce size of vent and drain pipes.

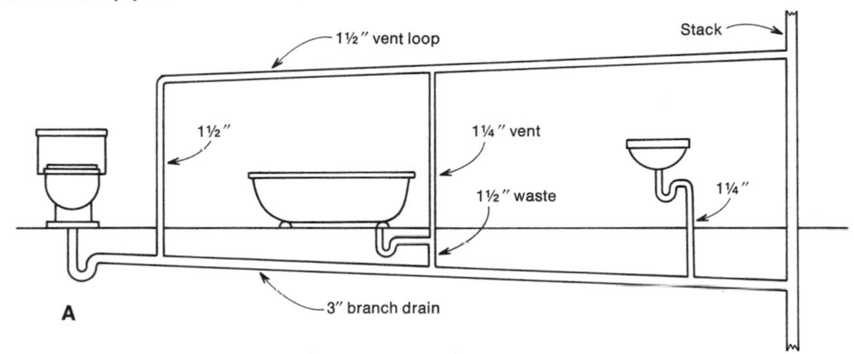

With the toilet at the end, the tub in the center and the lavatory close to the stack, a 3-inch drain is required for the entire distance. A 1½-inch pipe is required at the toilet and for the vent loop.

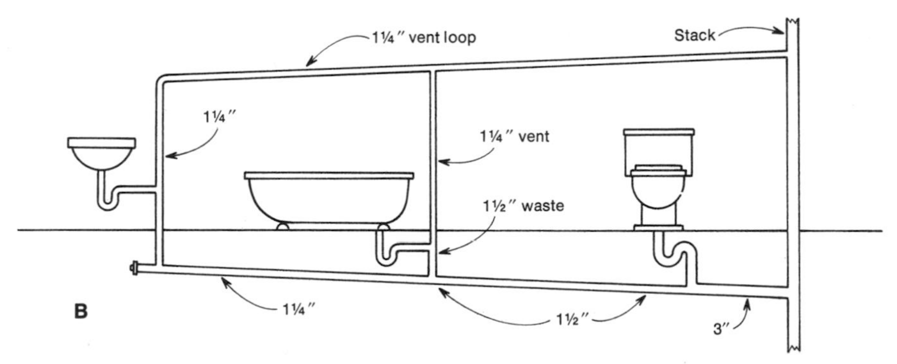

By moving the toilet close to the stack, placing the tub in the center and placing the lavatory at the end of the branch drain, the drainpipe may be reduced to 1¼ inches from lavatory to tub. From the tub to the toilet, the drainpipe may be 1½ inches. From the toilet to the stack, a 3-inch drain is needed. The toilet need not be vented and the vent for the lavatory and tub need only be 1¼-inch pipe.

There is one exception to the general 48-diameters rule. When the fixture to be drained has a flat bottom as some kitchen sinks do, the critical distance can be extended to 72 diameters. This is permitted because a lot of water drips down from the fixture after it has been emptied. The "after" drip refills the trap.

You cannot ignore the plumbing code and its local variations but you can, when practical, position your fixtures so that the code is satisfied and you use minimum diameter drainpipes.

The accompanying drawing shows how this may be done. In one arrangement the toilet is placed at the end of the branch drain. Doing so necessitates using 3- or 4-inch drainpipe (depending on local code and type of pipe) from the toilet all the way to the stack. In another arrangement the toilet is nearest the stack, the tub in the center and the lavatory is at the far end of the branch drain. With the fixtures in this sequence you can meet code requirements and use smaller drainpipe.

TRAPS

All plumbing codes require that a trap be installed in every drainpipe following all plumbing fixtures, with several exceptions: Two and three sinks may drain through a common trap. No trap is necessary following a toilet because it has its own built-in trap. And no trap is necessary with an indirect waste connection. An indirect waste connection is one that is open to the air. For example, a washing machine that empties into a mud tub by means of a pipe hooked on the edge of the tub is an indirect waste line. The washer drainpipe is vented at the tub.

By the same token it is contrary to code to run any drain through two traps. For example, if you had a wash basin in an attic

Table 36

Pipe Size Inches	Length Minimum (Inches)	Length Maximum (Feet)	Drop in Inches Minimum	Drop in Inches Maximum
1¼	2½	5	1¼	1¼
1½	3	6	1½	1½
2	4	8	2	2
2½	5	10	1¼	2½
3	6	12	1½	3
3½	7	14	1¾	3½
4	8	16	2	4
5	10	20	1¼	5

Minimum and maximum permissible horizontal drainpipe lengths and drops. The code prohibits connecting a fixture any closer to its stack or vent than 2 pipe diameters and any farther than 48 pipe diameters (critical distance). The table reflects this rule.

you are not permitted by code to run its drain down to another fixture drain and connect the basin drain behind the trap.

The large trap at the end of the house drain just before it passes through the foundation wall and becomes the house sewer is an exception to the no double-trap rule. It represents a second trap in the drainage system but is permitted. However, it is not a code requirement in all areas. Many plumbers dislike installing a house or running trap, as it is frequently called, because of the additional material and labor involved and because it is more likely to get plugged up than a straight run of pipe.

Whether a house trap is or is not installed, a cleanout should always be installed in the house drain close to the foundation wall.

Types of traps. There are two general types of traps used in most plumbing work—self-venting and nonself-venting. The first type is most often seen in the form of a drum trap, which is usually installed without a vent. The second type is manufactured in a variety of forms including the S, the P or half S and the U. Their names derive, more or less from their general shape. They must be vented or connected to a stack, which acts as a vent.

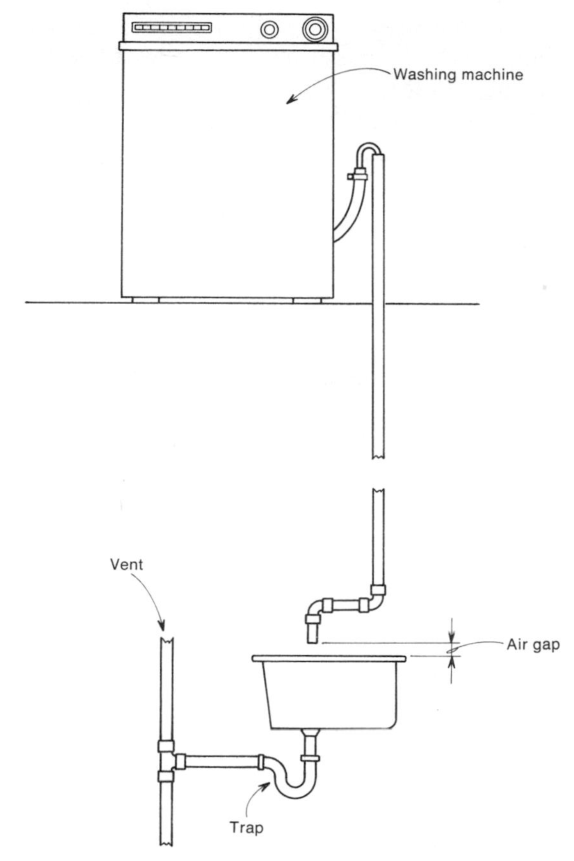

When a drainpipe is broken (opened to the air), as illustrated, there is no need for a trap in the pipe above the air gap. However, the pipe following the air gap must have a trap.

The letter-shaped traps operate by trapping a quantity of water in a bend. To some degree they are all self-scouring, meaning that the passage of water acts to clean the trap out.

Drum traps differ in two important ways. They do not require venting, and they are not self-scouring. A drum trap is a dual-ported container. Water enters one side, exits the other. As the entering pipe is low and the exit pipe somewhat higher, a plug of water always remains in the trap because of the pocket of air that forms above the water, breaking the siphon.

Drum traps are installed where it is impractical or impossible to install an ordinary trap and its venting pipe. Thus it would appear that drum traps are "position anywhere" traps, but it is not so. As drum traps are not self-scouring they are great for collecting debris. Therefore, a drum trap must always be positioned where its cover can be reached and removed without tearing a wall down. Drum traps cannot be "snaked" clean. Chemicals do not work very well either. Solid, non-dissolvable materials tend to collect in a drum trap and can only be removed by direct, physical means.

Trap size. There is no problem here. One simply selects a trap that accepts the size pipe one is running. Thus, when installing a trap in a 2-inch pipeline, a 2-inch trap is used, and so on.

The only thing to bear in mind when selecting a trap size is that reducing fittings should never be used. For although the capacity of a 3-inch trap is much greater than, let us say, the capacity of 2-inch pipe to which it is to be connected, the reducing bushings or fittings will obstruct the flow of hair, gunk and other matter so that a 3-inch trap, reduced to serve a 2-inch line, will clog up more quickly than a trap of the correct size.

Trap design. Outside of avoiding the older trap design which had a vent connection on top of the "hump" and which is no longer used because solid matter in the waste tended to get thrown up and into the vent, sealing it, one selects a trap design on the basis of installation convenience and joining method.

The trap selected for the house drain will depend on the position of the drain in respect to the building. If the drain runs beneath the cellar or basement floor, a U-trap with clean-

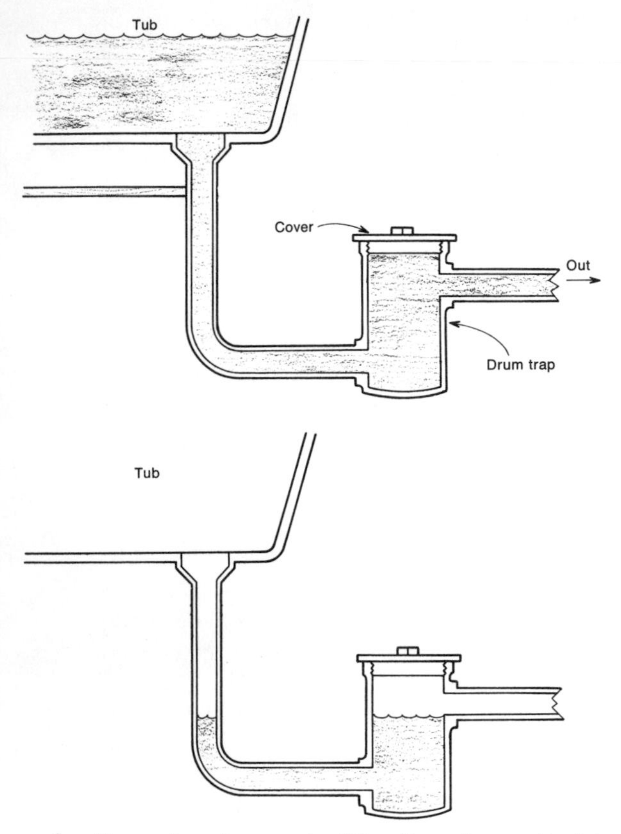

How a drum trap works. When the water level in the fixture being drained is higher than the outlet pipe on the drum trap (top), a siphon exists and the trap permits water to flow through. When the water level in the fixture or its drainpipe falls below the level of the drum trap's outlet pipe (bottom), the siphon is broken and flow stops.

out plugs at both ends of the trap is usually selected. If the house main runs close to the ceiling of the cellar, a trap with a cleanout plug on its bottom is best.

When the trap selected for one reason or another is such that it will be difficult to insert a snake into it and the following drain or sewer pipe, a cleanout must be installed.

And, as stated previously, every time a drainpipe makes a 45-degree turn or more, a cleanout should be installed. This is usually not a great increase in fitting costs because many 90-degree fittings can be had with cleanout plugs. And speaking of cleanout plugs, always use brass plugs. The iron plugs rust fast in time, even with a generous smear of pipe dope.

VENTING METHODS

As stated, the drainage system must be vented for proper operation and health. This may be done in one or more ways. No single method or group of methods is any better than any other. The choice of means is simply based on that which satisfied the code and holds labor and material to a minimum.

A number of venting methods are used. The most commonly used and the names by which these methods are most often called are as follows;

Direct venting. The fixture(s) are positioned within critical distance and vented directly by the drain or soil stack. There are no other fixtures connected higher on the stack. (In some areas a single lavatory may be connected higher on the stack.)

Back venting. The fixture(s) are connected beyond the critical distance. A vent pipe is run from the fixture drain to the soil or waste stack.

Wet venting. Fixtures exclusive of a toilet are connected to a common branch drain. The fixture farthest from the soil or waste stack is back vented by a vent pipe large enough to vent all the fixtures. Venting is based on using code-size drainpipe which is selected to keep liquid flow to no more than one-third the drainpipe. Thus venting for the fixtures between the stack and the single vent pipe takes place through the branch drain. This method is not permitted in all areas.

Some Basic Code Regulations

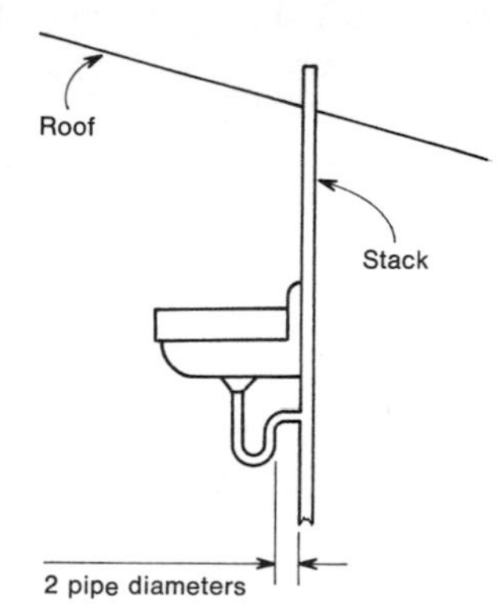

No fixture can be connected less than 2 drainpipe diameters from a stack.

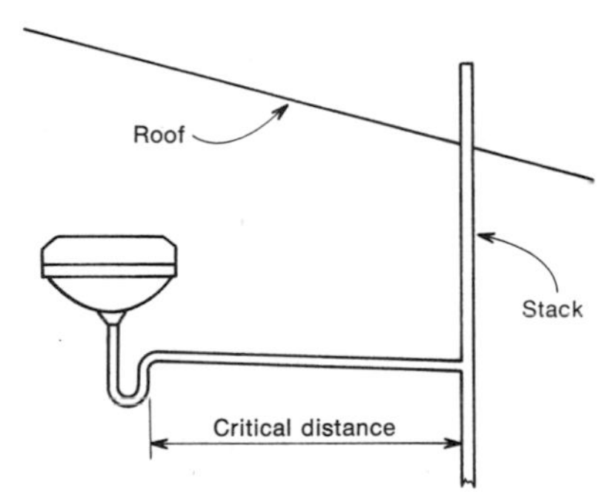

All fixtures must either be connected to a stack or vented at or less than their critical distance.

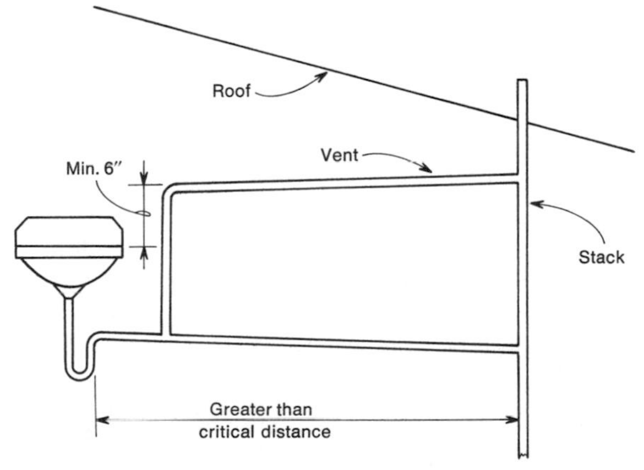

When there are no higher fixtures, that fixture may be back vented to the stack at a point at least 6 inches above the flood level of the nearest fixture.

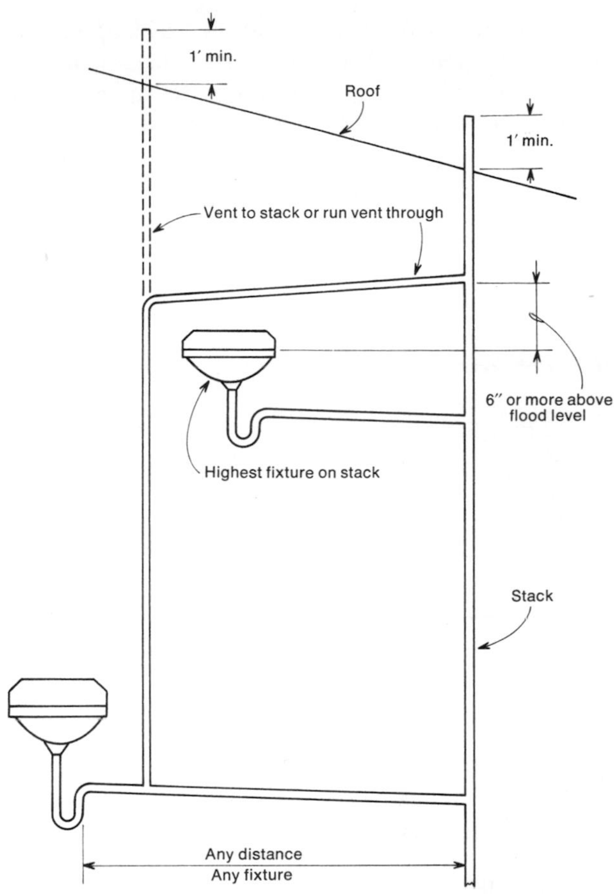

When higher fixtures are connected to a stack, the lower fixture(s) must be vented to the open air (vent stack) or back vented to the stack at a point at least 6 inches above the flood level of the highest fixture on that stack.

Individual venting. When one or more fixtures are installed beyond the critical distance and a vent pipe is brought up from each fixture to a common vent pipe which may join a stack or go on up through the roof, each fixture is sometimes said to be individually vented.

Loop or circuit venting. When one or more fixtures are back vented in contrast to being vented by a vent stack (described shortly), they are sometimes described as loop or circuit vented. The reason being that air moves from the stack up through the drain, enters the vent pipe and returns to the same stack.

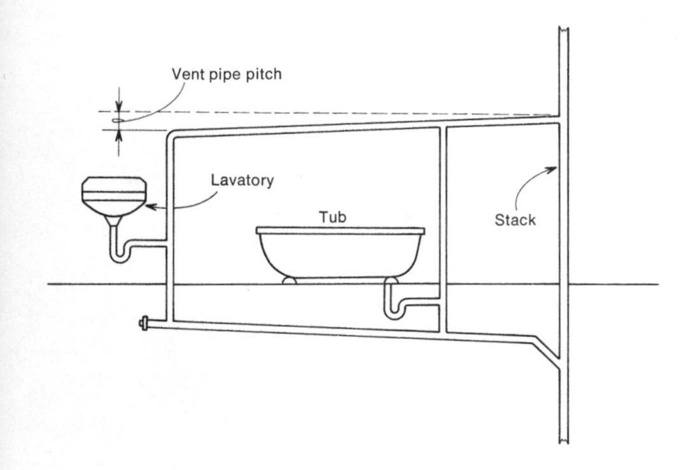

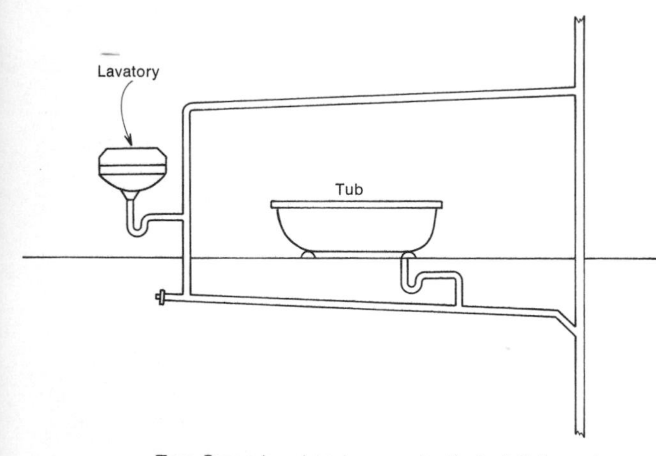

Top: Some local codes require that all fixtures beyond their critical distance and connected to a common branch drain be individually vented no matter what type of fixture they may be. Bottom: In some areas, when there is no toilet connected to a branch drain and proper size drainpipe is used, one or more fixtures connected between the end fixture and the stack may be wet vented. A single back vent connected to the last fixture vents all fixtures on branch drain.

BACK VENTS

As stated, any vent pipe that returns to the stack which drains the fixture is called a back vent. The point at which the back vent may be connected to the stack is code limited. Back vents, single or multiple, must be connected to the stack at a point 6 inches or higher than the flood level of the highest fixture attached to that stack.

Assume there is one fixture to be back vented on the first floor and a second fixture to be back vented on the second floor and no additional stack connections above the second floor back vent. The second floor back vent connects to the stack 6 or more inches above the fixture's flood level (water level when fixture is overflowing its rim). The first floor back vent must be brought up to join the stack at the point of *the second-floor back vent* or higher.

VENT STACKS

Any vent pipe that is run up through the roof of the building to the open air is termed a vent stack. According to the code the end of the stack is always left open and must extend at least 1 foot above the surface of the roof and 3 feet or more above any nearby window or similar opening.

A vent stack may be used to vent one or more fixtures. Like a waste or soil stack it may be used as a convenient central connection point for any number of vent pipes. When several vent pipes are connected to a vent stack the *entire length* of the vent stack must be large enough to properly vent all the connecting fixtures. The lower end of the vent stack must be connected to an as large a diameter branch drain or stack. In other words, if you require a 2-inch vent stack the entire length of stack must be of 2-inch diameter

Typical loop venting connections. Three individual vent pipes to the left of the plastic waste stack have been connected to a common pipe that joins the stack at an upward sweeping angle. The Tee to which the common vent pipe has been connected is an inverted drain fitting. (Compare this fitting with the fitting immediately below, which is used for drainage.)

Table 37

Diameter of Soil or Waste Stack	Total Fixture Units Connected to Stack	Vent Diameter in Inches 1¼	1½	2	2½
1¼	2	30			
1½	8	50	150		
1½	10	30	100		
2	12	30	75	200	
2	20	26	50	150	
2½	42		30	100	300
		Maximum Vent Length in Feet			

Minimum diameters and maximum lengths of vent stacks, the lower ends of which are connected to a waste stack, soil stack or building drain.

pipe and the lower end must be connected to a 2-inch or larger drain.

In northern areas it is advisable to terminate the top of the vent stack with 3-inch pipe to prevent closure by ice formation. Elsewhere, local code permitting, the size of the vent stack will depend on the number of fixtures it is venting, their total discharge volume in fixture units and the length of the vent stack itself. Minimum diameters and maximum vent stack lengths as governed by discharged fixture units are given in Table 37.

FRESH-AIR VENT

A fresh-air vent or fresh-air inlet is a short vent pipe connected to the house drain just ahead of the house trap. Its other end is generally brought directly outside the building by a long turn elbow. The pipe end outside the building is protected by a screen.

The fresh-air vent is not required by the national code, though many local plumbing departments do require it. Generally it is assembled of pipe of the same diameter as the house drain.

The fresh-air vent provides a lower opening to the ventilating system. Air entering the low vent can rise through all the drainpipes and pass out through the roof, thus providing much more ventilation than the opening at the top of the venting system alone. In addition, the fresh-air vent provides a very useful overflow release. Should your plumbing system plug up at the main trap, the excess will flow out the fresh-air vent, rather than back up into the tubs and kitchen sink.

BELOW SEWER-LEVEL DRAINAGE

Some buildings have drainpipes that exit through their foundation walls a goodly distance above the basement or cellar floor.

When fixtures are installed in such basements, their drains and traps are below the house drain and following sewer pipe.

In such cases there are two solutions. One is to install a pump and tank and drain the cellar fixtures into the tank, from which the pump lifts the water up and into the house drain. A check valve in the pump lines prevents house drain muck from running down into the tank.

The other solution involves constructing a dry well outside the building and leading the waste water from the cellar fixtures into the well. Many communities permit waste water to be dumped in a dry well.

CELLAR-FLOOR FIXTURES

Cellar-floor drains, toilets and other fixtures in the lower reaches of a building may be directly connected to the building drain at any point along its length 5 feet beyond the lower end of the stack. Connections are made according to the rules laid out for connecting fixtures to a branch drain. If the fixture can be held within the critical distances specified by Table 36, no separate venting is necessary. The stack does the venting. If the distance is greater the fixtures must be back vented.

12

Roughing-in a New Plumbing System

Previous chapters explained how to design a plumbing system in a new house. This chapter carries design into actual practice. Roughing-in is literally the rough work of measuring, cutting, joining and installing all the pipe in new construction. This can mean a new house or an addition on an existing one. Readers not planning a complete plumbing job from sewer to shutoff valve can skip to those portions of this chapter that are applicable to their problems.

LAYING OUT THE PIPELINES

The starting point is always the municipal sewer (or the septic tank), and if the architect is on his toes, this is the starting point for designing and positioning the building, for he must be sure the building drain (within the house) is higher than the municipal sewer to permit proper drainage by gravity. If the municipal sewer is higher than the house drain, a soil pump and soil tank must be installed. While the cost of soil-pumping equipment is not in itself prohibitive, one lives under the specter of power failure.

The location of the sewer in relation to your projected building, its depth beneath the road or surface of the ground, and whether or not there is a stub-out (branch pipe) prepared to accept your sewer pipe can all be learned at the office of the city engineer. Once you know where the sewer and stub-out is located, your next step is to transfer this point to the side of your foundation wall (or where your wall will be erected). This may be done with the aid of a transit and rule or a length of string, line level and rule.

If you have a transit (you can rent one by the day), position it above the center of the sewer main or at the termination of the stub. Have a helper hold a stick in a vertical position at the building line. Use a level to make certain it is close to vertical. Now sight through your transit, making certain the glass is level (check the bubble). As your transit is perfectly level, the position of the cross-hairs is a projection of that elevation. Have your helper mark this on your stick.

You have transferred the center of your transit to a mark on the stick. If the city engineer told you the sewer is, say, 9 feet below

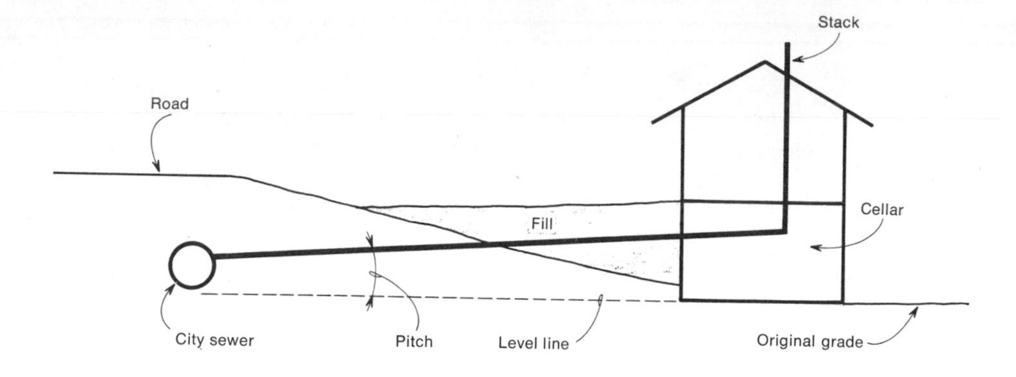

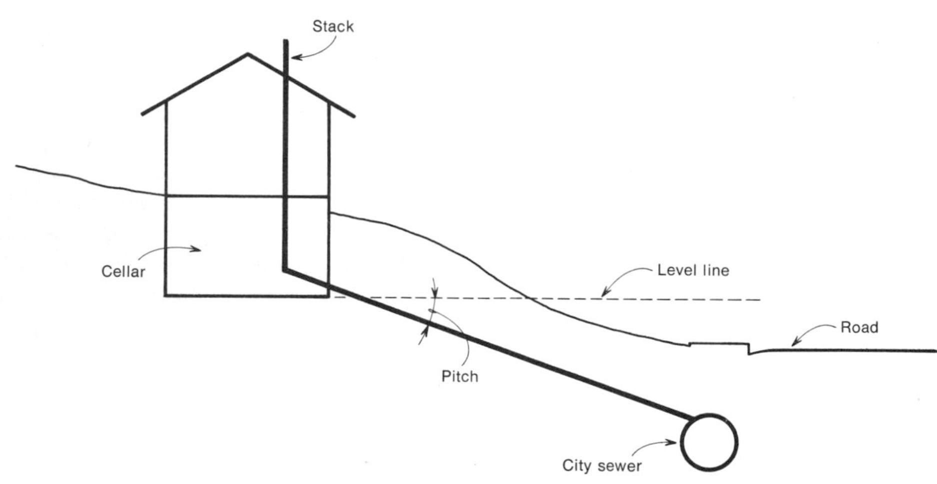

If the original grade on which you plan to build your home is *below* the city sewer, you will need to build a high cellar to raise the system above it (top). If the original grade is higher than the city sewer (bottom), you need go no deeper than the frost line.

the surface of your road, and the center of your transit is exactly 5 feet above the road, you know that the elevation of the sewer at the property line is 9 + 5 = 14 feet below the mark on the stick.

Now measure from your building line to the point directly under your transit, or the sewer's center. This will give you the distance from your foundation wall to the sewer. Actually, you should measure this distance at the angle the sewer pipe will run, but the difference can be ignored as it is very small.

Having measured from the stick to the center of the transit, which should be above the underground sewer, you know the length of your sewer line. Assume that the distance is 50 feet. If you pitch your sewer ¼ inch to the foot, the sewer at the building should be 50 × ¼ inch or 12½ inches higher at the building than at the city sewer main. If you make the pitch ½ inch to the foot, it will be 50 × ½ inch, or 25 inches, that you will have to raise the sewer at the building line. This figure is simply subtracted from the 14 feet we had in our example, thus lifting the house end of the sewer above the main.

The sewer should never be pitched less than ⅛ inch to the foot; ¼ inch is optimum, but ½ inch, where practical, is preferred because it reduces the number and extent of valleys in the sewer pipe caused by settling. If greater pitch is required to reach the sewer, run most of the pipe at ½ inch to the foot, and then use a 45-degree fitting and run the final few feet or yards at 45 or more degrees.

If you do not have a transit, you can use a length of string and an inexpensive line level. This can be a one-man operation if you drive the stick into the ground. To use the level properly it must be centered in the string, as the string sags. Also, turn the level

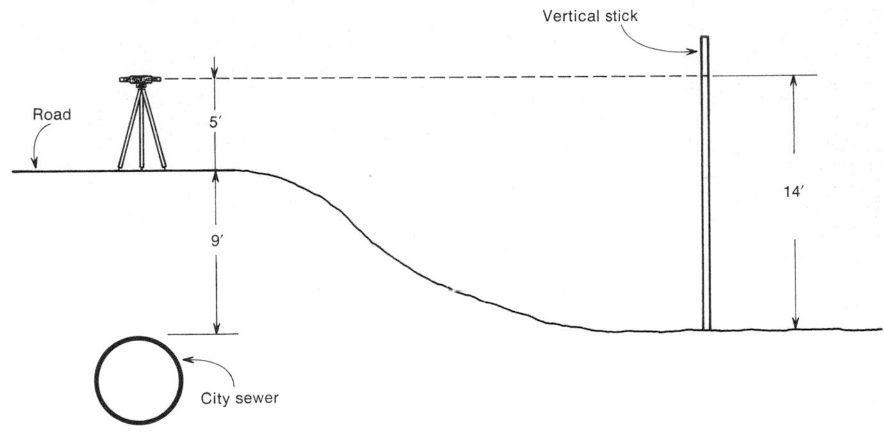

Transferring the depth of the city sewer to the foundation wall. Set the transit directly above the city sewer. Watch the bubble and level the glass. Then swing it around until you see the stick. Mark the stick at the center of the crosshairs. You thus have transferred the center of your transit to the stick. Measure down to determine the elevation of the center of the city sewer at your proposed building line.

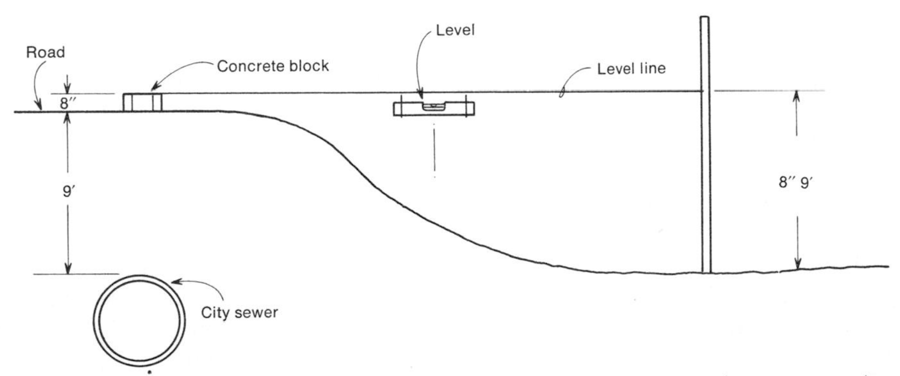

If you have no transit, set a concrete block atop the road and run a string across to the stick. Hang a line level in the center of the string. Move the string up and down until you center the bubble. The string is now perfectly horizontal and its "stick end" marks the top of the concrete block.

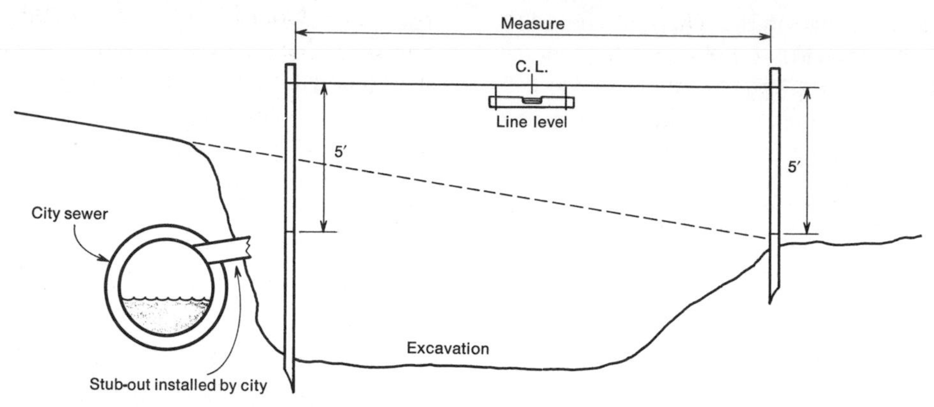

Most accurate method of transferring depth of sewer is to excavate the city's stub-out and set up your line from there. The shorter the distance, the more accurate are the string and line level.

around to make certain it reads the same hanging in either direction.

The string is not nearly as accurate as the transit. Estimate plus or minus 2 to 3 inches per hundred feet.

Should you feel uncertain when using the string, and there is a sewer stub-out entering your property, you can dig the sewer trench and measure from the stub-out to the building line. The shorter the distance, the smaller the error.

Once you know the elevation of the house sewer as it will enter the foundation wall you can plan your construction accordingly, bearing in mind that the sewer must be below the frost line. This is generally 2 to 3 feet below grade (surface of the earth when all landscaping is finished).

Locating the water service. Most local codes permit you to run your water-service pipe in the same trench as your sewer line if the water line is at least one foot above the top of the sewer line and the sewer line can pass a 10 psi test without leaking. In some instances this can be a savings of labor. However, if the sewer and the water connections are not close to one another at the property line, the additional pipe necessary to utilize one trench can be costly in material and lowered water pressure. As the water service must be below the frost line, lowering the sewer pipe one foot means you must dig more deeply, which is always more difficult, and increases your chance of encountering rock.

Laying out the trench. Start by installing a pair of batter boards and a guideline or grade string. A batter board is a horizontal board supported by a pair of stakes driven into the ground.

Position the first batter board close to the building with its stakes to either side of the line your sewer is going to take on its way to the city sewer pipe. Position the second batter board across the same line at a point close to where your sewer will connect to the city's sewer. Now adjust the heights of the two batter boards so that the board close to the building is higher by exactly the distance the sewer pipe is going to rise as you will install it working from the city sewer to the house. In our example the total rise or pitch is 25 inches, giving our pipe a pitch of ½ inch to the foot. Use the transit or line level to make this adjustment. Next, stretch the grade string from the top of one batter board to the next, making certain the string exactly follows the line the sewer pipe is supposed to follow.

Digging the trench. We are now ready to dig the trench. Move the string to one side and out of the way. Dig a little. Replace the string and measure down to the bottom of the trench. Repeat the digging and the measuring until you have a trench the bottom of which is exactly below the grade string and exactly parallel to it by the required distance.

Don't dig too deeply. Not only will you waste time and effort in removing soil, you will also waste more effort afterwards in blocking the pipe up on concrete block and stone. Do not place the pipe on soil that has been dug up and replaced. This is called fill and fill compacts with time. When it does so the pipe will move downward and possibly open at one or more joints.

INSTALLING THE SEWER LINE

If the sewer is to enter the building below the footings, the sewer should be installed before the footings are poured. If the sewer is to enter the building above the footings and the foundation wall is to be of poured concrete, a cardboard box or the like should be positioned before the wall is poured. If it isn't, be certain to remove the form and cut a hole while the concrete is still green. It gets very hard in a couple of days. If you are going to have a concrete block wall you can install the sewer line anytime. Concrete block is not too hard to cut.

Laying the pipe. The sewer run is always started at the sewer main. If the city has provided a stub-out, the connection is made to it. If the stub-out is larger than the pipe, install a reducer, bell side towards the building and then continue with the pipe. If the city hasn't provided a connection and there is nothing to greet you but the curved side of the city sewer, don't attack it yourself. Let the city fathers do it. It is much too easy to damage a section of clay pipe.

If you are required to make a pressure test on your sewer you can install a Tee, one connection side upwards, right after the reducer or your connection to the sewer main. When the sewer is in, place a cloth bag filled with wet clay into the Tee and plug the sewer. Following you make the test, then remove the plug and seal the Tee with concrete. If you do not install a Tee you can use a balloon device for the test. The uninflated balloon is floated down the sewer to its

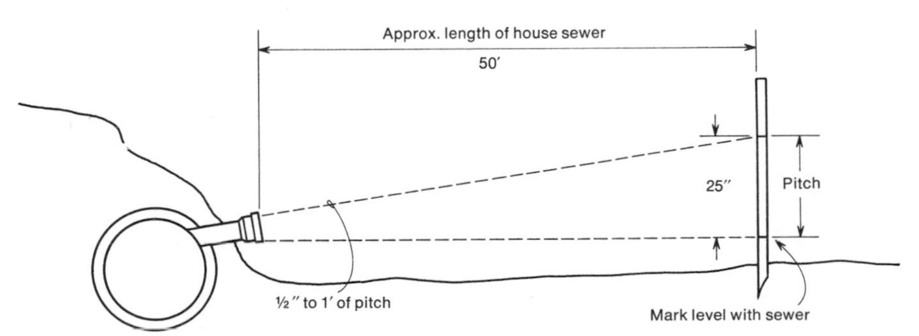

To pitch the sewer, measure from the main sewer to the stick. Then make a second mark on the stick, high enough above the first (city sewer level) to give the desired pitch.

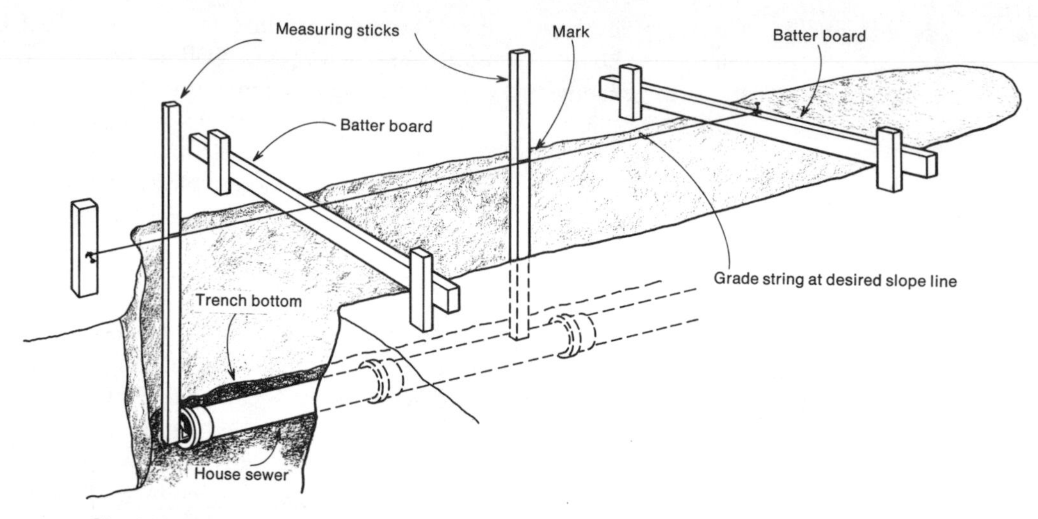

Checking on the pitch (slope) of the sewer line as you lay it. Set up batter boards as shown, with the board nearest the house higher than the other by the desired pitch. Stretch the grade string along the line you want the sewer to follow. Measure from the grade string to the bottom of the trench as you excavate. Measure again as you lay each section of pipe.

end. The balloon is inflated. It plugs the sewer end. The test is made. The balloon is then deflated and hauled out.

As you lay the pipe you must continuously check the pitch of each section in addition to making certain the pipe doesn't wander from its course. To do this, cut a stick to the proper length or mark it so you can constantly refer to your overhead guideline. Also, tape a wedge of properly angled wood to the bottom of your level. Use this to check on the pitch of each section of pipe as you lay it. When the bubble is centered the pitch is correct. If you are laying long sections of pipe you can rest your level against the barrel of the pipe without error. If you are running short length clay it is best to lay a straightedge over two or more sections of pipe to take a reading. In this way you can correct for minor inaccuracies at each joint and in the pipe.

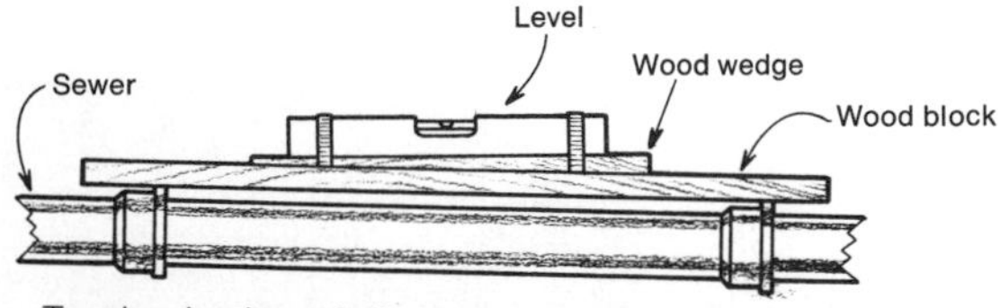

To check pipe pitch, tape a wedge of wood with the proper angle to the bottom of a spirit level. Place a straightedged board across two or more hubs and set the level on the board. If bubble is centered, pitch is correct. Do not depend on the level alone for correct pitch. Keep checking against grade string. On a long run a slight error each time you check pitch can lead you far astray.

Sewer pipe at the foundation. It is best to run sewer pipe directly from the city sewer into the building, laying your pipe on freshly uncovered earth. This is not always possible. In many instances the sewer pipe must enter the building a distance above the footings.

Sewer Pipe Outside Foundation Wall

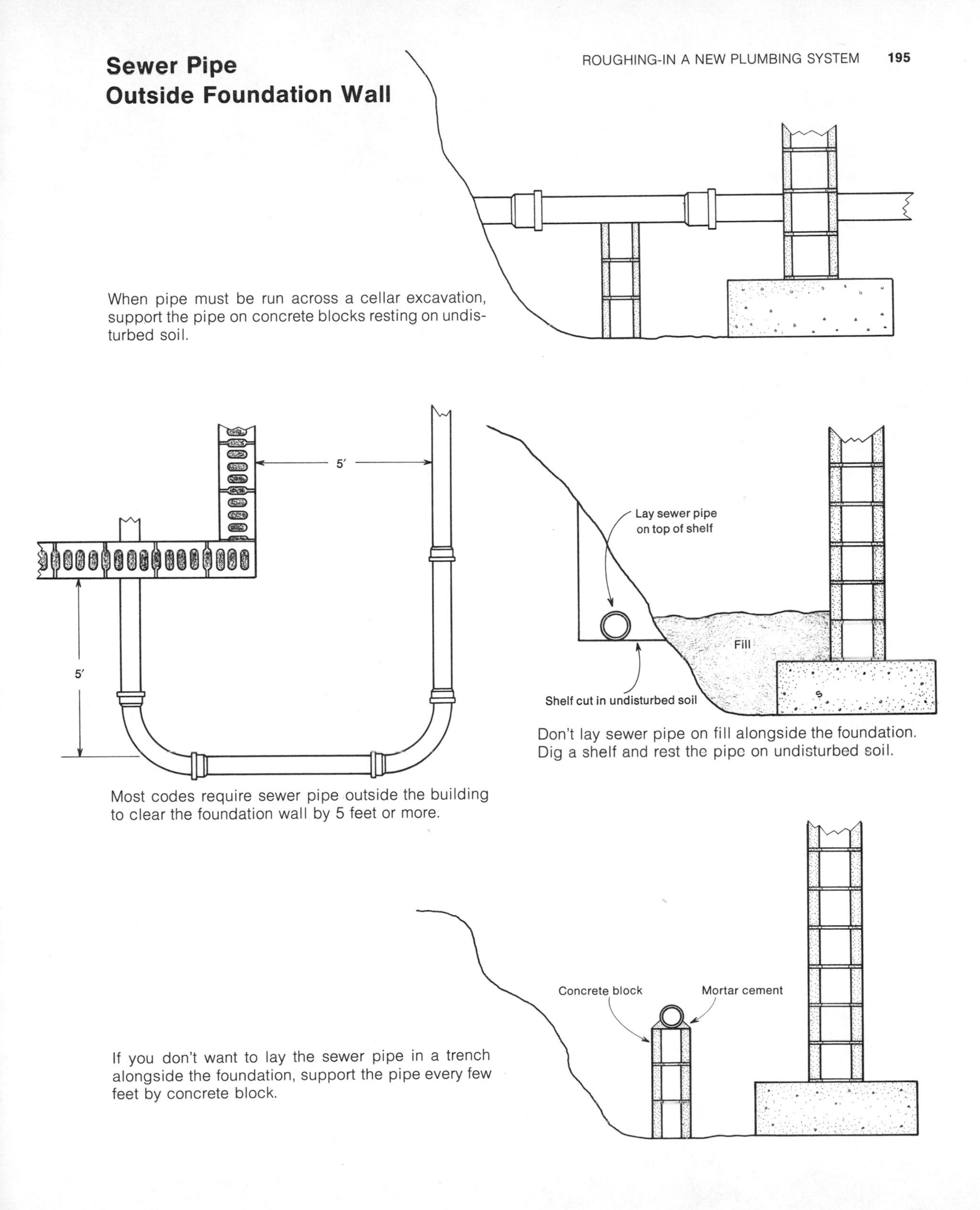

When pipe must be run across a cellar excavation, support the pipe on concrete blocks resting on undisturbed soil.

Most codes require sewer pipe outside the building to clear the foundation wall by 5 feet or more.

Don't lay sewer pipe on fill alongside the foundation. Dig a shelf and rest the pipe on undisturbed soil.

If you don't want to lay the sewer pipe in a trench alongside the foundation, support the pipe every few feet by concrete block.

In such cases be certain to support the pipe solidly where it crosses the open area next to the foundation. In some instances you may need to make a turn and enter the building from the side. If so, make certain you clear the foundation by the required code distance (usually 5 feet). Do not lay the pipe on fill alongside the foundation, but cut a ledge into the undisturbed earth and rest it there.

Covering the sewer pipe. This should not be done until the water test is completed and satisfactory. Back filling should be done with care. Don't put any rocks in the trench until you have covered the pipe with at least a foot of soil. And then, don't just drop the big ones in; ease them in. Clay and asbestos-cement pipe is easily broken.

Plumber caulking 3-inch cast-iron joint with oakum. Spirit level on pipe, to right of plumber's shoe, is used to check pitch.

INSTALLING THE WATER SERVICE

Use a single length of pipe. Run it as straight as possible. If you can't install it beneath the frost line be certain to insulate it with rolled up bats of rock-wool insulation. Use copper wire to tie the bats if you are running copper pipe. Don't use iron or galvanized wire.

Bringing the water service into the building is simple if you are running the pipe beneath the footings. In some areas, if you are coming in through the foundation wall you can run the pipe directly through and cement it fast. In others the code requires that the copper tubing be run through a section of galvanized pipe. If you have to do this, use much larger diameter galvanized pipe and seal and insulate the copper from the galvanized by a layer of caulking cement.

Once the water service is inside the building it is generally fastened to the nearest foundation wall and continued on from there to the main valve and then to the water meter.

Installing a cast-iron house drain. The pipe has been brought across the floor of the cellar in a trench. The cellar floor will be poured on top. The left-hand turn leads to the soil stack. The right-hand turn the plumber is working on will drain waste from the basement laundry tub and washing machine. Note water-service pipe to right of waste elbow. Pipe has been brought in under the footings.

INSTALLING THE STACK(S)

At this point we have brought our sewer pipe into the building, where it is now called the house drain. We know where our stack is going to go. What we need to know now is exactly where it is to go in relation to our drain so that we can install an elbow and follow with the stack.

Plumber melts lead in a propane-fired pot (left), then pours it to form a caulked joint (right). Plumbers usually pack a number of joints with oakum, then pour them all at one time.

To locate the drain elbow, cut two holes in the floor above the end of the house drain. One hole will accommodate the stack. The other hole will accommodate the toilet flange (assuming there is to be one near the stack). Temporarily connect an elbow to a short length of the same type and size of pipe that is to be used for the stack. Position the two pieces of pipe (or more if you need a section between the elbow and the stack) exactly where you want them to finally go. Drop a plumb bob and line through the center of the vertical section of pipe. The tip of the bob marks the exact center of the elbow that must be connected to the horizontal house drain. Now you can properly mark and cut the house drain and connect the elbow pointing upwards.

Supporting the stack. If you are running cast iron up through two floors you can easily approach half a ton in dead weight resting

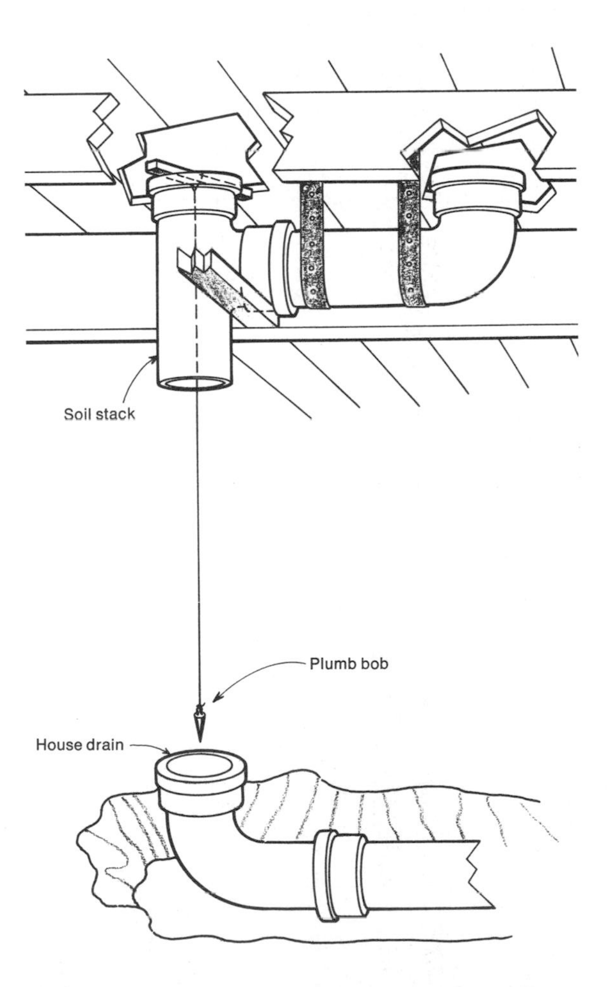

To make certain the elbow of the house-drain is directly beneath the stack, position the elbow exactly where it should be, and drop a plumb bob on a line down the center of the stack. The bob marks the exact center of the elbow.

on the house drain. Obviously this weight must be properly supported. If the soil pipe is coming up through a poured concrete floor and the concrete is below the elbow hub, use a trowel and mortar cement to make a cement collar under the hub. If the house drain rests or will rest on top of the concrete cellar floor, use cement and concrete block if necessary to make a firm support. Cement the block to the floor and the pipe to the block.

If you are running cast-iron soil pipe up near the cellar ceiling, use heavy strap iron lag-bolted to the sides of the joists to carry the weight.

In addition to providing secure support at the base of the stack, it is advisable to transfer some of the stack's weight to each floor the stack passes through. This is done by using special fittings that fit beneath the pipe's hub and are lag-bolted to the joists or studs.

Copper and plastic drainpipe, being far lighter, do not require as sturdy support. However, both copper and plastic pipe must be protected from damage. If the house drain and stack end rest on the cellar floor, it is advisable to cement several concrete blocks around the foot of the stack so that it cannot be accidentally pushed out of position. If the copper or plastic drainpipe is exposed to foot traffic it should be protected by a covering of some sort.

You may use any kind of metal to support a house drain of plastic pipe that has to be hung beneath the cellar ceiling. Do not use anything but copper, plastic or wood to support a copper drain (or any copper pipe).

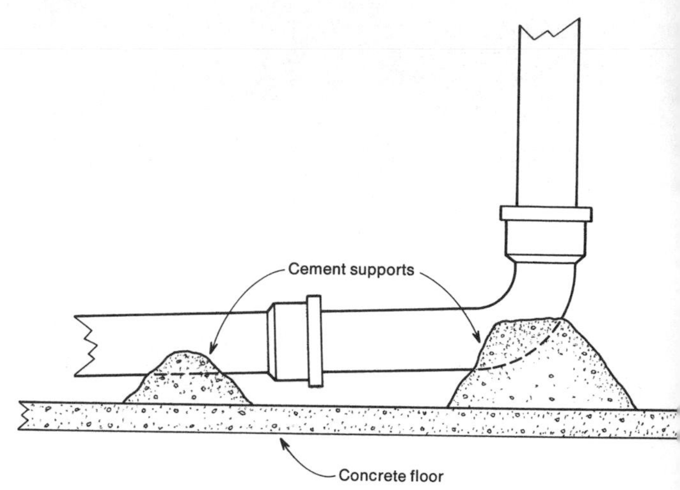

The bottom end of a stack and attached house drain should be supported and locked in place by cement.

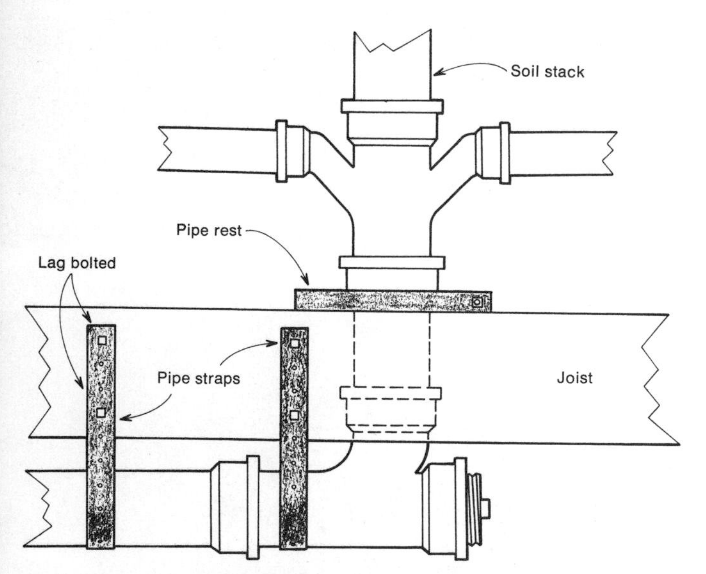

How a stack and house drain may be supported by a pipe rest and pipe straps.

Assembling the stack. When you have finished joining the upward-facing elbow to the end of your house drain you are ready to assemble the stack.

At this point you have already positioned the next-above section of pipe. Measure upwards from the in-place elbow and cut whatever length(s) of pipe you may need to make the connection. Install the pipe "dry" to

Installing a Copper Stack and Fittings

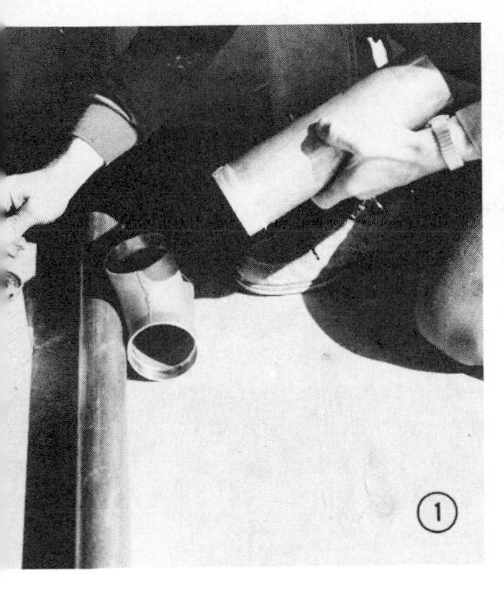

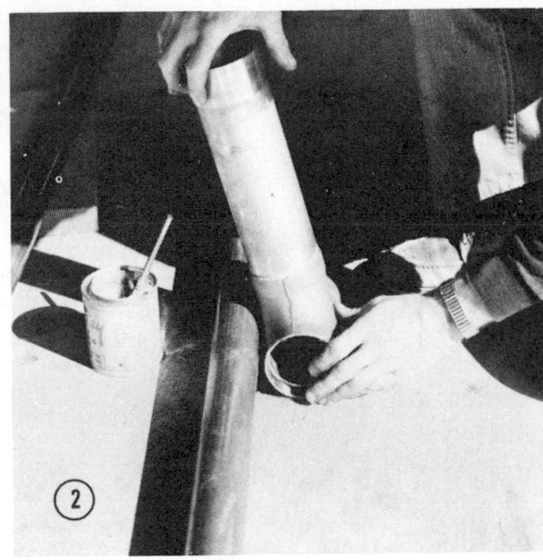

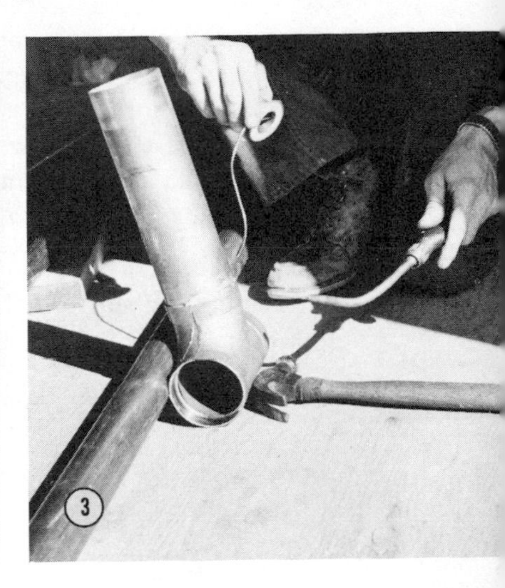

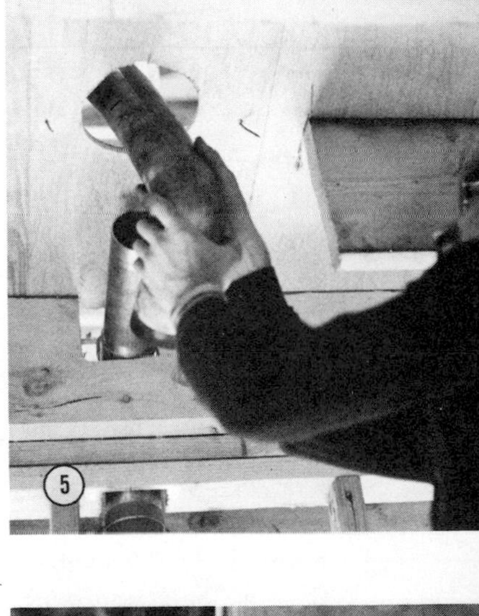

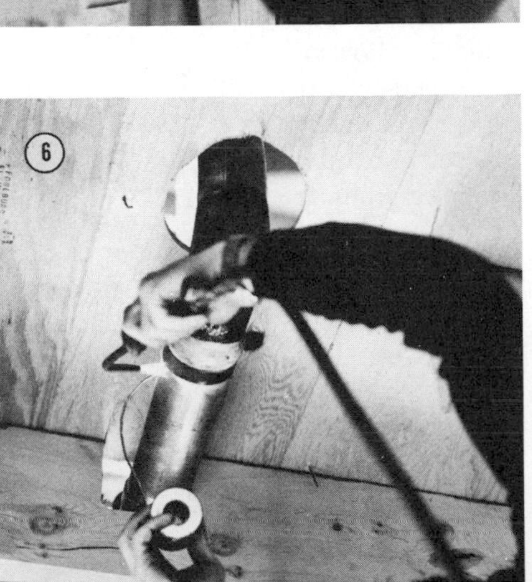

1. The end of the cut pipe is coated with flux.

2. The end of the pipe is inserted in its fitting.

3. The joint is soldered.

4. The joint is soldered to the copper stack.

5. A lead toilet bend joined by a wiped-lead joint to a brass ferrule is connected to pipe section.

6. The brass ferrule is soldered to the pipe section which is connected to the stack.

7. View of the lead toilet bend projecting into the bathroom. Note that end of lead pipe has been pinched and soldered closed.

check on its size. If correct, join the pipe to the next section or fitting and continue upwards.

You work a section at a time, from joint to joint or fitting to fitting, making certain all the fittings point in the correct direction away from the stack.

Framing the stack wall. Stacks and other pipes are usually hidden within walls. If you are running 3-inch copper soil pipe, you can use standard 2x4s for the wall studs. The width of the 2x4 is comfortably greater than the diameter of the 3-inch copper plus the usual drain fittings.

If you are running standard 4-inch cast-iron soil pipe, you need to use 2x6s to secure the necessary space. If you don't have 2x6s on hand you can use 2x4s and fur them out by adding strips of wood.

When framing the stack wall take the time to position the studs where they will not interfere with fittings that angle off to one side, and make certain your sheetrock does not rest on the pipe. If it does, pipe noise will be amplified and the pipe itself may be outlined because of temperature differences.

RUNNING BRANCH DRAINS

Branch drains are the slightly pitched, horizontal drainpipes that run into the stack or house drain. The same procedure is followed. Start at the fitting connected to the stack or drain. Position the following fitting exactly, but temporarily where you want it. Measure from the stack fitting to the following fitting. Cut and connect the pipe.

Cutting through. When roughing in stacks and branch drains you will run into joists and studs and will have to cut your way through. When cutting through joists try to keep cuts near the end of the joist. As the accompanying drawing illustrates, you can do a lot of cutting without seriously reducing the strength of the joist if you cut at the right place or drill smooth, round holes. Don't chop holes in the joists; use a slow-speed electric drill and a large bit. Although no one has published data on what and where you can drill, smooth holes up to 1¼-inch clearance can be drilled anywhere, almost, in a 2x8 joist, and even 1½-inch clearance holes near the ends of the joists and in 2x10 and larger boards. The point to watch out for is drilling near loose or large knots and drilling too many holes close together.

When you have to notch a joist somewhere along its middle, nail a length of 2x4 alongside to take up some of the load. Make the board at least 4 feet long, and use at least 6 10-penny nails. A short brace is worthless.

When the drainpipe has to come up right through the width of a joist, cut the joist and box it. This is simple enough, but you have to make a fairly snug fit on the boards if they are to have any holding strength.

When you have to go sideways through a number of studs and cannot drill, for one reason or another, cut a series of slots, no deeper nor larger than you have to. Some plumbers alternate the slots, so two slots do not follow on one side of the studs. This provides a little extra strength, but is hardly worth the effort. Instead, if you are worried about having weakened the studs, or if it is a bearing wall (there is another wall directly above it carrying a floor or roof), try slipping some more studs in sideways. That is to say, place the additional studs with their widest side against the pipe.

Vent pipes originate near a fixture trap or at the end of a branch drain. They run vertically for the required distance and then they slope *upwards* towards the stack, which may be a soil, waste or vent stack. In order to connect the upward sloping section of the

Installing a Copper Branch Drain

1. A series of holes are cut through the joists. The holes are staggered vertically so the pipe has the proper pitch.

2. To provide more clearance for the pipe, a saber saw is used to enlarge some holes.

3. One section of pipe is slipped into place and a measurement is taken to ascertain the correct length of the next piece of pipe.

4. The next section is positioned and coated with flux.

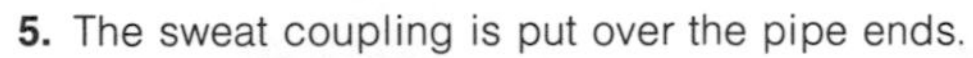

5. The sweat coupling is put over the pipe ends.

6. The coupling is soldered fast, completing the joint.

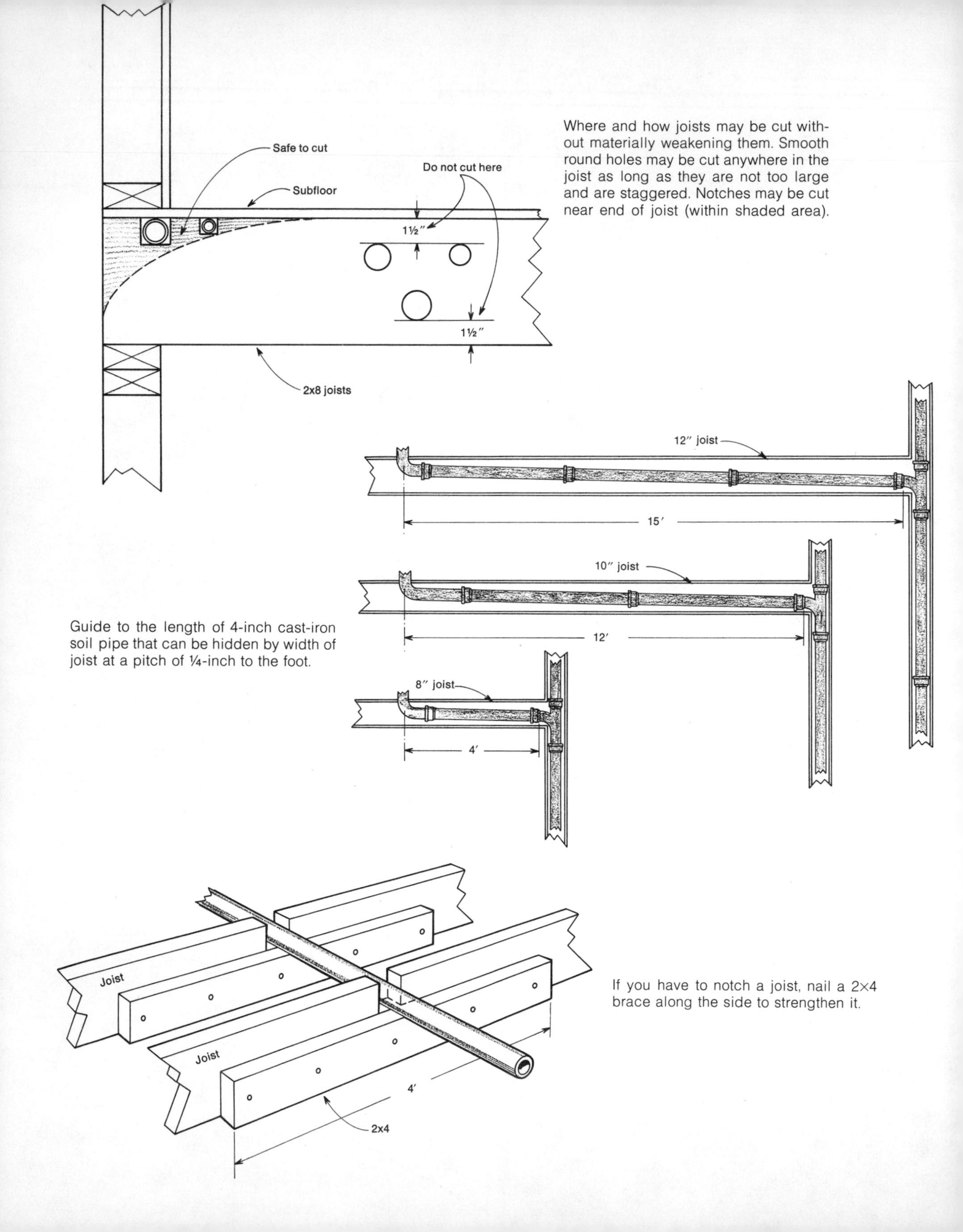

Where and how joists may be cut without materially weakening them. Smooth round holes may be cut anywhere in the joist as long as they are not too large and are staggered. Notches may be cut near end of joist (within shaded area).

Guide to the length of 4-inch cast-iron soil pipe that can be hidden by width of joist at a pitch of ¼-inch to the foot.

If you have to notch a joist, nail a 2×4 brace along the side to strengthen it.

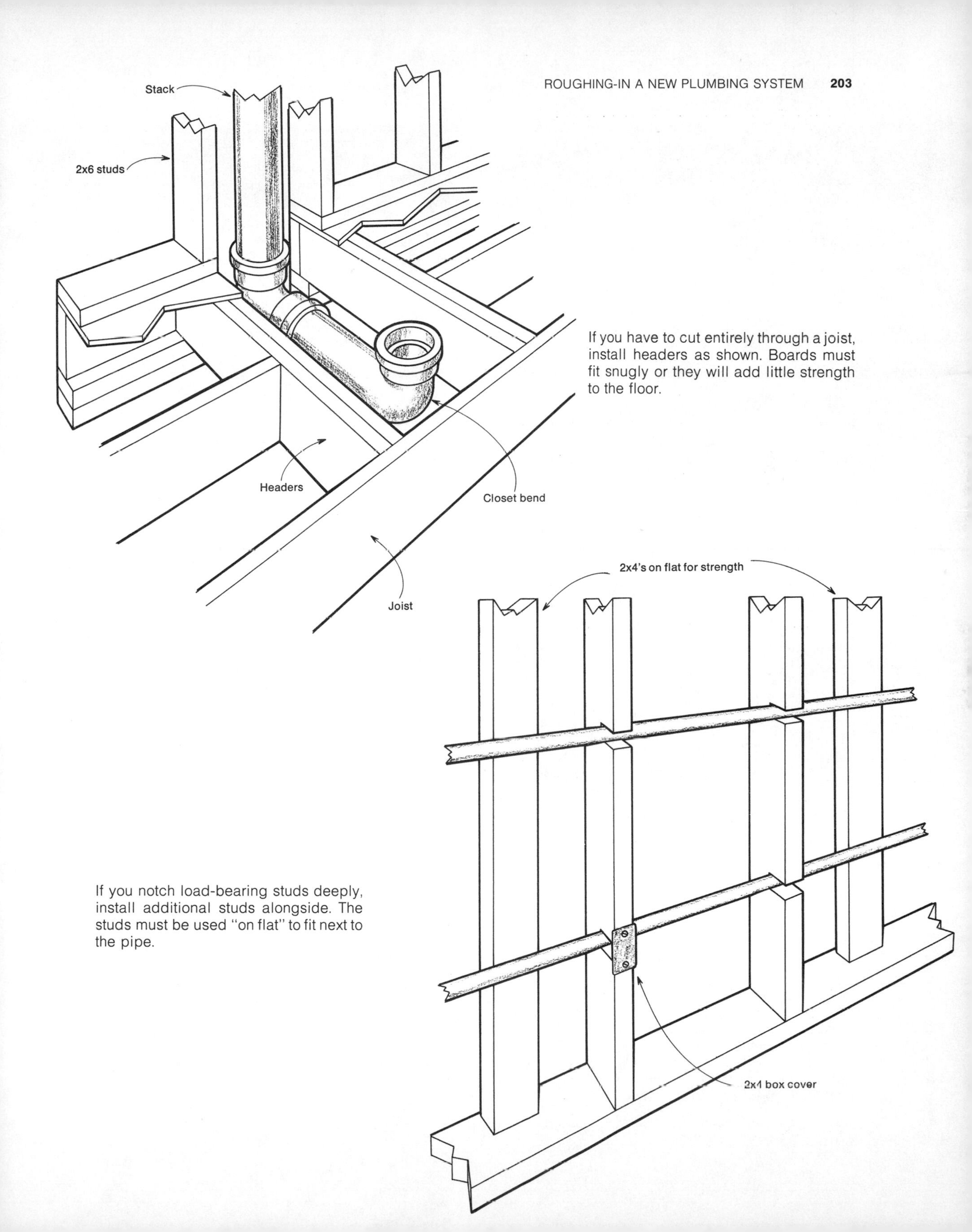

If you have to cut entirely through a joist, install headers as shown. Boards must fit snugly or they will add little strength to the floor.

If you notch load-bearing studs deeply, install additional studs alongside. The studs must be used "on flat" to fit next to the pipe.

Installing a Plastic Toilet Flange

1. A length of drainpipe is measured.

2. The plastic pipe is cut with a crosscut saw.

3. The "shine" is taken off the part to be cemented with a strip of sandpaper.

4. Cement is applied to the pipe.

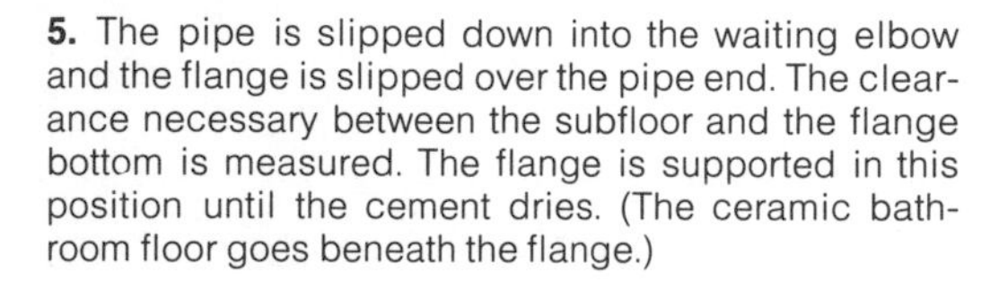

5. The pipe is slipped down into the waiting elbow and the flange is slipped over the pipe end. The clearance necessary between the subfloor and the flange bottom is measured. The flange is supported in this position until the cement dries. (The ceramic bathroom floor goes beneath the flange.)

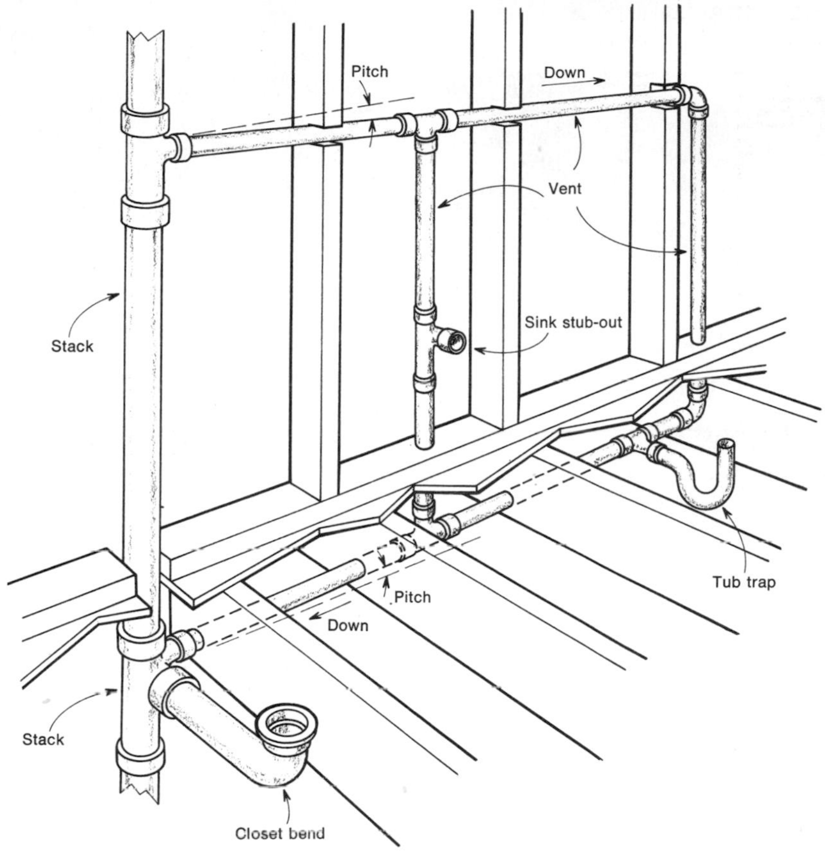

Typical vent and drain piping. Both the tub trap (shown) and the lavatory trap (not shown) are back vented to the soil stack.

vent pipe to the stack, the stack fitting must slope downward. Usually a standard drain fitting installed upside down is used here when rigid pipe is used.

RUNNING WATER PIPE

The same procedures just discussed may be used for running water pipe. The only difference is that it is easier to measure and cut water pipe because the pipe itself, even the rigid types, have a little give so that you can err a bit without problems.

When you run pipe through a series of holes cut through the joists, it obviously needs no support. When you run pipe beneath a ceiling or up a wall the pipe must be supported every so often, even though the rigid types of pipe appear to need no support. This is because without support a long length of pipe vibrates and may eventually open at the joints.

Galvanized pipe should be supported at least every 10 feet. Copper pipe should be supported at least every 6 or 8 feet and by copper clamps and copper screws and nails only. Plastic pipe should be supported every 4 to 6 feet. Any type of loose-fitting clamp may be used.

Protect soft pipe. Whenever you run copper tube or plastic pipe always consider the possibility that the pipe will be later attacked

The water pipes being soldered are supported by the holes in the joists through which they run. No additional support or fastening is required.

Copper tubing must be strapped down with copper straps and copper nails. It is important that no other metal should be used.

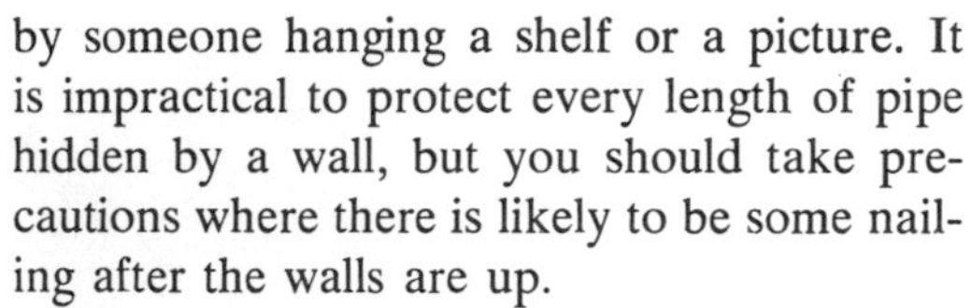

by someone hanging a shelf or a picture. It is impractical to protect every length of pipe hidden by a wall, but you should take precautions where there is likely to be some nailing after the walls are up.

You can use any kind of strip of metal that will slow a nail down. The protective strip can be nailed atop the studs, the fraction of an inch the strip will add to the stud width will not be noticeable when the sheetrock is in place.

Laying out water pipe. The usual arrangement is to connect the water service to the main valve and then to the water meter. From there the cold-water line may turn up to the cellar ceiling and cross the length or width of the building. Although in our calculations we showed the cold-water pipe taking an immediate upturn, there is no special reason for doing so. You can run cold-water pipe in any direction you wish. The goal, of course, is to use as little pipe as practical.

Professional plumbers make neat square joints when they cross a ceiling. This is unnecessary. You can cross at an angle if you wish. You can install your riser at any point that is convenient. You do not have to use a single riser for two floors. You can use two risers or a riser for every fixture. We used a single riser in our example because it was simple and made the associated math easy to explain. No other reason.

In most instances you will find it convenient to run the hot- and cold-water lines side by side. This is fine but just keep them 6 or 8 inches apart so the hot water doesn't warm the cold and vice versa.

While the position of the furnace is primarily dictated by the location of the chimney, heating requirements and the usual desire to leave as much open space in the basement as possible, bear two facts in mind before you finally settle on where the furnace will go. You need clearance on all four sides to get at the domestic hot-water coil and the furnace itself for service and repairs. And, the closer the furnace is to a cellar wall, the more heat that is lost to the wall and surrounding earth. Try to place the domestic hot-water heater, whatever type it may be, as close to the fixtures that use hot water or a central feed pipe as you can. The shorter the pipe from the heater to its faucet the less heat you will lose in the pipe.

13

Installing Fixtures

This chapter assumes that most of the roughing-in work is done, and that with the exception of the last section or two of pipe, all the domestic hot- and cold-water lines and drains are in. This chapter further assumes that the room or rooms in which you plan to install the fixtures are unfinished; the walls are not covered with sheetrock or plaster and you are standing on the subflooring.

Thus, this chapter concentrates on the problems you will encounter in selecting, mounting, fastening and connecting fixtures to newly installed pipe, so-called "new work," which may be a single fixture or a house full.

The next chapter, Remodeling Plumbing, covers replacing existing fixtures, updating systems and adding appliances (dishwashers, etc.) to old plumbing systems.

CARE OF FIXTURES

Fixtures are usually packed with pads of folded paper or foam supporting the unit within a crate of rough boards. If any of the boards are broken or cracked it is obvious the fixture was not handled with care during shipment. But even if the crate is in good condition, take your time and examine each piece carefully before you sign for it. Once you have accepted delivery it is all yours.

Lighter fixtures can be carried by one or two people and placed out of harm's way. Heavier fixtures, such as a tub, cannot be easily moved. If you cannot find three or more helpers, leave the tub in its crate. Lift one end and slip a length of iron pipe underneath. The pipe will act as a roller and you can more easily move it without damage. Do not remove the tub or any other fixture from its crate until you are ready to install it. With a tub, move it the final few feet by placing it atop some smooth boards and sliding it into place. Then remove the boards. Don't push the tub on its bare bottom across a rough, nail-studded subfloor.

Protect the inside of a tub or other fixtures with a dozen layers of wet newspaper glued in place with any kind of water-soluble glue. The paper will protect the tub's inside from scratches and prevent debris from falling into the drain.

INSTALLING STUB-OUTS

Stub-outs are used wherever a pipe has to enter a room through a visible, finished wall. They are used with lavatories, kitchen sinks, toilets, washing machines and dishwashers. They are not used with bathtubs and showers because the water and drain connections are made inside the wall or floor. They are not used for basement washers because the piping is exposed.

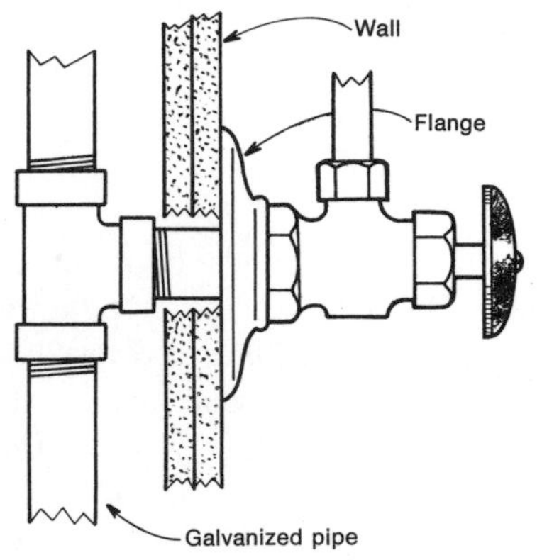

Short nipple stub-out. Here a nipple, cut to the proper length, eliminates the need for plated pipe. Stub-out is hidden by flange and wall.

Water-supply stub-outs. Water-supply stub-outs are preferably positioned about 1½ feet away from the faucet to which they are connected. A shorter distance makes it difficult to bend and install the tube connecting the stub-out to the faucet or valve. A greater distance is a waste of tubing and pressure.

For the sake of appearance, water supply stub-outs are positioned directly beneath the faucets they serve and out of direct sight.

Choice of the stub-out pipe material depends on the feed pipe, use of shutoff valve and whether or not the stub-out is directly visible. If you are running galvanized you can use a threaded nipple for the stub-out.

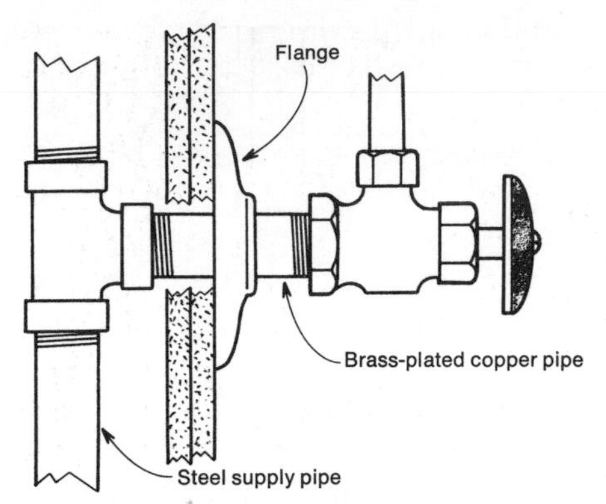

Projecting stub-out. This is an alternative when no nipple is available. A length of plated pipe is used for the stub-out.

The shutoff valve can be screwed directly onto the end of the nipple.

If you are running plastic or copper you can use a similar material for the stub-out, but in order to attach a valve to the end of the stub-out you will need a transition fitting. This is either cemented or sweated to the end of the stub-out. The valve screws onto the threaded end of the transition fitting.

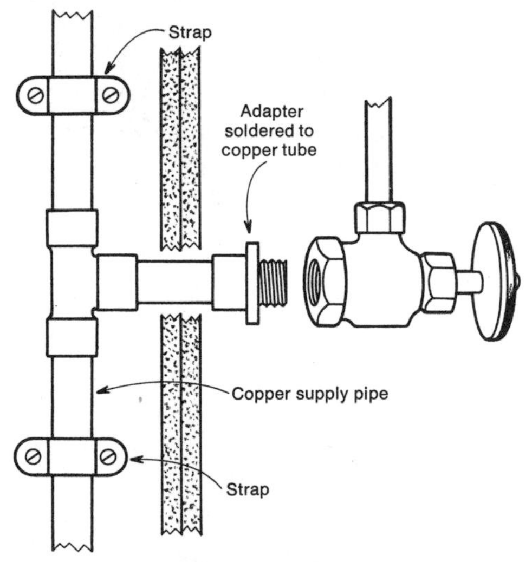

Copper tube stub-out. To connect tubing to a screw-thread valve, solder a transition fitting (adapter) to the end of the tubing.

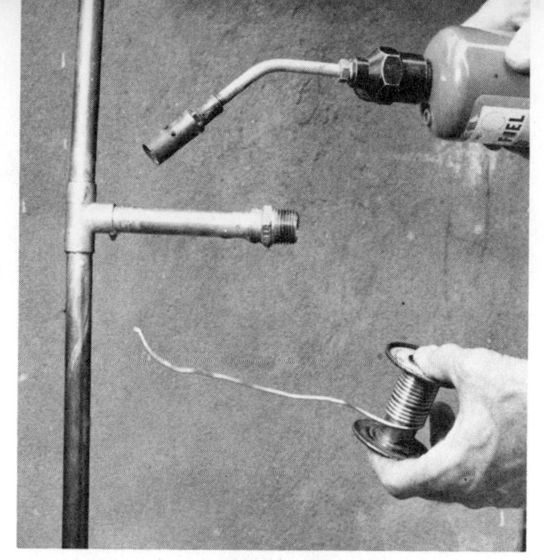

Soldering a copper tube stub-out to a copper Tee. This has to be done before the wall is up.

If you don't plan on using a valve you can simply continue the feed pipe a distance beyond the studs. Later, after the wall is up, you can continue the run.

If the stub-out appears beneath a kitchen sink or inside a cabinet supporting a lavatory, it doesn't really matter what finish the pipe may have; it is out of sight.

If the stub-out is in plain sight you can hide it by making it just long enough to go through the wall and into the valve. The valve body itself or a flange under the valve body hides the edges of the hole and the stub-out is invisible. Alternately, you can use a nickel or chrome-plated pipe for the stub-out.

All stub-outs must be securely strapped to the studs. If a stud isn't handy, nail a few blocks of wood to the floor or a distant stud. If you don't fasten the stub-out down you may have trouble attaching or removing the valve.

Always use a copper strap and copper nails with copper tubing. Use any metal with plastic pipe and iron with galvanized. If you use iron on hot-water copper tubing there is a good chance the copper will corrode through at that point.

Drainpipe stub-outs. Whereas the stub-out for a water feed pipe can be made with a standard Tee or elbow, drain fittings only

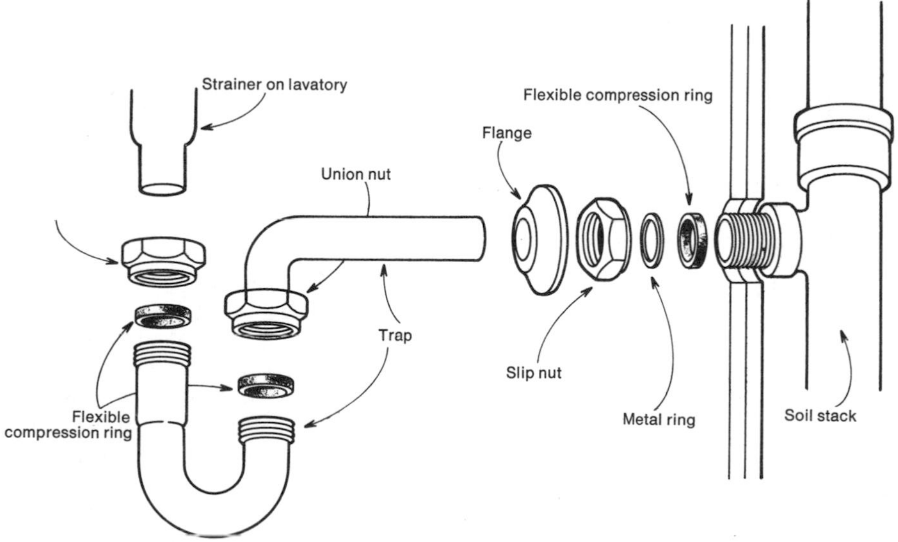

Out-of-line fixture drain. When the fixture's drain opening is not in direct line with the drain stub-out, use a two-piece trap to make the connection.

should be used to connect a drainpipe stub-out to the drainpipe. And, the drain fitting must be installed so that the stub-out angles upwards towards the fixture.

If you can position the drainpipe stub-out exactly in line with the fixture's drain you can get by with one section of drainpipe and a single flexible-ring compression joint on the end of the trap. If you have to offset the drain you may need more than one section of pipe and suitable couplings and a two-section trap which will permit you to swing one end a few inches from side to side.

The height of your stub-out from the floor is not too critical because you can make considerable vertical adjustment by means of a flexible-ring compression joint either at the end of the trap or between two vertical sections of drainpipe connecting the trap to the fixture's drain.

Several methods are used to make a neat connection between the drain stub-out and the drainpipe. When the drain fitting is threaded, a close nipple followed by a flexible washer and a flange nut can be used to connect a length of thin-wall brass drainpipe. If the hole in the wall is just right, the flange nut covers the opening and makes a neat joint.

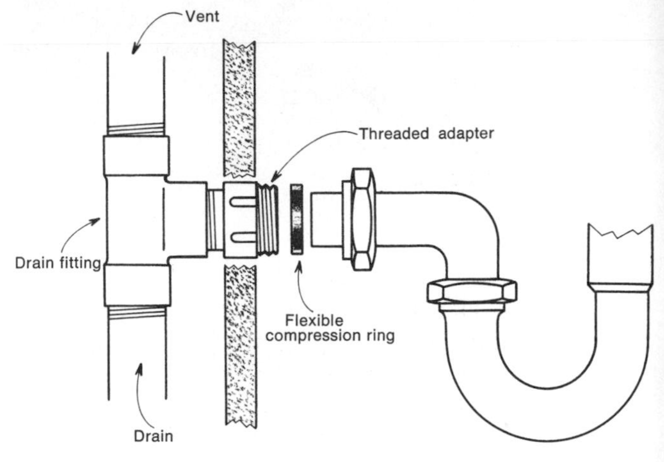

Connecting a trap to a copper stub-out. A transition fitting is sweated onto the end of the copper stub-out. A flexible-ring compression joint is added to complete the connection.

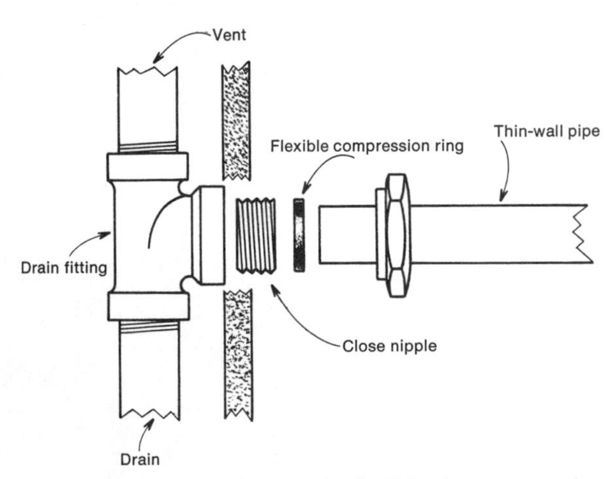

Using a close nipple and a flexible-ring compression joint to connect a thin-wall brass drainpipe directly to a Tee. If the hole in the wall is not too large, the flange nut will neatly cover the hole.

When the drain fitting is of sweated copper, a threaded adapter or transition fitting is soldered to a short length of copper tubing, which is then sweated to the drain fitting. The flexible ring goes over the end of the thin-wall brass drainpipe, which enters the adapter and is held in place by a flange nut.

INSTALLING WALL-HUNG LAVATORIES AND SINKS

Start by studying the manufacturer's literature or measure the fixture itself to secure roughing-in dimensions, faucet and drain locations. Draw a reasonably accurate, full-size outline of the fixture on the studs. Position the drawing so that the top of the basin is 31 inches above the finished floor. Allow

1 inch if the floor is to be of wood, 2 inches if the floor is to be ceramic tile on mortar cement.

Some fixtures can only be hung from brackets supplied by their manufacture. Some can be hung from a pair of individual brackets which can be spaced as necessary, and some are hung from a bar bracket. Hold the bracket(s) up against the studs where it will be mounted when the fixture is in place. Mark the position of the bracket. Now cut a length of 1x6 or 2x4 lumber long enough to cross two or more studs. Rabbet this cross brace into the studs and nail it fast.

Install stub-outs, water and drain, in proper relation to the outline of the fixture and where you know the faucets and drain are going to be.

At this point you can finish the entire room if you wish—walls, floor, trim and paint. There is nothing more you need do to the pipes. You work from the stub-outs from this point on.

Next, mount the fixture bracket with good-sized screws and fasten the bracket(s) atop the sheetrock or plaster wall and to the cross brace you installed.

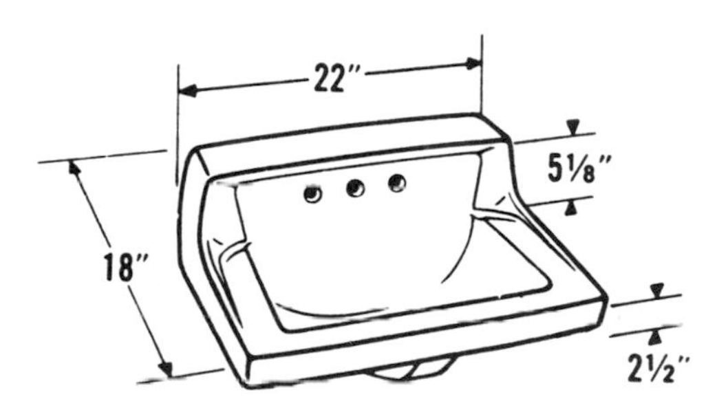

Manufacturer's sketch and roughing-in dimensions for a wall-hung lavatory. *Courtesy American Standard.*

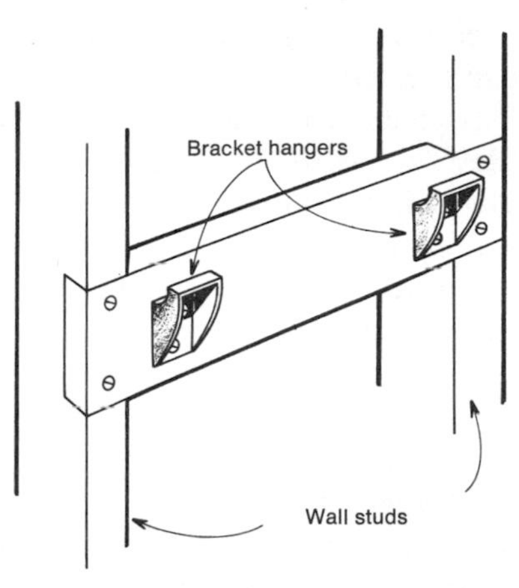

A pair of fixture-support brackets mounted on a cross brace rabbeted into two studs. Brackets must be suited to fixture and must be spaced properly. Bracket height from floor establishes fixture height.

Now hang the fixture on its bracket. If it has legs, these may be attached at this time. The water connections are usually made by means of Speedees, preformed copper tubes

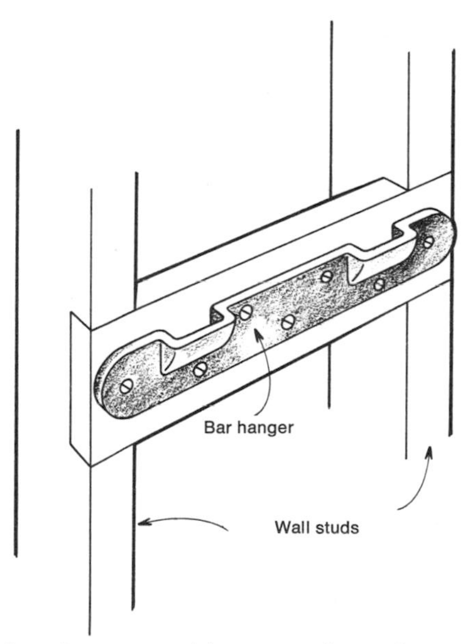

Bar-type hanger used for supporting a fixture. Bar dimensions must match those of fixture to be hung.

made for the purpose. They are discussed along with other methods used to connect faucets to their water supply in Chapter 5, Repairing and Installing Faucets and Valves.

There should be a washer beneath the top and bottom of the drain fitting or strainer where it makes contact with the porcelain bowl. If there is no washer put a thin layer of plumber's putty behind these parts. The putty helps prevent cracking when you tighten the drain fitting and seals the joints against leakage. Just make the drain parts snug rather than tight.

Following, the drain stub-out is connected to the lavatory drain. Nickel-plated thin-wall brass pipe is used along with flexible-ring compression joints. Again, beware of making these joints too tight.

Connecting a plastic stub-out drain. The stub-out is made a little longer than necessary and permanently cemented to the drain fitting within the wall. The connection between the plastic pipe and the thin-wall brass drainpipe is made at the end of the stub-out if a metal trap is used, or between the trap and the tail-pipe (drainpipe) on the fixture if a plastic trap is used.

The joint between the plastic pipe and the metal is made with a pipe trap adapter, which consists of a short length of plastic pipe threaded at one end. With the addition of a

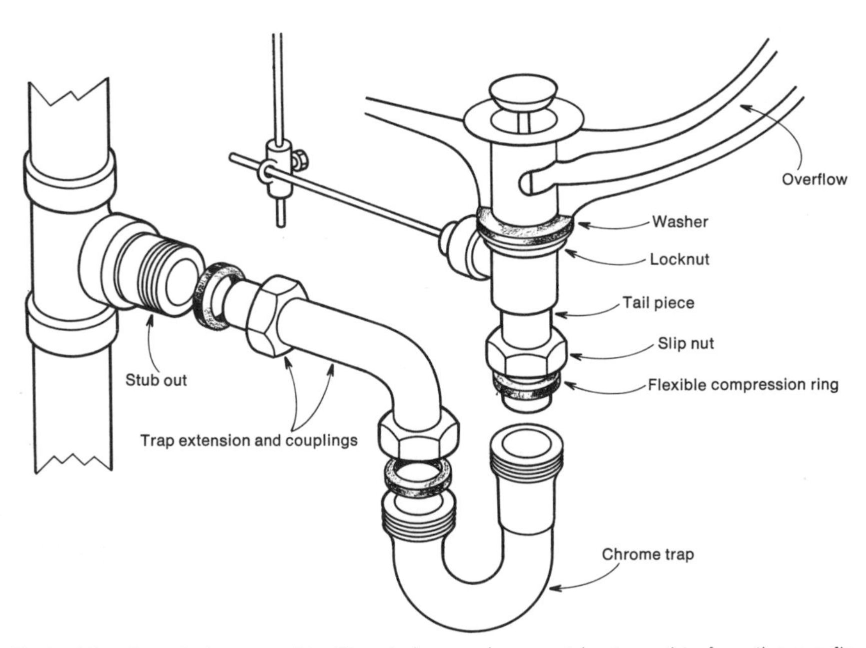

Typical lavatory drain assembly. The drain opening must be turned to face the overflow channel in the basin. Washers or putty must be used between the drain fitting and the basin. The trap is connected after the tail piece and the trap extension have been connected.

Typical kitchen sink stub-outs. Drain stub-out consists of a galvanized nipple threaded into a copper Tee which is sweated to the DWV tubing. Water-feed stub-outs are galvanized nipplės screwed into copper elbows, the lower ends of which have been sweated onto the copper tubing. Copper reducing Tees are used to connect the ½-inch stub-out tubing with the ¾-inch feed pipes coming up from below.

plastic flexible ring and a plastic flange nut this end forms a flexible-ring compression joint. The metal pipe is slipped inside and the flange nut simply tightened to make the seal. The other end of the pipe trap adapter is cemented onto the plastic stub-out end. Some trap adapters go over the end of the stub-out, others fit inside. For fixture drains, use the type that fits inside.

Plastic traps are constructed like the metal traps with slip joints and flexible-ring compression joints. The trap is cemented to the plastic stub-out. The fixture's tailpipe slips into the trap and is held by the trap's compression joint.

Since you cannot disassemble a cemented plastic joint it is best to try the fittings and trap for size before you cement them together or to the stub-out.

INSTALLING SELF-SUPPORTING LAVATORIES AND SINKS

The steps necessary to install a self-supporting lavatory or sink are almost identical to those used for wall-hung lavatories and sinks. The difference is there is no need to install a bracket to support a hanger. And you need not be quite as meticulous about the stub-outs as they will be hidden by the supporting cabinet.

Start by outlining the fixture on the wall it is going to adjoin. Position and install stub-outs in proper relation to the fixture's faucets and drain. Install and finish the wall and floor. Move the fixture into place and connect the pipes. Fasten the fixture permanently to the wall and/or floor.

INSTALLING COUNTERTOP FAUCETS

If the holes have already been drilled, apply a thin layer of plumber's putty beneath the faucet to seal it to the countertop before you insert it and install the nuts.

If you have to drill a hole or holes for the faucet, make certain you allow clearance between the underside of the faucet and the bowl as well as between the faucet and the wall or rear of the supporting cabinet.

Some modern kitchen sinks and lavatories come equipped with single-lever faucets. As they do not accept Speedee fittings but have

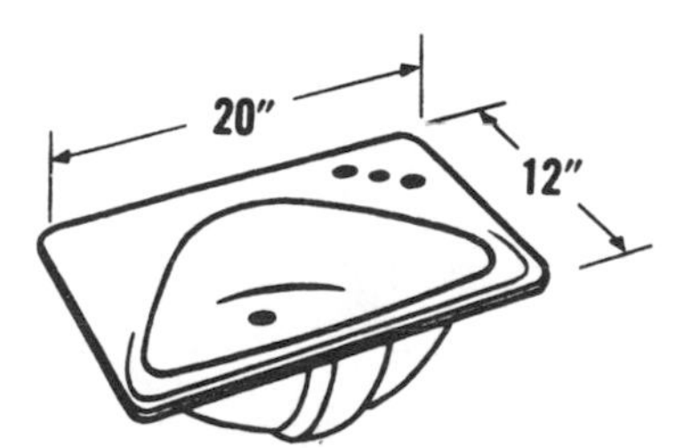

Manufacturer's sketch and roughing-in dimensions for a countertop lavatory. Type shown overlaps edge of hole in countertop. *Courtesy American Standard.*

two downward-protruding lengths of ⅜-inch copper tubing, connections are made somewhat differently.

If you have already installed stub-outs terminating in shutoff valves made to accept a Speedee fitting, you can use a sweat coupling and solder the end of a length of ⅜-inch copper tube to the end of the faucet tube. Cut the other end of the tube perfectly square, remove all the burrs and insert it into the top of the shutoff valve. The metal-ring compression fitting atop the shutoff valve will accept ⅜-inch copper tubing and make a watertight seal if the tube's end is perfectly straight and smooth.

As an alternative to using rigid pipe stub-outs and angled shutoff valves, you can let your ½-inch copper cold- and hot-water pipes project into the room about 1 foot, after which the wall is installed and finished. Then the ½-inch tubing is sweated to a pair of valves with "ears." Lengths of ½-inch tubing are sweated to the other side of the valves and joined by means of a sweated-on reducing coupling to the ⅜-inch tubing leading to the single-lever faucet. Next the valves are screwed fast to a board fastened to the wall underneath the fixture.

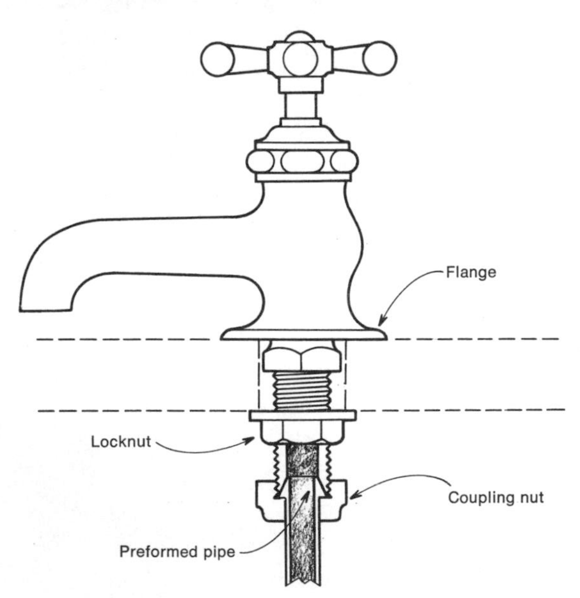

How an individual faucet can be mounted in a countertop. Connection between faucet inlet and feed pipe depends on inlet design. In faucet shown, a rigid length of pipe containing a preformed end (ground joint) is used.

When doing this you will find it helpful to do most of your soldering with the fixture turned away from the wall so you can easily reach the faucet pipes from the rear of the cabinet. Also, there is no special need to keep the pipes under the cabinet short. A couple of extra inches will ease the task of getting the valves and the board into position on the wall.

INSTALLING OLD-FASHIONED LAVATORIES

The major difference you will encounter with old fixtures is in the way the faucets are connected to their feed pipes. On really ancient sinks and lavatories, the faucet is screwed right down on a rigid, threaded feed pipe. On the merely old lavatories and sinks, a rigid feed pipe with a prefabricated end (sometimes called a ground joint) is held in place against the bottom of the faucet inlet with a flange nut.

When installing either of these two types of vintage fixtures you must be very careful to make the stub-out pipe length just right as the elbow you place there must point straight upwards to the bottom of the faucet. And the length of the final section (or sections) of pipe must also be quite accurate as it must just reach the faucet and not exert a stress on the fixture.

The best way to handle these connections is to temporarily make up the entire run from feed pipe to faucet before you finish the wall and floor.

INSTALLING BATHTUBS

There are presently three general types of tubs: cast iron, pressed steel and plastic. The cast iron and pressed steel, which also have a porcelain coating, look almost exactly alike and wear almost exactly alike. However, the cast-iron tub requires four strong men to carry it upstairs. Two men can handle the pressed steel. The plastic tub is lighter still. It will probably last as long as the other two but as it is of plastic it will change with time.

Tub design features such as size and shape and outlet hole position will depend on taste and the configuration of your bathroom. From a plumber's point of view, all tubs are installed exactly alike.

You are well advised to secure roughing-in dimensions and construct the tub's framing before you position and connect the tub. If you position the tub first and then build the stud frame around it, you stand a good chance of damaging the tub. They chip and scratch very easily.

A point to bear in mind before framing the bathroom or the tub enclosure is that some tubs are broader than the bathroom door opening. If this is true of the tub you select, just hold off framing the door until you have installed the tub.

Load considerations. A full-size iron tub with a fat person inside can weigh a full ton. Therefore, you cannot place a tub anywhere. Be certain the floor will take the load.

If you are working from plans approved by your building department, you can be certain you are safe in placing the tub in the bathroom as indicated on the plans. However, this presupposes that you will not cut any joists without replacing the support they provided one way or another. The simple way is to double the thickness of the remaining supports. If you cut right through a joist, double the thickness of the joists on either side of the cut section.

If you are building your own home, or an adjoining room, and you have any doubts about the strength of the joists you are using, check with an architect or a local builder. There is no point in overbuilding, but using joists too long or thin isn't wise either.

Preinstallation work. The tub sits on the subflooring and presses right up against the framing studs. Therefore the subfloor must be clean, level and free of protruding nails. If you are installing a pressed steel tub, you must nail 1x4 strips of wood around the perimeter of the tub's stall at a height that will place them just under the rim. If you have a cast-iron tub, these rim or ledge supports aren't necessary.

Next, cut a clearance hole for the drainpipe. Use the roughing-in dimensions or measure directly from the tub itself and locate the position of the tub's drain hole on the subfloor. Cut a generous clearance hole in the subfloor to permit you to later connect the drainpipe and associate trap and pop-up valve assembly.

To provide trap and valve access, cut and install whatever framing is necessary to permit you to later install a trap door or other means of getting to the trap and pop-up valve for cleaning and servicing at some later date.

You must also install the faucet and shower-head assembly. The faucet assembly consists of two water valves, a diverter valve, a connection to take the spout stub-out and a connection to take the pipe running upwards to the elbow that carries the shower head stub-out. The valves are strapped to a cross brace rabbeted to the studs. The elbow at the top is similarly supported. However, before you start hacking away, there are a number of things to consider.

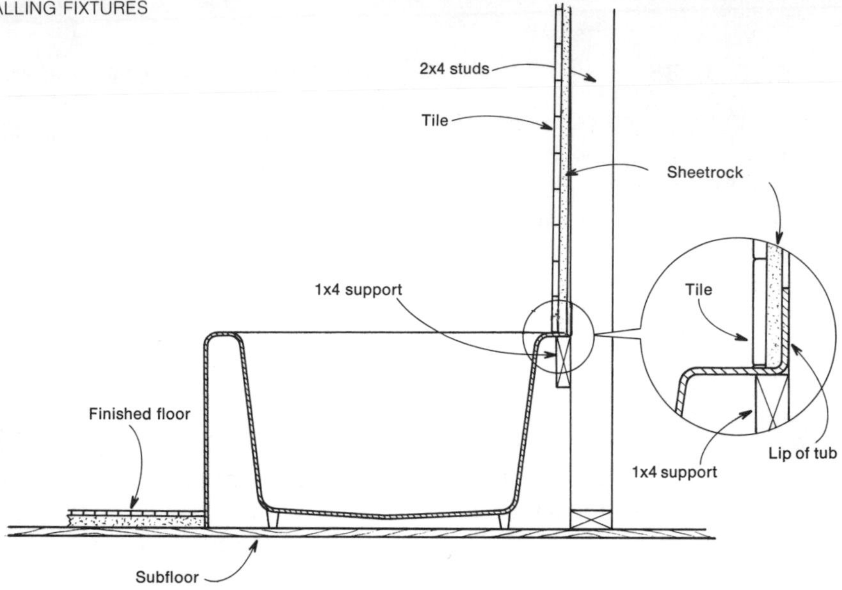

Typical bathtub shown in cross section. This is a pressed steel tub; therefore a 1 × 4 support is used under tub's rim along wall. Note that tub's lip goes beneath sheetrock and tile. Also, tub rests on subflooring and finished flooring butts up against the side of the tub.

First there is height. How high shall you position the supporting cross braces? This is easily answered. Position the lower brace at least 2 inches above the tub's rim. This will permit one full tile between the stub-out and the rim. (A standard tile is 4x4 inches in size.)

Position the second brace as high as you wish. Just remember it is advisable to run the tile above the shower head, so the higher you place the shower stub-out the more tile you may need.

Secondly, there is horizontal placement to consider. This too is easy. Place the tub spout stub-out and the shower head stub-out anywhere you wish so long as the water is directed into the tub.

Next there is brace depth—distance from the finished wall surface. This is fairly critical. The sure way to find this dimension is to assemble the valves, flanges and handles.

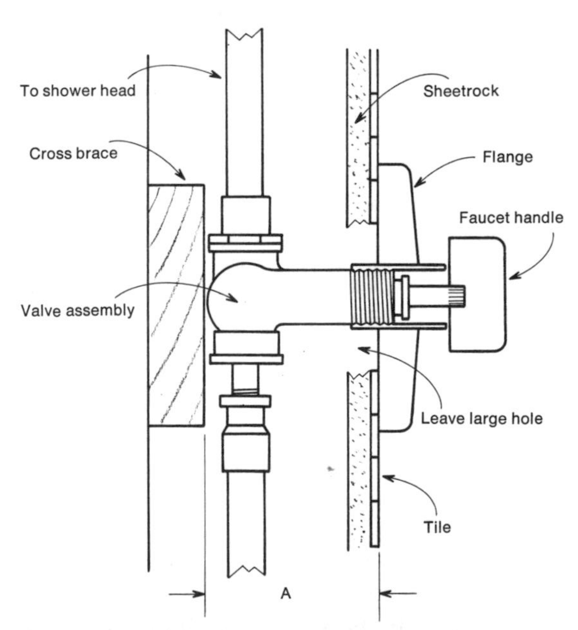

Cross-brace depth, dimension *A*, determines just how far the faucet stems will project beyond the finished wall. To find this dimension before you mount the cross brace, place the flange and handle on the faucet stem and measure from the underside of the flange to the rear of the faucet assembly or fitting.

Close-up of a stub-out for a bathtub spout. Feed pipe is copper tubing, so a transition fitting has been sweated to a copper stub-out. Stub-out must be long enough so spout screws down tightly and ends up pointing downward. Rear of spout covers hole in wall.

Now measure from the rear of the assembly to the underside of the flange. The flange rests on the surface of the finished wall. The valve assembly or fitting rests on the cross brace. Position your cross brace this distance from the *finished wall's* tub-side surface and your valve flanges and handles will work out just right. Some manufacturers provide collars that can be adjusted to accommodate brace depths. Some do not. With these you have to work carefully.

When you have determined these dimensions you can cut your braces into the studs. Use 2x4s or 1x4s and use straps and not bent nails to hold the assembly and pipe in place.

With the shower and faucet assembly firmly strapped down, connect the pipes and stub-outs.

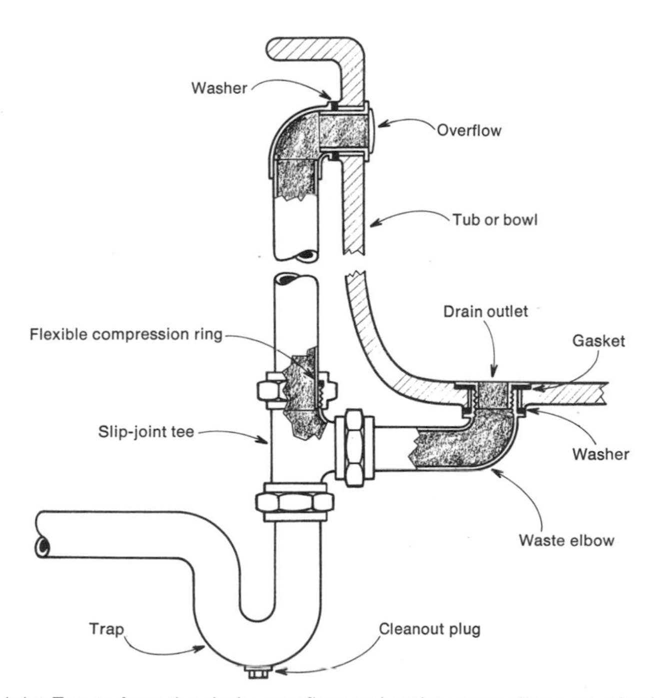

Use a slip-joint Tee to form the drain, overflow and outlet connections on a bathtub. Brass pipe needn't be cut to exact length to assemble the joint since there is some leeway in Tee.

Installing a Prefabricated Bathtub and Enclosure

1. Frame the tub's "pocket" according to the manufacturer's sketch, which is reproduced below. Note that all the studs are not spaced the standard 16 inches on centers, and that a pair of studs (black) are recommended for the ends of the pocket.

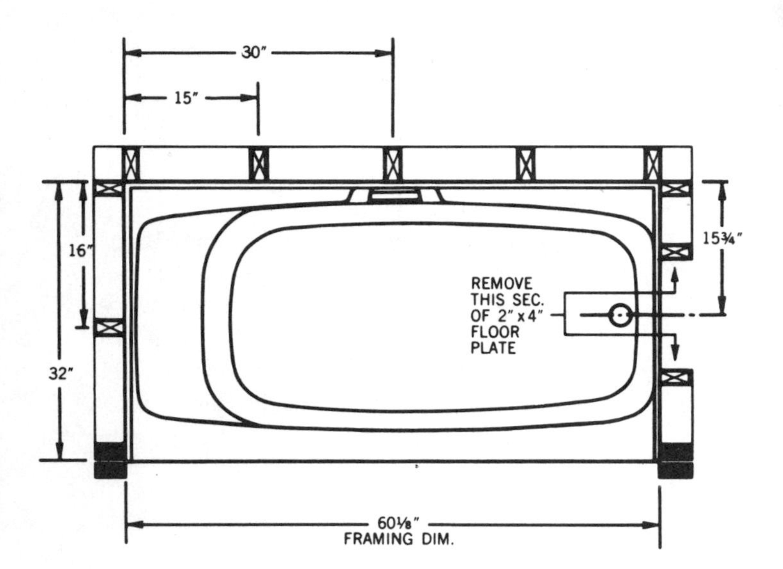

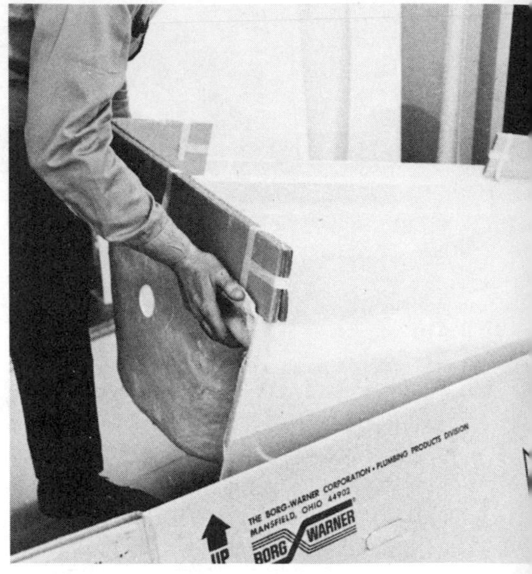

2. With the pocket complete, install the necessary stub-outs at precisely the locations given. Then uncrate the tub and carefully slide it into place.

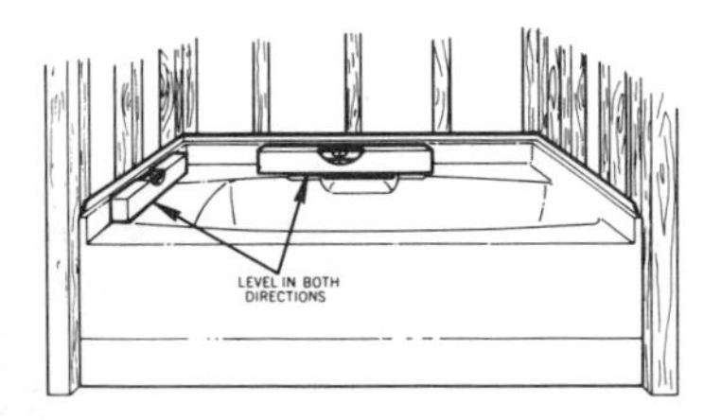

3. Level the tub in both directions. However, a slight pitch toward the drain hole is permissible. If the tub slopes in the other direction, level it by building a false subfloor of ⅜-inch plywood. Do not use shims to level the tub. A plastic or pressed-steel tub cannot be supported at only a couple of points; it will crack when someone stands in it.

4. After the tub has been leveled, drill and countersink holes through the lip and fasten it to the adjacent studs with flathead brass screws.

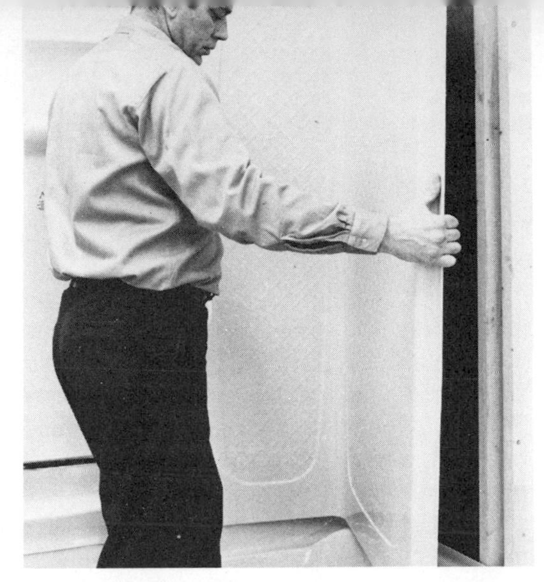

5. Install prefab plastic walls that come with the tub.

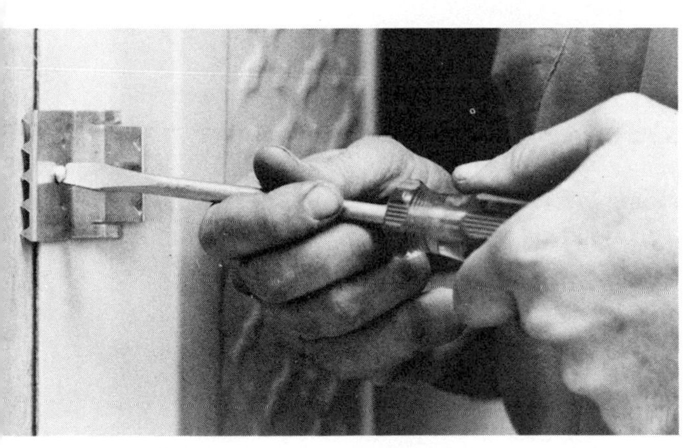

6. Fasten plastic walls to studs with special clips supplied by the manufacturer.

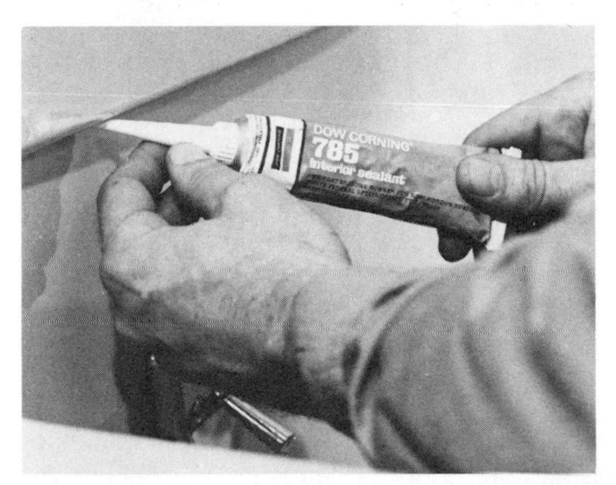

7. Seal joint between the tub and plastic walls with Dow sealant.

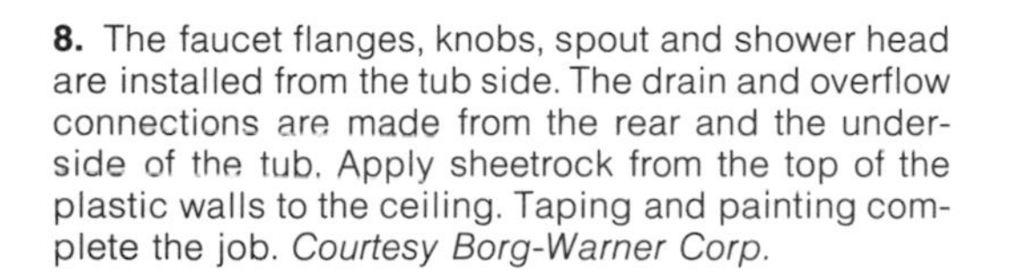

8. The faucet flanges, knobs, spout and shower head are installed from the tub side. The drain and overflow connections are made from the rear and the underside of the tub. Apply sheetrock from the top of the plastic walls to the ceiling. Taping and painting complete the job. *Courtesy Borg-Warner Corp.*

Take care with the spout stub-out. It must be just long enough to enable you to screw the spout fast with its spout pointing downward.

Installing the tub. When preinstallation work is finished, move the tub into place and connect the drain, trap and pop-up valve assembly. Use plumber's putty if you don't have soft washers to place above and below the drainpipe-to-tub connection. This done, protect the tub with a dozen layers of wet newspaper held in place with water-soluble glue.

As shown in the accompanying drawing, the lower edge of the wall goes over the lip of the tub's edge. Leave sufficiently large holes in the wall for the faucet stems and the diverter valve handle to permit you to later insert a socket wrench to disassemble these valves for service and repairs.

When the wall is completed, install the stem flanges, faucet and diverter valve handles and the shower-pipe head. The floor can now be laid down. It butts against the side of the tub and helps hold it in place.

Prefabricated bathtubs. Many manufacturers are now offering complete, lightweight tubs and enclosures. Whether or not the total material price is less than that of tub and walls (tile, etc.) is something you will have to check yourself. However, the complete package is definitely much easier and faster to install.

The steps necessary to install a typical prefab tub and enclosure are a portion of the manufacturer's drawings and photos are shown in accompanying illustrations.

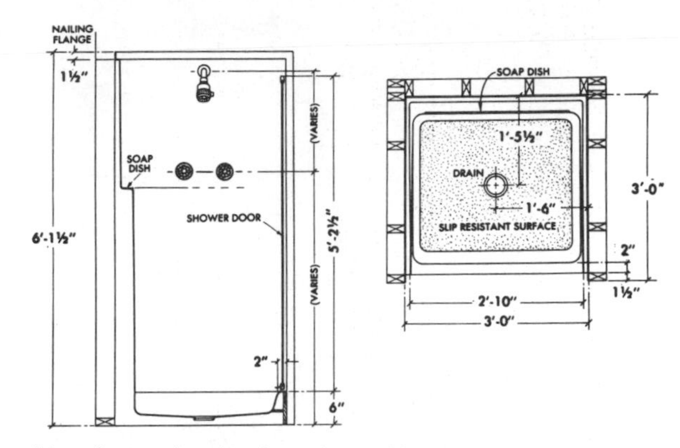

Manufacturer's sketch and roughing-in dimensions for a shower stall. Note framing details; also, note that the drain hole is not exactly in the center. *Courtesy American Standard.*

INSTALLING SHOWER STALLS

For convenience of discussion, shower stalls can be divided into two general types. One type is constructed in place. You lay down a waterproof pan, build a wall around it, and mount a shower head and valve on one side. The other is a complete unit which is slid into place and the necessary pipes connected.

The frame for a shower stall is constructed somewhat like that for a tub. The stall is dimensioned so that the bare studs rest against the sides of the waterproof pan that forms the bottom of the shower. The shower walls go over the lip at the edge of the pan. Thus, the descending water runs easily from the walls onto the pan.

Follow the manufacturer's literature or measure the pan itself to locate the position of the drain hole in the subfloor on which it rests. Cut a generous clearance hole for the drainpipe. If you are going to use a drum trap, cut whatever floor members are necessary and make provisions for later access to the top of the trap. Drum traps have to be cleaned out from time to time.

The shower drainpipe or strainer slips down through a soft washer and through the hole in the bottom of the pan. If there is no washer use a layer of plumber's putty. The strainer is then locked into place from below, again using putty or a washer. From here you connect the drainpipe to a standard trap or drum trap and then lead the pipe on to the branch drain or stack, as the case may be.

The shower head and valve assembly come next. Rabbet two cross braces into the studs to support the shower-valve assembly and head connection. Strap the faucet assembly and head firmly to the braces with brass or copper straps and copper nails. Connect the hot- and cold-water pipes.

Finally install and finish the walls and floor. As with the walls around the tub, you must leave sufficient clearance around the valve stems to enable you to work on them in years to come. Finish by installing the shower head, valve flanges and handles.

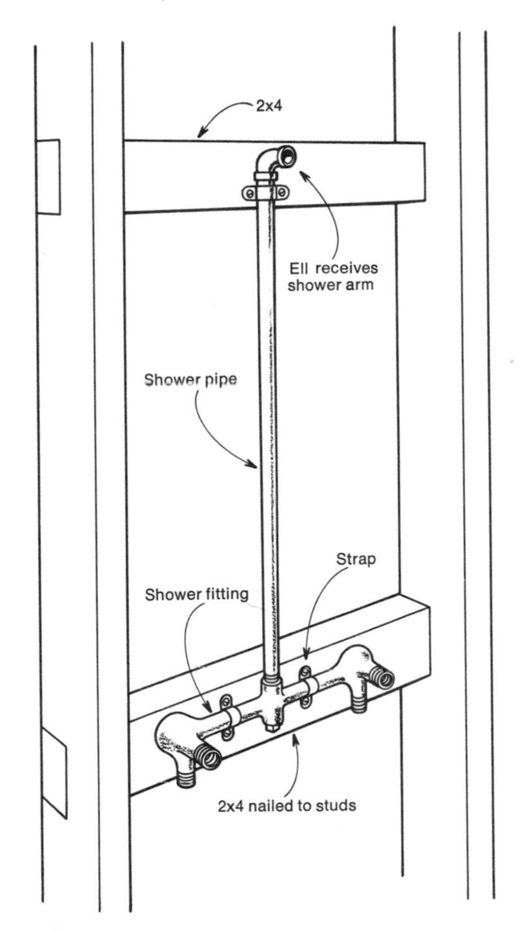

Cross braces used for valve and shower-head assembly. The same arrangement of parts is used for a tub. Difference would be in the valve assembly. Tub would use fitting having a diverter valve in the center and a spout connection as well as the two water valves.

Single-unit shower stalls. The stall's framing is prepared from the dimensions taken from the manufacturer's roughing-in literature or by measuring the stall itself. A clearance hole is cut in the subfloor to accommodate the drainpipe. The shower stall is slid into place. The drainpipe and fittings are connected from below. The water connections are made from the side. With some designs stub-outs are installed before the stall is slid into place.

INSTALLING WASHTUBS

Washtubs are large, deep sinks provided with hot and cold water and a drain. They differ from one another only in size.

Old tubs were made of soapstone cut into slabs and cemented together. They had individual faucets mounted in holes cut in the rear of the tub. Modern tubs are made of porcelainized pressed steel, porcelain, asbestos cement and plastic. They have dual faucets which may be mounted on a nearby wall or bolted or clamped to the edge of the tub.

To install a tub of this type, start by connecting its drain outlet to a trap and nearby drainpipe using standard thin-wall brass pipe and trap, or plastic pipe and trap. Generally the tub stands free on its own legs. However, it is good practice to either daub a little mortar cement around the legs, if the tub is on concrete, or to nail them in place if the tub is on wood. Tubs have a habit of walking away in response to vibration.

Fasten a 12-inch 1 x 6 board to the surface of the wall immediately above the top of the tub. Run hot- and cold-water pipes along the surface of the wall and connect them to a plain, dual faucet, fastened to the board. So long as the spout of the dual faucet projects over the rim of the tub it is properly positioned. Generally, ½-inch copper feed pipes are used along with an inexpensive, sweated-on faucet assembly.

INSTALLING TOILETS

There are two general types of new toilets manufactured today. One is the familiar floor-mounted design. The other is hung from the wall. The floor-mounted design is much less expensive and easier to install. The other, being clear of the floor, makes the room easier to clean and appear to be larger.

The two basic configurations are manufactured in several variations. The most expensive, least noisy, most effective in operation and efficient in use of water is the siphon jet. During discharge a jet of water enters the bowl and speeds the soil out of the bowl and down the drain.

A somewhat less expensive and less effective design is the siphon whirlpool. A portion of the water from the tank is directed to produce a whirlpool in the bowl and speed discharge. Like the siphon jet it uses comparatively little water, and is quiet.

Another is the reverse trap. It is less costly and less effective because it has no jet. It can be used with a flush valve or tank.

The least expensive and slowest to discharge the waste is the wash down. It requires the most water and because its action is slower than the others, it is most liable to stoppages.

Still another type is called the blow out. It can only be used with a flush valve, which required 1-inch or larger supply piping all the way from the water meter, and fairly high pressure. Thus it is uneconomical for most installations.

Floor-mounted Toilets

The horn (discharge pipe) on the bottom of a floor-mounted toilet sits inside the toilet drainpipe. When it does, the rear side of the flush tank's cover should be about 1 inch away from the finished wall. Since the toilet is installed after the finished floor has been laid down, which locks the toilet drainpipe firmly in place, it is obviously most important that your first step is to make certain the drainpipe is the proper distance from the rear wall.

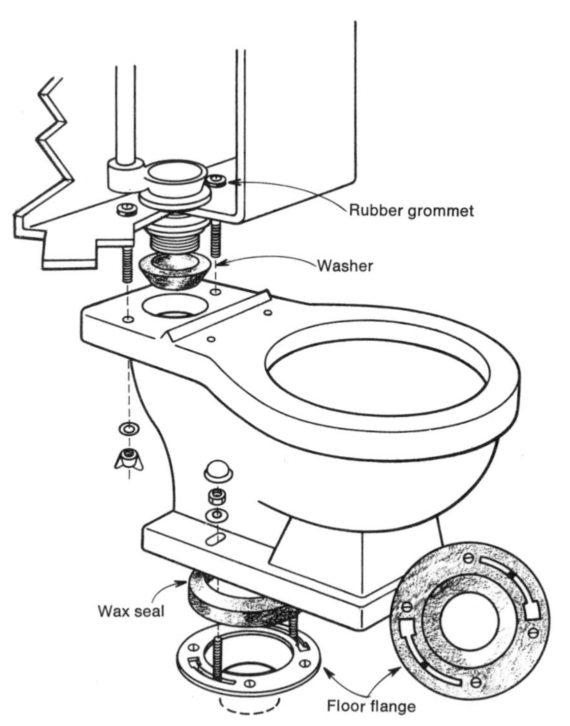

Major parts of a modern floor-mounted toilet. Horn at bottom of toilet can be seen in photo in Chapter 6.

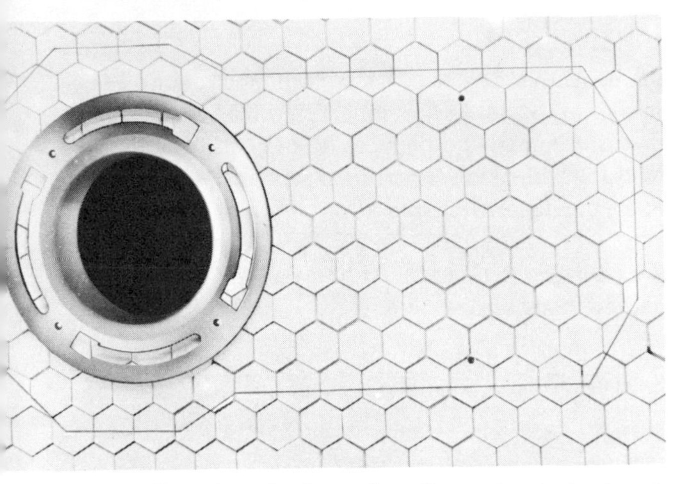

Top view of a brass floor flange. Lead pipe is soldered to inside beveled edge of flange. Toilet horn fits inside lead pipe. Wax ring seals the joint.

If the wall to the rear of the toilet is of bare studs, as it would be in new construction, allow ½ inch if you are going to use sheetrock alone, allow 1 inch if you are going to set tile atop the sheetrock and allow 1½ inches if the finished wall is to be tile on plaster. Draw a line along the subfloor next to the bare wall, indicating the distance the finished wall will project into the room.

Measure from the wall line to the center of the toilet drainpipe. Refer to the manufacturer's literature to get the distance from the center of the toilet horn to the rear of the toilet, including the cover's overhang. Or, measure along the bottom of the toilet to a point directly under the end of the cover. The difference between these two dimensions should be at least 1 inch; 2 will do no harm. Three inches of clearance is a waste of floor space.

If the distance from the center of the toilet drain to the wall is not right, now is the time to move the toilet drainpipe.

Check the toilet horn against the toilet drainpipe and flange. If the drainpipe is of lead or copper, it fits *inside* the toilet floor flange. The toilet's discharge horn fits *inside* the drainpipe.

If you are connecting to a plastic drainpipe, the toilet horn fits inside the toilet floor flange, which fits *inside* the drainpipe. So, before you solder or cement away, make certain how the toilet horn should fit the drainpipe. If the drainpipe is of lead or copper, check that the floor flange you are going to use fits outside the drainpipe and that its holes can be lined up with the holes in the bottom of the toilet. With a plastic floor flange, remember that the lower end of the floor flange slips *inside* the plastic drainpipe.

Also check that the center of the toilet's drainpipe is at least 15 inches from the nearest side wall or lavatory. A few inches more will do no harm; less will limit the toilet's use to skinny people.

Begin by outlining the toilet tank on the rear wall studs. This will be a double check to make certain you have sufficient side clearance near the tank for appearance and to remove the cover. Also, it will enable you to mark the position of your cold water stub-out.

Next, install the cold water stub-out. Position the stub-out a good foot below the toilet-tank connection and in line with the feed pipe. Bring the stub-out no more than an inch or two into the room. There is no need for it to project farther. Connect the stub-out and strap it to a nearby stud.

If you install the floor flange after the finished floor is in, whoever installs the floor will have an easier job. However, if you are going to use asphalt or rubber tile and are going to solder the floor flange to a lead or brass pipe, you can't lay the tile down first. In such cases, nail wood shims as thick as the floor tile near the drainpipe. Then install the flange. When the floor is to be laid, remove the wood shims.

Soldering floor flange in place. Assuming you have installed a ceramic tile floor or have shimmed the area around the drain to the correct height, position the floor flange with the bevel side up. Make certain it is level and screw it fast to the subfloor with brass screws. If the tile setter hasn't already drilled holes, drill your own. Then use a hacksaw to cut the top of the lead toilet pipe flush with the top of the flange. Use a file or sandpaper to clean the outside top inch of lead pipe of oxide. Make it shine. Apply flux to the flange. Heat the flange with a propane torch. As you do so keep touching the tip of a length of wire solder to the flange a distance away from the torch flame. Continue heating until the solder melts. Continue applying solder until the entire beveled surface is covered with solder. Shut off the torch. Use a hammer and gently spread the top inch of lead pipe until it smoothly and evenly rests against the beveled edge of the flange. Apply flux to the joint. Relight the torch. Heat the flange until it again is hot enough to melt solder. Now, very carefully, heat the lead until it comingles with the solder. *Do not overheat the lead.* The melting point of lead is very close to that of the solder. If you apply too much heat the entire top section of pipe may collapse. Let cool. File the top of the flange smooth.

Cementing a plastic floor flange. This is much easier than the previous job. The floor is laid or shims are used as described previously. Sandpaper the end of the flange fitting —the section that enters the drainpipe. Insert the flange fitting into the pipe and try it for size. If all is well, apply an even coating of plastic cement to the pipe end of the fitting and slip it into place. When the cement hardens in a minute or two, the job is done.

Finish walls and floor. With the toilet feedpipe stub-out in place and the floor flange permanently connected, the wall and floor (if not in) should be installed and finished—that is, trimmed and painted. The correct way to do a bathroom is not to install the walls and floor in sections but to go as far as you can with each fixture until the walls and floors are needed, and then to do all the walls and the floor at one time.

Install bowl and tank. Disconnect the toilet bowl from its tank and place the bowl on its

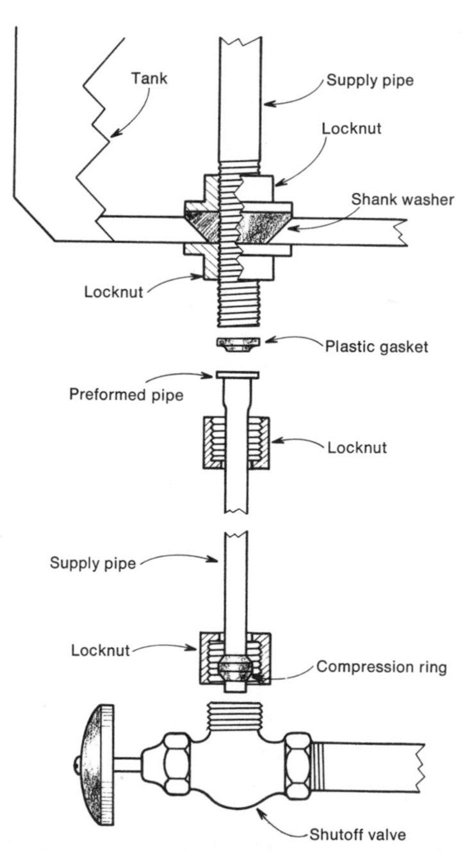

Connect the water-supply pipe to the bottom of the ballcock valve with a preformed Speedee fitting. A locknut and gasket seal the joint.

side. Position a prefabricated wax ring toilet seal around the toilet's discharge horn. Press it firmly against the bottom of the bowl to make it stick. Insert the toilet hold-down bolts, head down in the floor flange. Turn toilet bowl right-side up and lower it over the hold-down bolts and drain opening. Press the bowl down to expand the wax and bring the bottom of the bowl against the finished floor. Run nuts down on hold-down bolts.

Now you can bolt the tank to the bowl. Make certain there are rubber grommets with each tank bolt and that the large "spud" washer points downward into the hole in the rear of the bowl. Do not overtighten the bolts as you can easily crack the tank.

With the tank in place, rotate the bowl to make certain the tank is parallel to the rear wall. Tighten the bowl hold-down nuts. Just make them snug; the wax seals the joint. Fill the porcelain caps with a little wet plaster and invert over the hold-down nuts.

Finally, assemble and install the ballcock valve and associate parts, and the seat. Use a Speedee fitting to connect the cold-water stub-out to the bottom of the ballcock valve assembly.

Installing Old-style Toilets

If you are building a vacation home, or an addition to your present home, you may want to save money and buy a used toilet. Investigate the used building material yards in your area; you can often pick up a working toilet for $5 to $10. As long as the toilet isn't cracked it will function. The very old designs —those with the tank near the ceiling—are museum pieces. Later designs are also in two pieces, but the tank is fairly close to the bowl. The procedure used to install an old-timer is similar to that used to install a one-piece toilet, with a number of exceptions that will be noted.

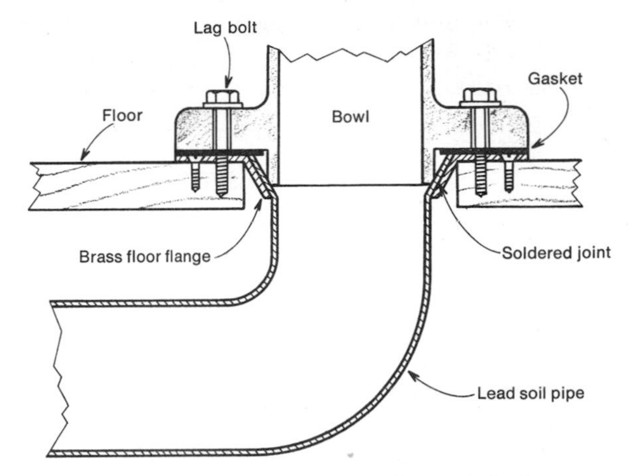

Cross section of an old-fashioned toilet-to-drainpipe connection which is sealed by a gasket. Therefore it is important that the horn does not rest on the lead pipe and that there is solid wood beneath the bottom of toilet bowl so that the lag bolts can take hold.

First of all, check the horn-to-flange sealing requirement. Make certain the diameters are correct and determine whether or not a wax ring seal can be used. Some of the older designs use a flat gasket between the top of the flange and the bottom of the toilet. With these you have to be very careful to provide a solid "footing" for the lag bolt that goes through the flange and into the subflooring. You also have to make certain there is sufficient clearance (a fraction of an inch will do) between the toilet's horn and the inside of the drainpipe. The bottom of the toilet should rest on the gasket and not on the inner surface of the drainpipe. Otherwise you will not be able to seal the toilet to the flange without danger of cracking the bottom of the bowl.

Next, check the water-closet elbow. On some models you can use a standard elbow between the water closet (tank) and the bowl. On others nothing but the elbow supplied by the manufacturer will fit. So check

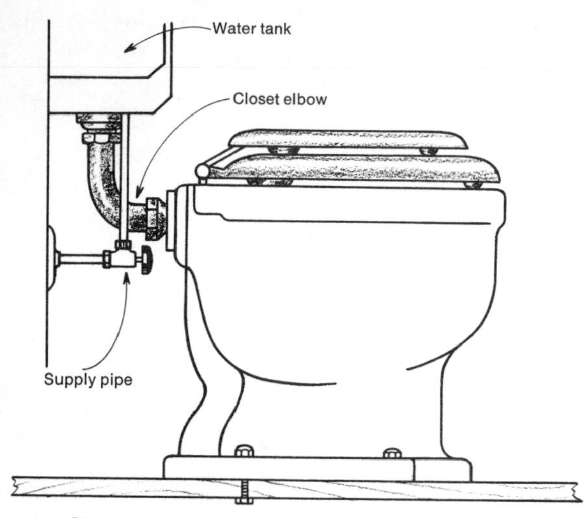

Side view of an old two-piece toilet. If you must use water-closet elbow furnished with the toilet, make certain to measure rear-wall clearance carefully (horn center to rear of supply tank). Otherwise the closet elbow won't fit.

to make certain the elbow is in good condition, and if not, that you can replace it.

Temporarily assemble the toilet by hanging the tank from two screws inserted in the holes in its rear wall and connecting the water-closet elbow to the tank. Slide the rear of the bowl toward the elbow. If necessary, raise or lower the bowl. Connect the elbow to the bowl. Measure the distance from the bottom of the tank to the floor on which the bowl stands and record this dimension.

Check rear and side clearance. To do this, draw a line on the subfloor parallel to the bare stud wall behind the toilet drainpipe. This line should be ½ inch from the studs if you are going to use sheetrock, 1 inch if you are going to add ceramic tile and 1½ inches if you are going to install tile on plaster. Measure along the bottom of the bowl from the center of the horn to the line you have drawn marking the surface of the finished wall-to-be. This is the rear-wall clearance. If the water-closet elbow is such that you cannot vary it, you must make certain the center of the toilet drainpipe is exactly this distance from the rear wall. If it is not, you will find yourself either shimming out the wall or chopping some of it away.

When the rear-wall clearance is satisfactory, check the clearance from the center of the drainpipe to the nearest wall or obstruction. There should be about 15 inches here for comfortable sitting.

You'll find it easier to lay the floor if you install the flange afterwards, but if you are going to lay down asphalt or rubber tile and are going to use solder, shim the flange to the

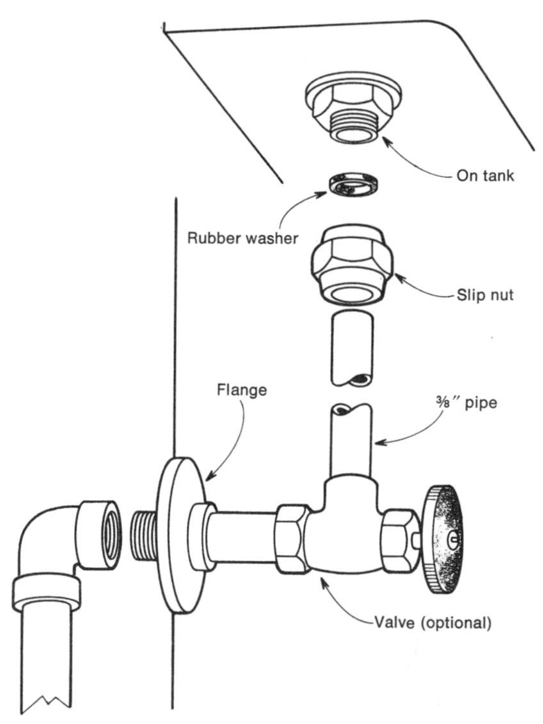

Some old-time toilet tanks are connected to a rigid ⅜-inch feed pipe by a special slip-joint nut and washer. You can replace the entire ballcock assembly and use a standard Speedee *if* the replacement ballcock will properly fit the hole in the tank.

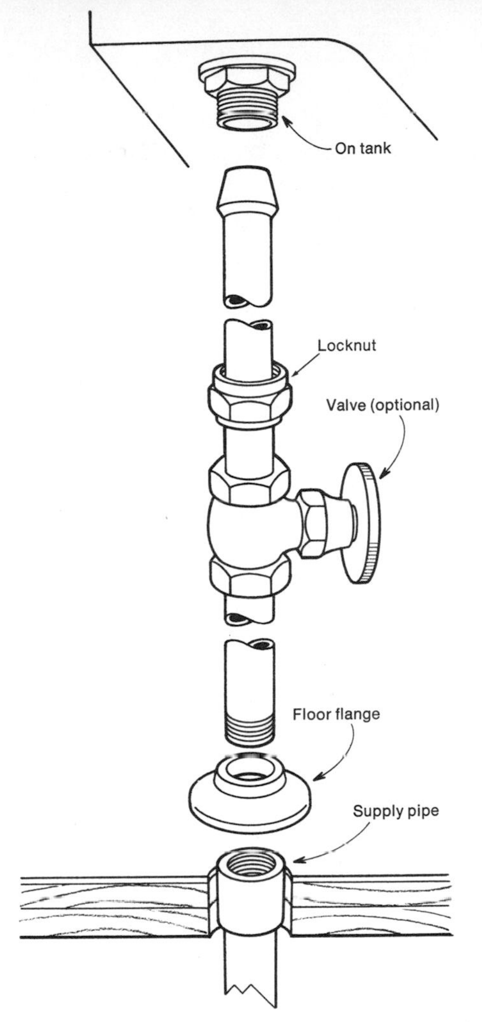

Some of the old ballcock-valve assemblies are designed to accept a preformed feed pipe. These can be brought up through the floor as shown, or replaced when the tank will accept a new ballcock valve.

proper height and solder it in place before going any further. When the solder is cold, file the top of the flange free of solder and remove any small lumps of solder that may have formed inside the end of the drainpipe.

Temporarily hang the tank, using shims equal in thickness to the wall you are going to install. Position the tank in exactly the position necessary to connect the elbow. Use the tank-to-floor measurement to put the tank at the correct height.

Examine the ballcock inlet connection. If it is designed to accept a rigid pipe with a preformed end (ground end) you must be very careful to position the angle shutoff valve directly beneath the inlet connection. And, you will need to use a valve with a screw connection outlet to accept the rigid, threaded feed pipe. If you can install a new ballcock valve assembly, one that accepts a Speedee, your task is considerably simplified, as you can bend the copper tube as much as necessary. Either way, install the stub-out at this time. Try a dry run following, if you are going to use rigid pipe for the connection, to make certain all is well before you install the wall.

Install and finish the wall.

Install the toilet bowl.

Hang and connect the toilet tank. Use the tank-to-floor dimension you recorded. Subtract the thickness of the finished floor and use this figure as a guide to hanging the tank. Use large brass wood screws, after first checking to make certain their shanks will pass through the holes in the tank's rear wall.

Installing Wall-Hung Toilets

In order to properly install a wall-hung toilet it is necessary to plan and prepare for it before you start roughing-in—for several reasons.

Wall-hung toilets are almost always positioned close to their soil stack. Therefore the position of the soil stack just about determines the position of the toilet. Of course, this means you have to consider the type of toilet you want when deciding where to install the soil stack.

Depending upon the particular design you select, the toilet may or may not hang directly on the soil stack. If the model you select hangs on the stack, the stack must be

4-inch cast iron. (Obviously, a toilet can't be hung from a copper or plastic DWV pipe.)

Since a hung toilet drains through its rear directly into a short section of pipe connected to a drain Tee in the soil stack, the height of the Tee and the direction in which it points is highly critical. There's positively no room for correction between a drain Tee leading to a hung toilet as there is between a Tee leading to a lead elbow connected to a floor-mounted toilet.

Since the wall-hung toilet hangs from its bracket, you must take special care to make certain the stack—if that is what it is holding on to—or the building members are securely fastened. The position of the bracket is critical as it and not the toilet determines the toilet's height above the finished floor.

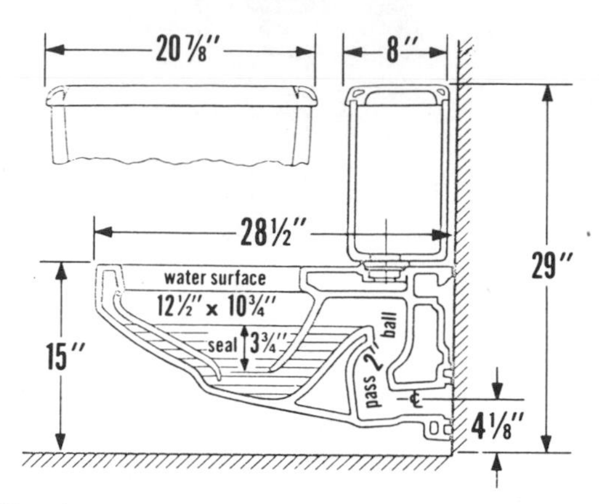

Sketch and roughing-in dimensions for a modern wall-hung toilet. Note that upper edge of bowl is only 15 inches above the floor. *Courtesy American Standard.*

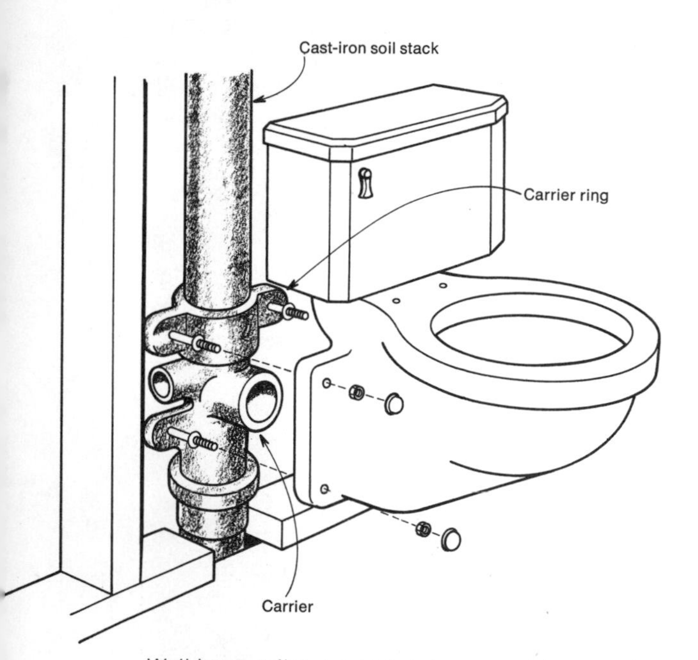

Wall-hung toilet supported by a cast-iron carrier attached to the soil stack. This type of carrier forms a portion of the stack and must be installed with the stack. Center of toilet stub-out is usually 4¼ inches above the finished floor.

General procedure. Study the manufacturer's data to determine the support method, dimensions and height of the drain Tee outlet. Generally the center of the drain Tee inlet is positioned about 4¼ inches above the finished floor.

The following sequence should be followed for best results when installing a wall-hung toilet:

1. Install the stack and the drain Tee.
2. Install the nipple from the drain Tee (which will engage the toilet's discharge horn).
3. Install the supporting brackets.
4. Install the cold-water stub-out.
5. Finish and install the wall.
6. Hang the toilet from its supports.
7. Install the cold-water feed pipe.

BEWARE OF CROSS CONNECTIONS

A cross connection exists when waste or soil water mixes with the domestic hot or cold water. As has been emphasized previously, waste and soil are alive with disease, and any time sink and toilet discharge gets into the drinking water there is better than a good chance illness will follow.

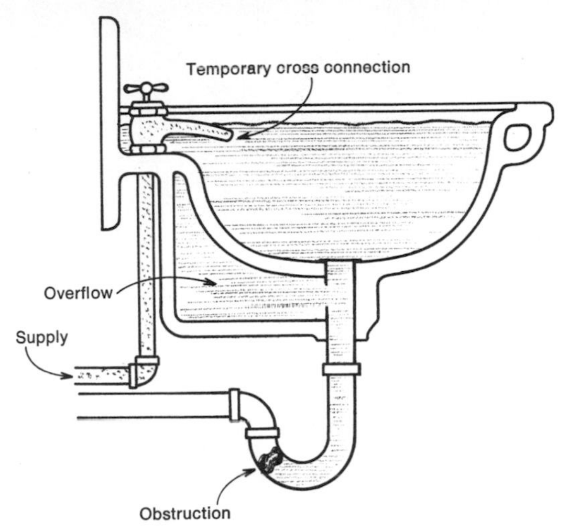

Unacceptable bowl design can lead to a cross connection. Faucet spout is beneath water level when bowl's drain plugs up.

Now it would seem that the only way soil and waste flowing down the drain could get into drinking water would be for the plumber to make a bad mistake, but this isn't so. It may also appear strange that a warning against cross connections should appear in the chapter on connecting fixtures, but again this isn't so, as shall be explained.

Waste water most often enters the domestic hot- and cold-water system by one or two means. Most commonly the condition necessary for a cross connection exists in the form of an old, improperly designed sink, lavatory or tub. In such fixtures the faucet(s) are very low or the overflow outlet is very high so that any slowdown in the overflow causes the water to rise above the faucet spout. This condition will of itself not cause a cross connection, but should the pressure in the domestic hot or cold line be removed, as for example the main valve is closed, the pressure of the soiled water against the immersed faucet is greater than the pressure in the pipeline. If a faucet on a lower level is opened at this time, dirty water in the sink or tub will flow down and out the lower faucet.

This possibility may be prevented by making certain when you purchase and install an *old sink or tub* that none of the faucets is mounted so low that should the sink or bowl overflow the faucet end will be immersed. Modern fixtures cannot be cross connected in the manner described.

Another fairly common cause of cross connections occurs when there is no permanent shower head and a tube and spray head are clamped to the bathtub water spout. Should the shower head be permitted to rest on the bottom of the tub while the tub is full, and the building's water pressure is cut off one way or another, the dirty water in the tub will flow down through the pipes and emerge from any open faucet on a lower level.

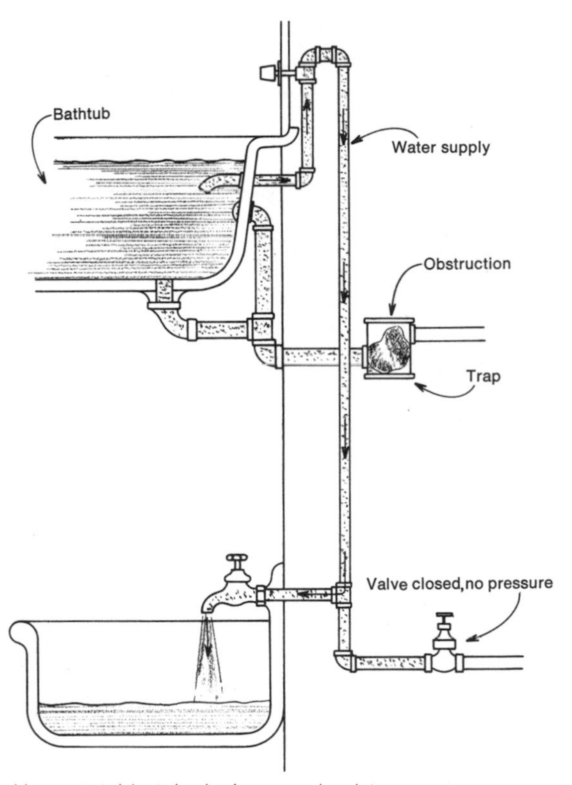

Unacceptable tub design can lead to a cross connection. In this example an obstruction in the tub's trap has stopped drainage. Water level in the tub is above the supply spout. If the valve leading to the tub and sink at a lower level is closed, opening the faucet on the sink will release dirty tub water.

14

Remodeling a Plumbing System

The basic procedures and methods of roughing-in were described in Chapter 12. The following chapter dealt with completing the job—that is, connecting the fixtures. In a sense, this chapter covers the same ground. However, there is one major difference. Whereas in our previous chapters we began our work with an open frame and no piping, in this chapter we start with a complete, in-place plumbing system and concentrate on the problems of finding the old pipe in the wall, connecting the new pipe to the old, adding additional fixtures and replacing old fixtures with new. In addition, this chapter covers the work involved in connecting appliances such as dishwashers, washing machines, water softeners and food disposers.

While the planning involved in adding a fixture or altering an existing system is little different from that required to plan a job from scratch, the actual work differs in that it is much more difficult, physically, to do "old work," as it is called in the profession, than new work. When a building is still incomplete there is an open frame. You can see where the pipes are going. You can get at the studs and joists. When the walls are up there is a lot of blind groping and chopping.

FINDING PIPES WITHIN A WALL

Sometimes you can connect your new pipe to the old pipe where it is exposed in the cellar or attic. Frequently, however, it is best to connect your new pipe to old pipe hidden in a wall. The first step in such cases is finding the pipe without opening up an entire wall.

There are a number of methods and aids you may use in finding hidden pipe. One is to locate the riser in the basement and have a friend bang away on the bottom end of the pipe while you listen for the sound upstairs. If that doesn't help, try stopping and starting the flow of water through the pipe. If you still aren't certain where it is, try measuring from the bottom of the pipe to two side walls and then transferring these dimensions to the same walls upstairs.

Drain and vent pipes can sometimes be located by examining the roof of the building. Generally the drain and vent stacks rise straight up so their top ends indicate their position.

When you have narrowed your search and there is no copper or plastic pipe in the wall, you can sometimes pinpoint the pipe by driving long finishing nails into the wall. You can easily tell when you hit galvanized or cast-iron pipe.

Still another "trick" is to use a magnetic stud detector. This device has a sensitive magnet that responds to the presence of iron. The detector is, of course, no good with copper and plastic pipe.

Don't be too surprised if the pipe eventually turns up in a perfectly ridiculous location. All plumbers aren't logical and in some old houses the plumbing has been changed several times over.

CONNECTING TO EXISTING WATER PIPES

If you have to join your new pipe to an existing galvanized pipe and there is no union in the line, you have no choice but to use a

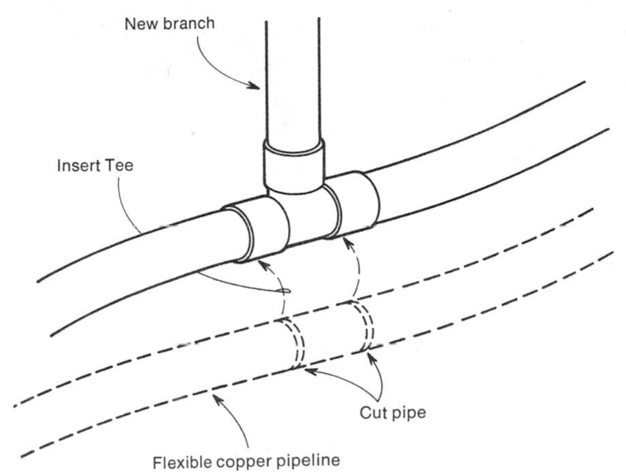

To insert a Tee into a flexible copper pipeline, cut the tubing, bend it to one side, insert the Tee and sweat it in place.

hacksaw and cut through the pipe. Then unscrew the ends and replace them with two sections of pipe, a union and a Tee. If you have the equipment or can have a shop do it for you, cut and thread the old pieces of pipe and use them again.

If you are working with soft copper and have enough space and extra tubing, cut the

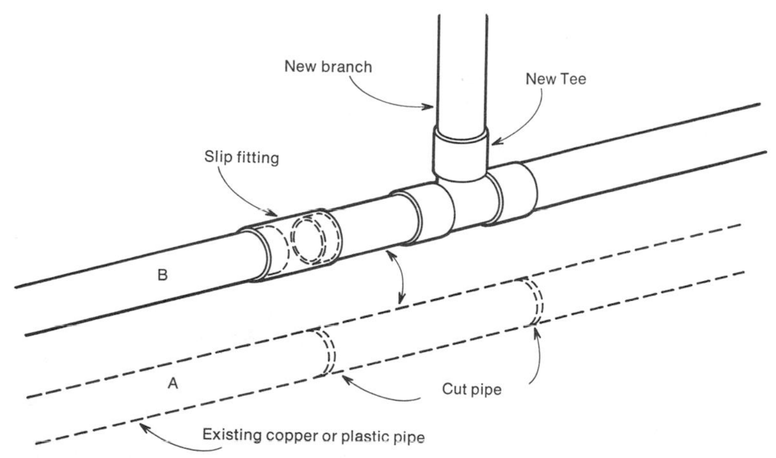

To insert a Tee into a rigid plastic or copper pipline, follow these steps: Cut a section of the pipe out and shorten it by about 1 inch. Position the Tee over one end of the pipe. Sweat it or cement it fast. Slide a slip fitting over the other end of the pipe. Position the shortened section of tubing within the Tee. Sweat it or cement it fast. Move the slip fitting over the one remaining open joint. Sweat or cement the slip fitting fast.

pipe and remove just enough to permit you to insert the fitting and sweat it in place.

If you are going to tap into a rigid copper line, or if you are working with soft copper in a cramped space, cut out a section of pipe. Sweat join this to the fitting. Shorten the section just enough to permit it and the fitting to go back into the pipeline. Then use a slip fitting over the cut end to rejoin it to the pipe.

You can do the same with plastic pipe, using solvent cement in place of solder to form the joints.

To connect a small-diameter copper tube to a larger diameter tube, use any of the "tap-on" fittings sold under a variety of trade names. Clamp the fitting onto the pipe to be tapped. Shut off and drain the water. Remove the screw valve. Drill through the valve hole into the tube. Replace the screw valve. Connect the smaller copper tube by means of the metal-ring compression joint that is a part of the clamp.

CONNECTING TO AN EXISTING CLEANOUT

Almost every cleanout is a potential drainpipe connection. Remove the plug, screw in or caulk a nipple or a section of pipe and add a second cleanout and elbow plus the new drainpipe. This is a lot easier than tearing up a floor or wall and cutting into a drainpipe.

Often it is easier to install a new stack and connect its bottom to an existing cleanout than it is to open a wall or floor to connect to a nearby branch drain. You may use more pipe, but you may also save yourself the work of cutting and patching the floors and walls and possibly opening a joint or two while you work on the top section of the drain.

CONNECTING TO AN EXISTING STACK

It is much more difficult to connect a vent or drainpipe to an existing stack than it is to

Installing a "Tap-on" Fitting

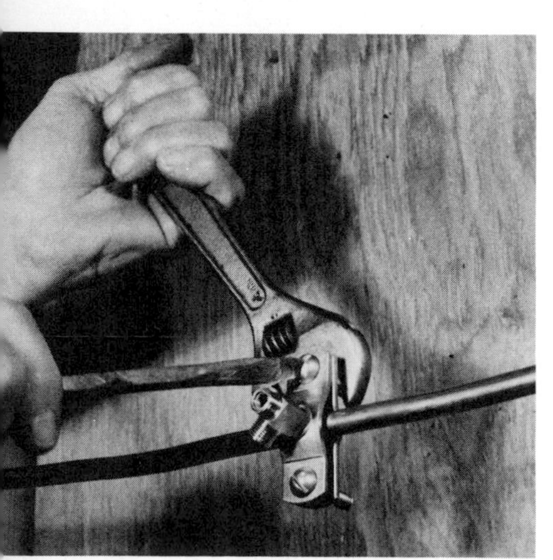

1. Remove the screw valve and clamp the fitting onto the pipe or tube.

2. Drill a hole through the wall of the pipe or tube.

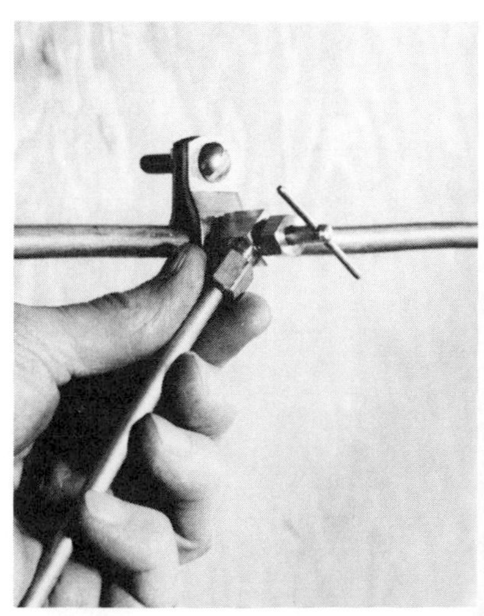

3. Replace the screw valve and connect the smaller tubing to the fitting.

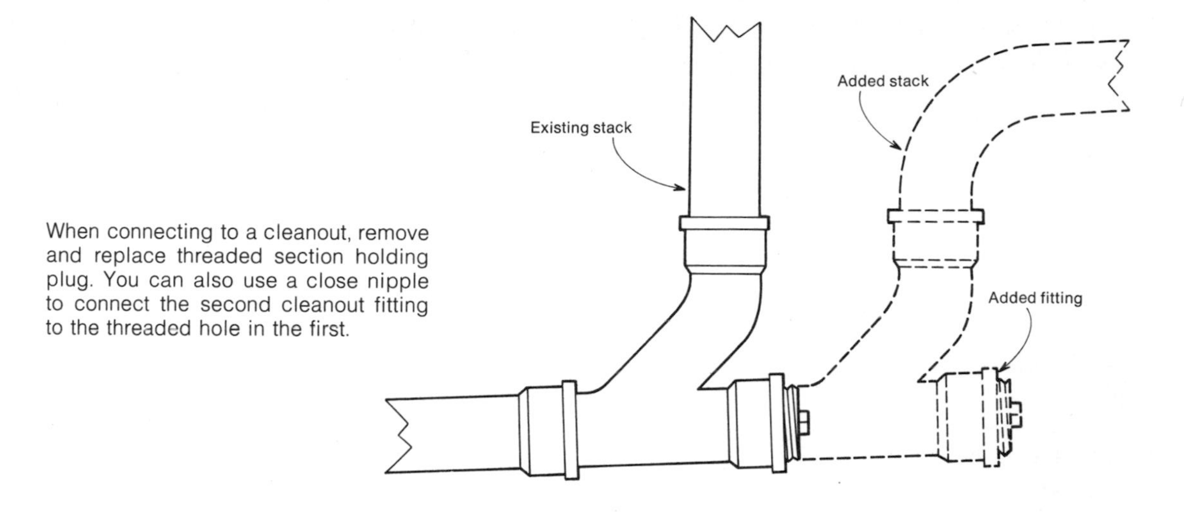

When connecting to a cleanout, remove and replace threaded section holding plug. You can also use a close nipple to connect the second cleanout fitting to the threaded hole in the first.

connect a water pipe to an existing, in-place water pipe. Stacks are always larger in diameter and are usually buried in the walls, except for where they may be exposed in the attic and the cellar.

The suggestions that follow are applicable to all cast-iron stacks, be they soil, waste or vent. There are several methods. Your choicc will depend on the local code, the position and diameter of the stack, and access.

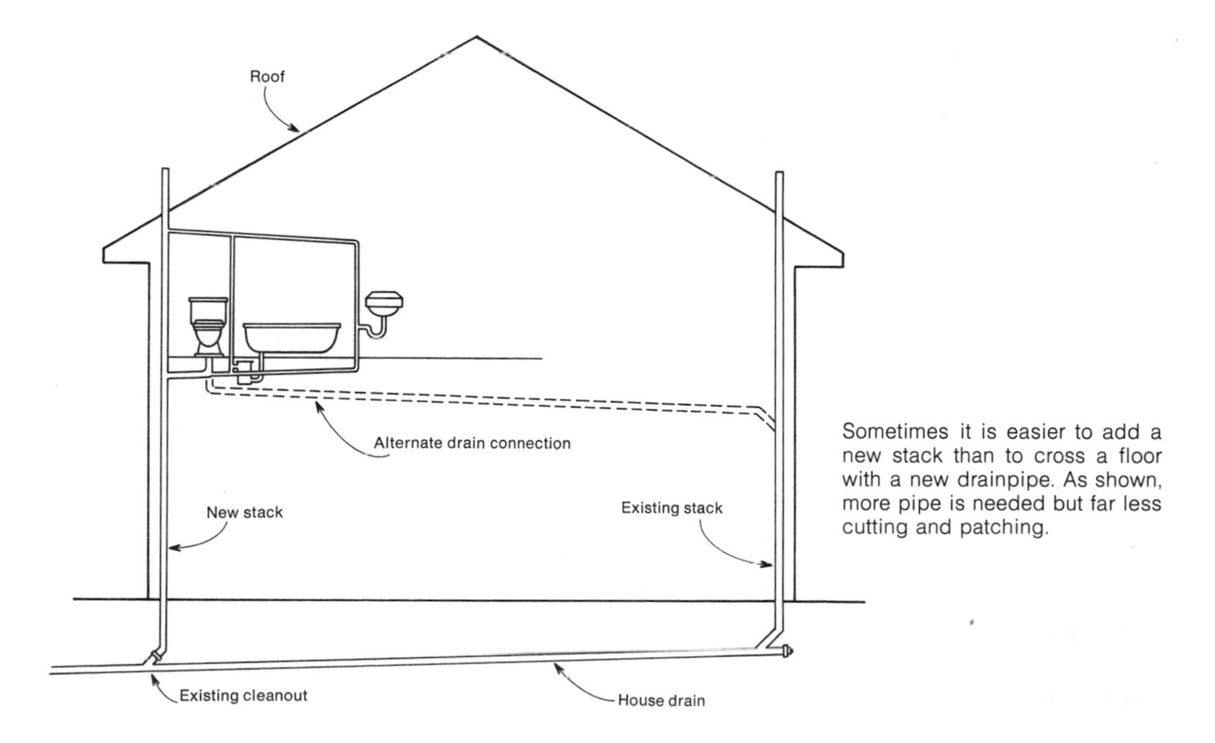

Sometimes it is easier to add a new stack than to cross a floor with a new drainpipe. As shown, more pipe is needed but far less cutting and patching.

If you are close to the roof and there are no connections above the location of your proposed tap, open the nearest lower caulked joint by removing the lead and oakum with a chisel. Lift the upper sections of stack and insert your fitting in place. Then caulk the joint closed.

(This sounds easy because we have ignored the cutting and chopping which may be necessary to get at the pipe and make room for the drain fitting. But this is a plumbing book, not a carpentry book, so we stick to pipe work.)

You can leave your stack a couple of feet above the roof. Just recaulk the joint between the pipe and the roof flashing.

If, for one reason or another, you cannot lift the stack you will have to cut out a section of the pipe large enough to permit you to install the new fitting.

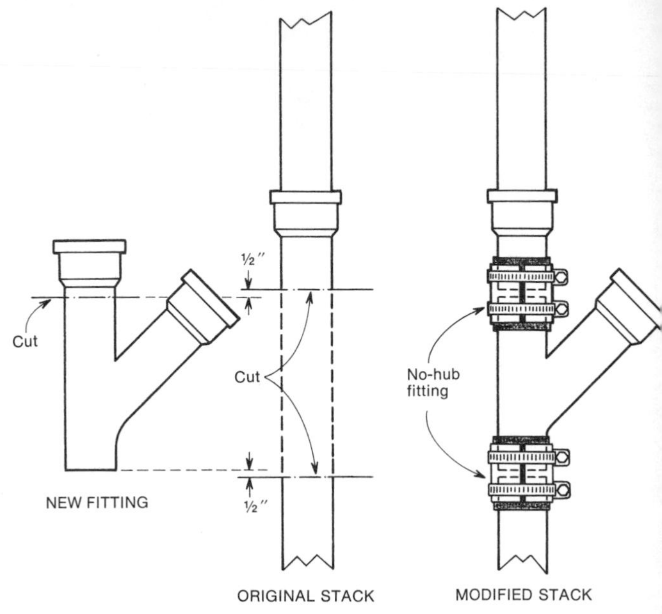

Using No-hub fittings to insert a drain fitting into the middle of an existing stack or drain. Make the two pipe cuts about ½ inch farther apart than the length of the new fitting.

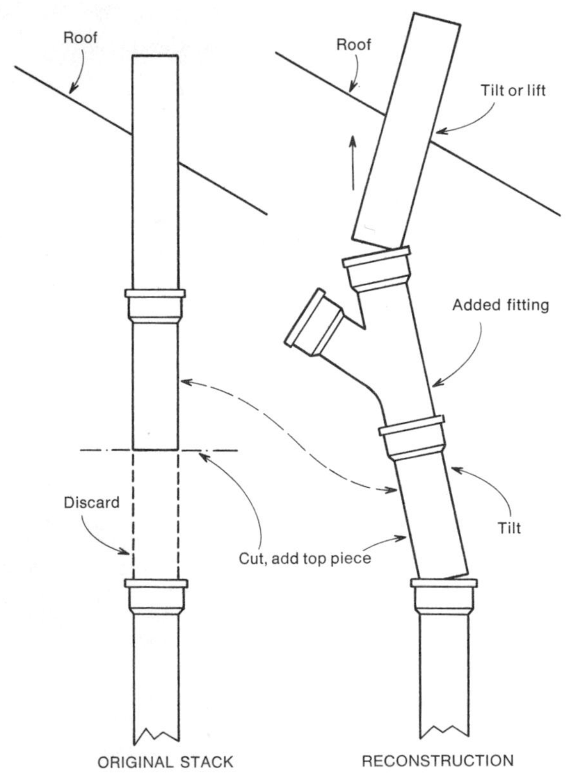

To add a stack fitting, lift and tilt the top section of pipe, allowing it to protrude farther above the roof.

Before you attack with your hacksaw, there are two things you must do. One, make certain the upper sections of pipe will not slip down. The best way to do this is to install a number of heavy iron straps snugly under two or more pipe hubs. Don't depend on the original plumber having done this. Two, decide on the method you are going to use to connect the new fitting and get the fitting in hand before proceeding.

Two methods are commonly used. The first uses No-hub joints. The second uses a special slip fitting called a sisson fitting. The No-hub system is the better way, but some codes do not as yet permit it.

To install a sisson fitting it is usually necessary to remove an entire length of soil pipe. Then the new fitting and the sisson fitting are positioned and the sisson fitting is expanded.

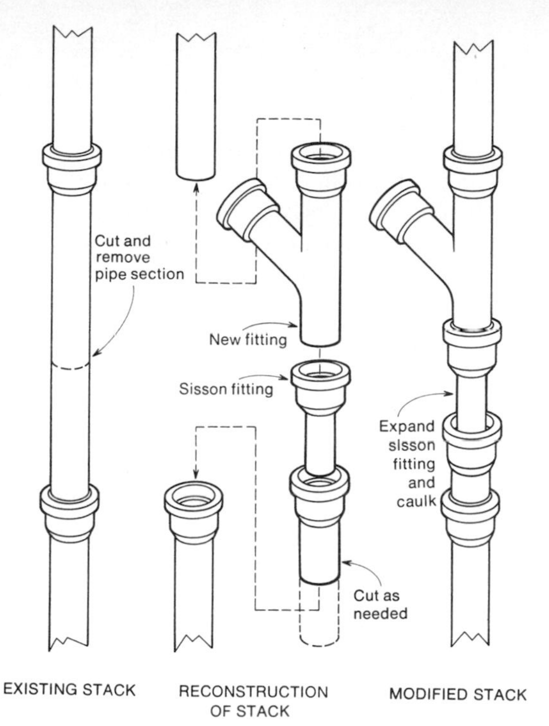

Using a sisson fitting to insert a drain fitting into the middle of an existing stack or drain. The sisson fitting expands after it has been positioned. With this fitting you need to caulk four joints.

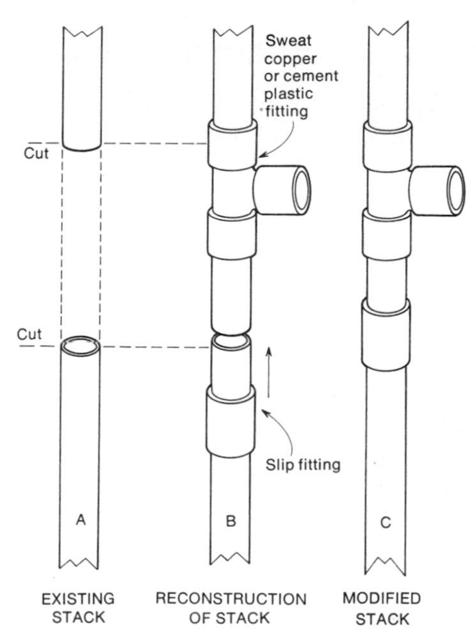

Using a slip fitting to add a drain Tee to an existing plastic or copper DWV pipe. Steps are exactly the same as those used with rigid plastic or copper water-supply pipe.

All the joints are then caulked. Both fittings are illustrated.

Connecting to a copper stack. Use the same procedure as for connecting to a rigid copper tube. With a hacksaw or a large tube cutter, cut a section out of the stack where you wish to install the new fitting. Slip the fitting over the open end of the pipe and sweat it in place. Cut a length of tubing just a fraction shorter than that necessary to go from the new fitting to the end of the lower end of the pipe. Place a slip fitting over the lower end of the pipe and sweat the short tube you just cut to the lower end of the new fitting. Move the slip fitting up and sweat it into position over the joint.

Connecting to a plastic stack. Follow the same procedure. The only difference is that you cement the new fittings and section of pipe into place instead of sweating them fast.

Connecting plastic pipe to cast-iron pipe. Insert the plastic spigot into the cast-iron bell and seal it with a compound made for the purpose.

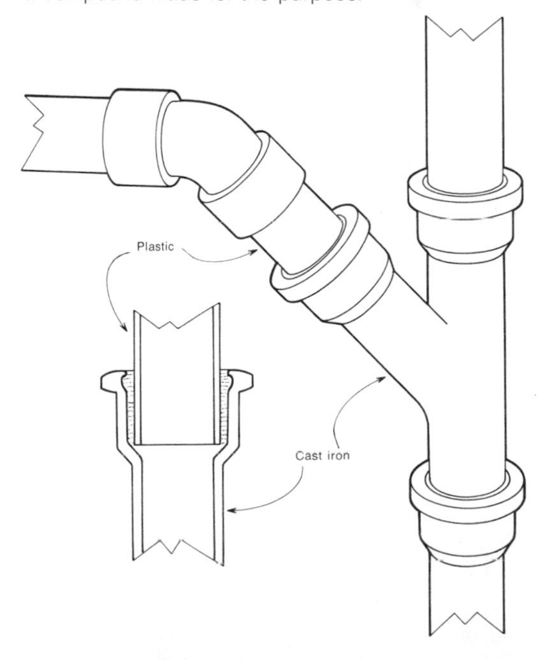

BRANCH DRAIN CONNECTIONS

The problems encountered in connecting a fixture drainpipe to a branch drain are similar to those just discussed. The only difference between making a connection to an existing stack and an existing branch drain is that the latter is made of smaller-diameter pipe and is horizontal instead of vertical. Cutting and joining techniques are exactly the same.

Bear in mind when you add a drain to a branch drain that you cannot overload the existing drain without a penalty. And you must make certain your new drain is properly pitched and vented.

Generally, you can slip by with a little overloading—for example, letting a 1¼-inch pipe remain when the new total load in fixture units calls for 1½-inch pipe. But don't carry it too far, especially when adding a kitchen sink which is used a lot and which discharges relatively "soiled" water as compared to a lavatory.

A common fault of uninhibited plumbing remodelers is to continue to tie on fixture after fixture, as the need or desire arises, to an original branch drain. The final result is a string of fixtures on one undersized drainpipe. Unclogging the drain becomes a weekly entertainment, especially if there isn't sufficient pitch and the drain isn't vented.

Should you encounter this condition, try a temporary vent just before the last fixture on the line. Don't trouble to run it through the roof; a short, test length of pipe will do. If that doesn't help, replace the undersized line with pipe of the correct size, properly vented, properly pitched.

MIXING PIPING MATERIALS

To connect a copper DWV (drain, waste, vent) pipe to a cast-iron pipe, caulk a transition fitting into the cast-iron pipe and solder the copper to the other end of the fitting.

To connect a plastic pipe into a cast-iron pipe, slip the spigot end of the plastic pipe into the cast-iron bell and seal the joint with a compound made for the job. It is available at plumbing supply shops under a variety of trade names.

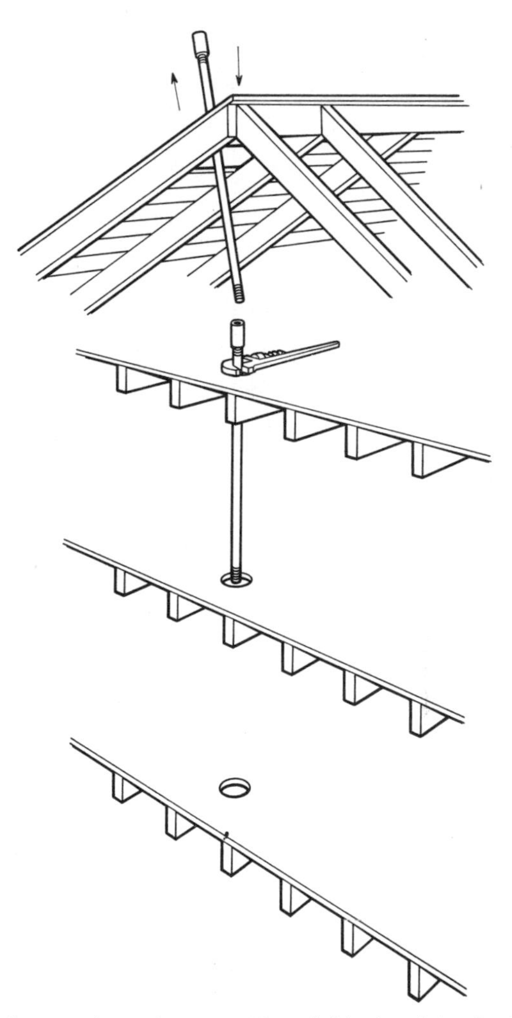

As an alternative to cutting rigid pipe into short lengths and rejoining it, consider opening a hole in the side of the house or the roof. Hold one pipe in place with a wrench while making the joint. Then lower two pipes, add on a third, etc.

GETTING THE PIPE IN

Obviously, you are going to have to cut holes in the walls as well as holes through the various frame timbers that may lie in the path of your pipe. Perhaps not so obvious is that you can reduce the labor of cutting and later patching by careful planning. For example, in order to cut a hole from one floor level to another you have to cut a hole in the wall to get the drill and saw between the studs. If there are no obstructions you will do better making the hole in the wall near the ceiling rather than near the floor because there is no trim at the ceiling, and if you have to extend your hole into the room, it is a lot easier to cut and patch sheetrock than two layers of flooring.

If you are going to cut more than a pair of holes it is advisable to rent power tools. Get a heavy-duty angle drill and a large saber saw with a good supply of drills and blades, as you must expect to damage the cutting tools. Purchase a cat's paw, which is a nail puller and invaluable for the purpose. If you go at it with hand tools alone you will spend more time cutting than plumbing.

Check before you join. It is difficult enough to cut pipe to the correct length when most of the pipe is visible. Doing so when most of the pipe is hidden by a wall is even worse. As the pitch on your branch drains are critical, and branch-drain pitch depends in part on the elevation of the drain fitting on the stack, it is advisable to assemble each section of the stack "dry" and check the elevation and angle of the drain fitting (toward the branch drainpipe) before you permanently join the pipe and fittings. Observing the same precaution with branch drains will also save you a lot of tears.

Tight quarters. If you are working with rigid pipe and haven't enough room you have two courses of action. You can cut the pipe into short lengths and rejoin the pieces as you feed them into the hole. Or you can cut a hole in the roof or wall and feed longer pipe lengths in that way.

Short-length pipe means more cutting and joining. A hole in the roof or side wall means cutting and patching the hole. The choice of evils is yours.

Running flexible pipe. To run flexible pipe vertically down through a wall, tie a weight to a rope and the rope to the end of your pipe. Lower the weight to a helper below and have her haul away as you guide the pipe down.

To run flexible pipe horizontally between a ceiling and floor, use an electrician's snake in place of the rope. Tie the end of the snake to the pipe and then have your helper pull while you guide the pipe into the hole.

Tie a rope to the end of flexible pipe and have a helper assist you in running the pipe through the wall.

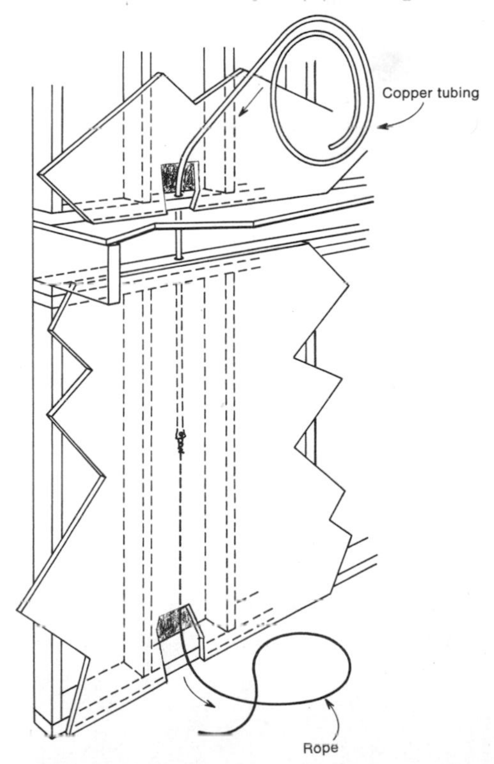

Some Suggestions for Hiding Pipe

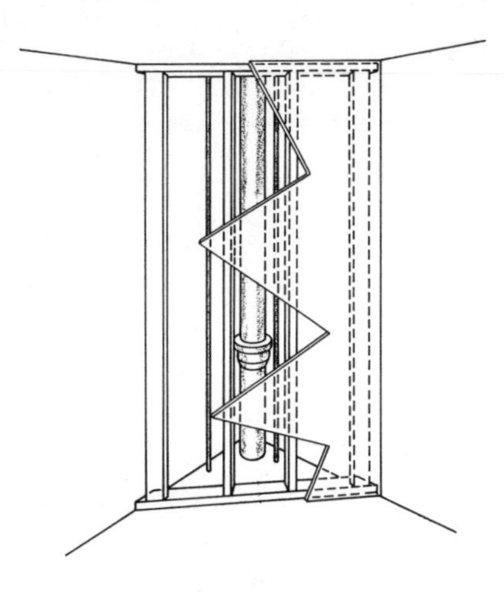

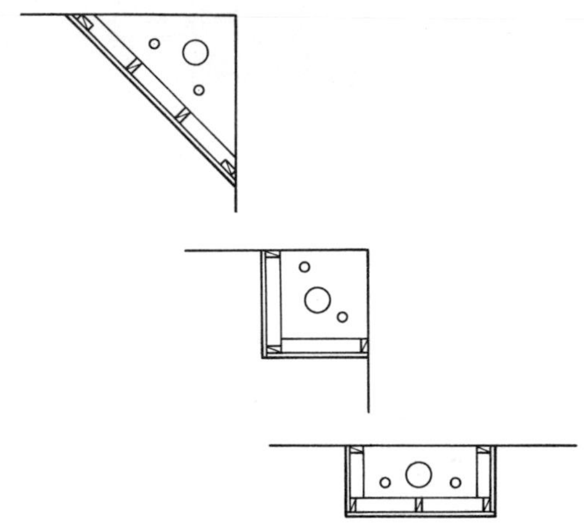

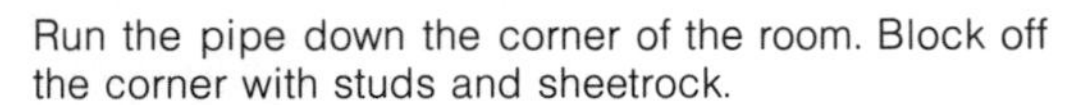

Run the pipe down the corner of the room. Block off the corner with studs and sheetrock.

Top view of three ways to box pipe in corner of room or along the wall.

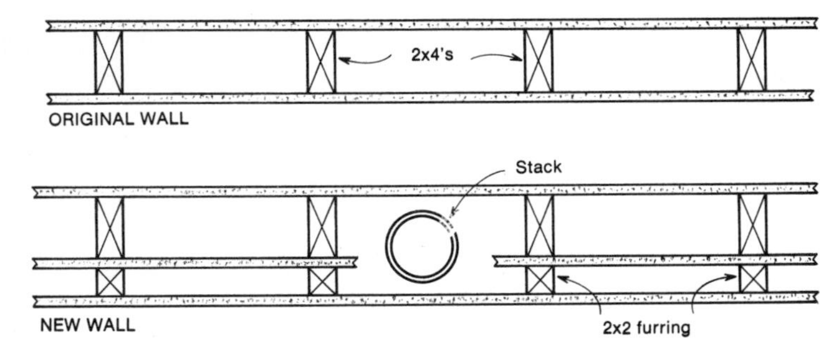

Build a false wall and hide the pipe between the studs.

Build a false floor on top of the old floor and run the pipe between or through the joists.

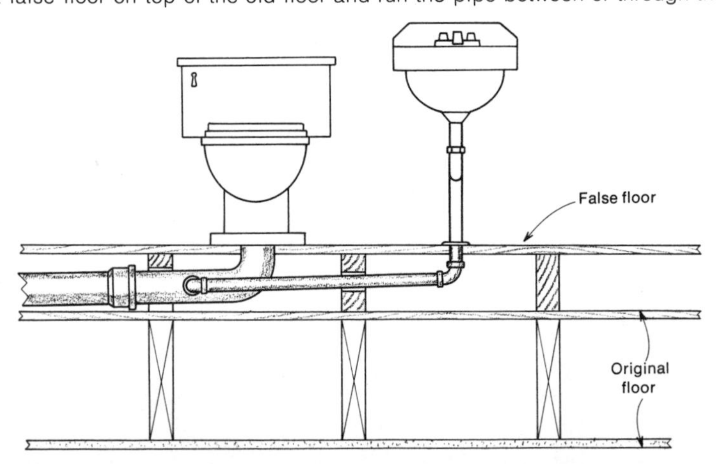

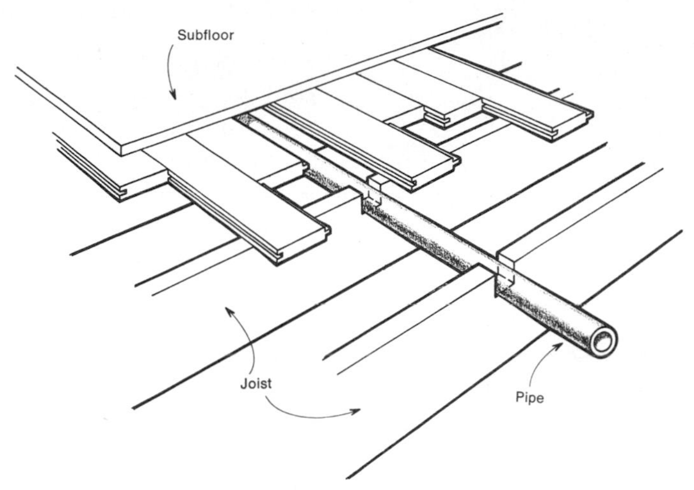

Cut slots in the joists supporting the old floor and run the pipe in them; then lay a plywood floor on top.

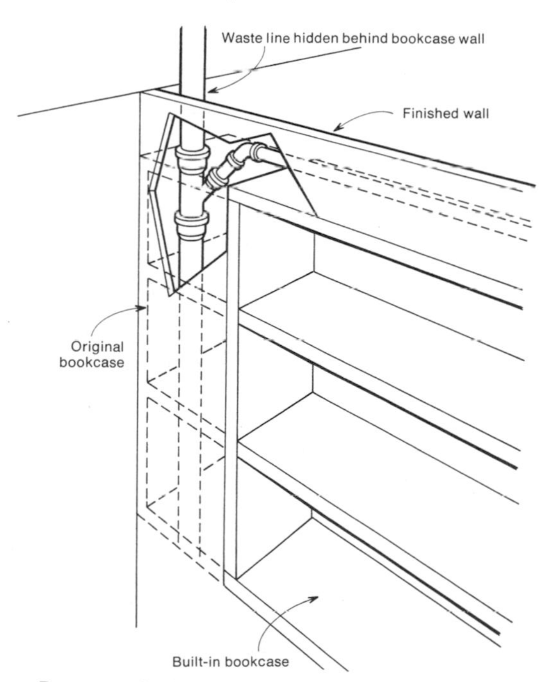

Box part of a bookcase and run the pipe inside.

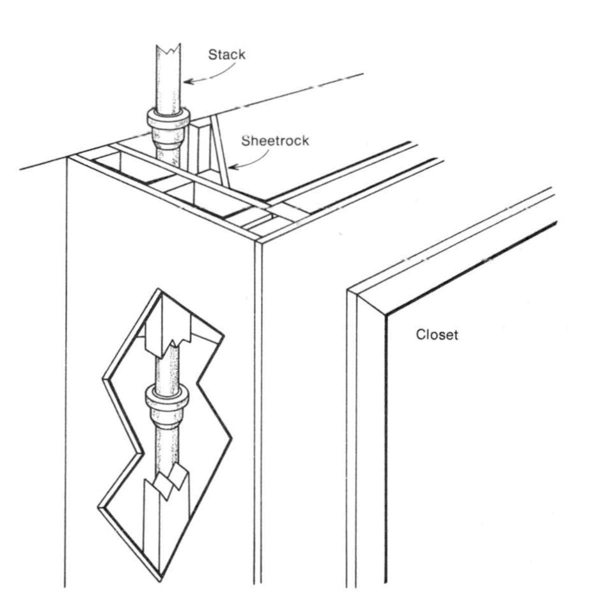

Run the pipe inside a closet and conceal it with a length of sheetrock.

Hiding the pipes. In many instances when a new bathroom or perhaps an additional kitchen is added, all the pipes cannot be run inside the walls or floor. Some of the pipes have to pass through the rooms, an unsightly intrusion that most women will not abide. The accompanying illustrations show some ways of hiding pipes.

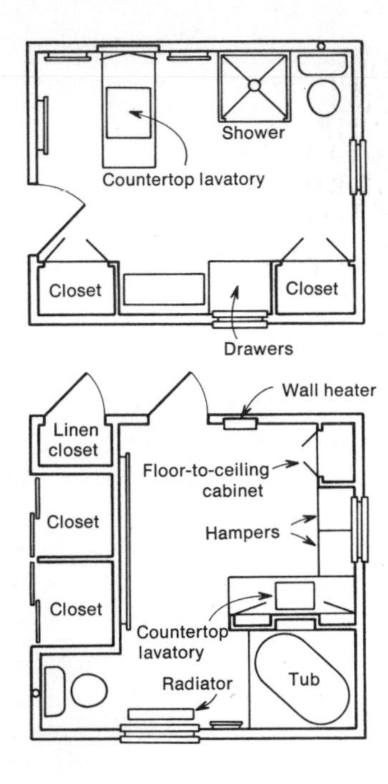

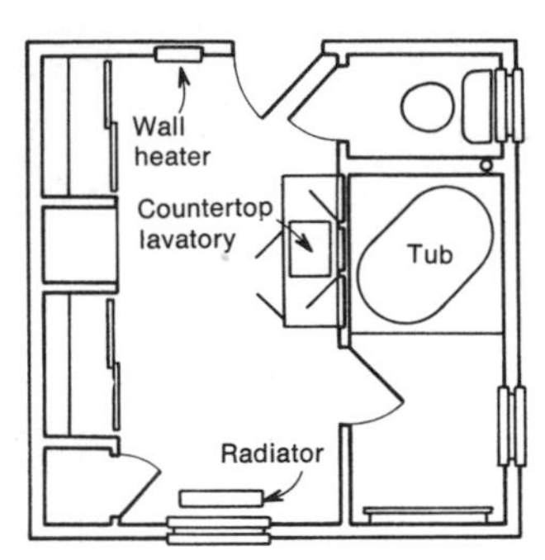

Suggestions for large bathrooms.

BATHROOM SUGGESTIONS

Sometimes when you remodel you will be converting a fairly large room into a bathroom. In their way, large rooms can require as much planning as very tiny rooms. Some suggestions for handling large rooms are illustrated.

Sometimes an existing bathroom can be made more useful to a large family by dividing it so that several individuals can use it at one time in privacy. One approach is to replace a bathtub with a shower stall and provide it with an individual entrance. Another is to add a second shower so there is a complete bathroom—toilet, lavatory and shower—plus a second shower. As a toilet can be loosened and rotated about its drainpipe, after a new water supply is provided, it is sometimes possible to do just this, add a shower stall and a second toilet and make two complete bathrooms without a great deal of additional piping.

REPLACING OLD SINKS

If your kitchen contains one of those quaint wall-hung sinks with the large faucets, you can replace it with a modern sink, either wall-hung or standing.

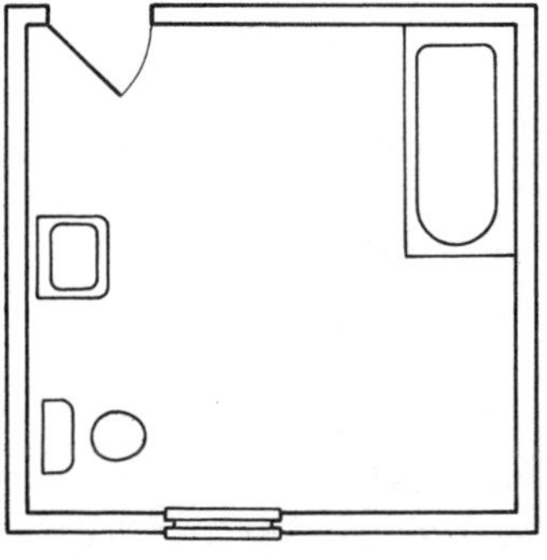

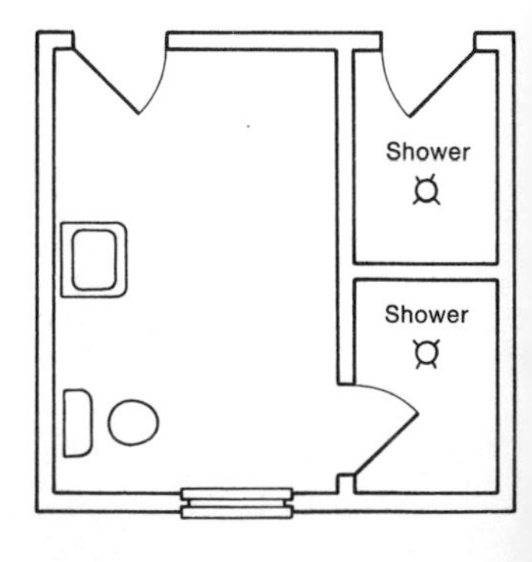

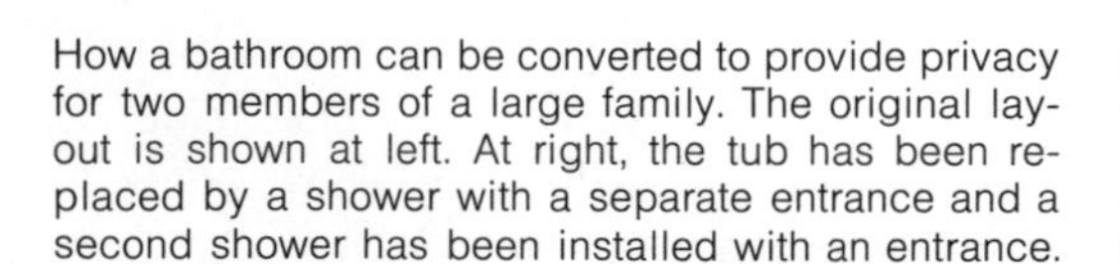
How a bathroom can be converted to provide privacy for two members of a large family. The original layout is shown at left. At right, the tub has been replaced by a shower with a separate entrance and a second shower has been installed with an entrance.

Two Ways to Replace an Old Sink

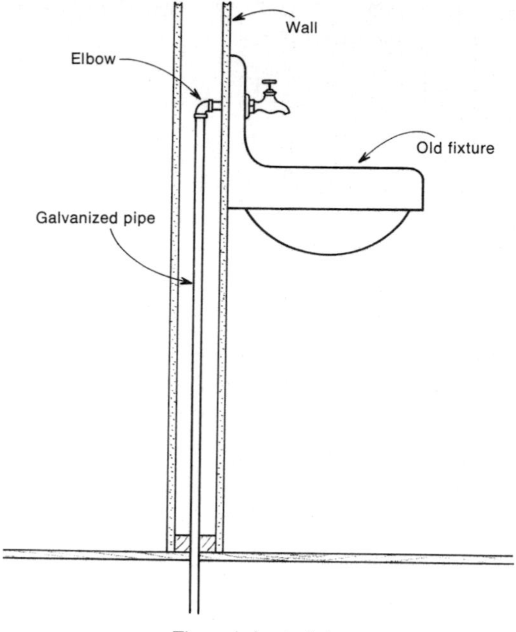

The original sink.

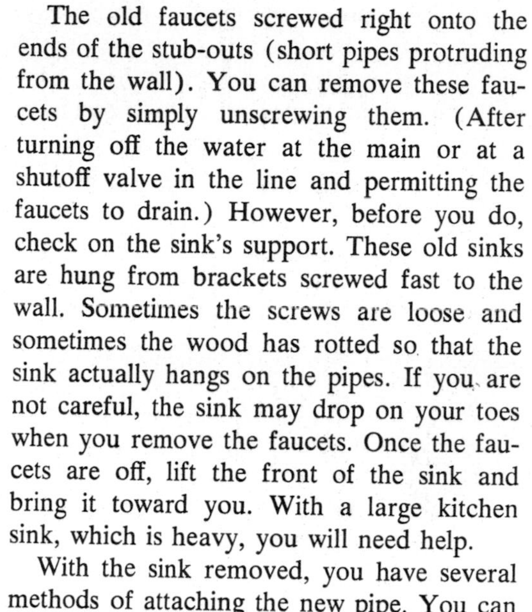

The old faucets screwed right onto the ends of the stub-outs (short pipes protruding from the wall). You can remove these faucets by simply unscrewing them. (After turning off the water at the main or at a shutoff valve in the line and permitting the faucets to drain.) However, before you do, check on the sink's support. These old sinks are hung from brackets screwed fast to the wall. Sometimes the screws are loose and sometimes the wood has rotted so that the sink actually hangs on the pipes. If you are not careful, the sink may drop on your toes when you remove the faucets. Once the faucets are off, lift the front of the sink and bring it toward you. With a large kitchen sink, which is heavy, you will need help.

With the sink removed, you have several methods of attaching the new pipe. You can

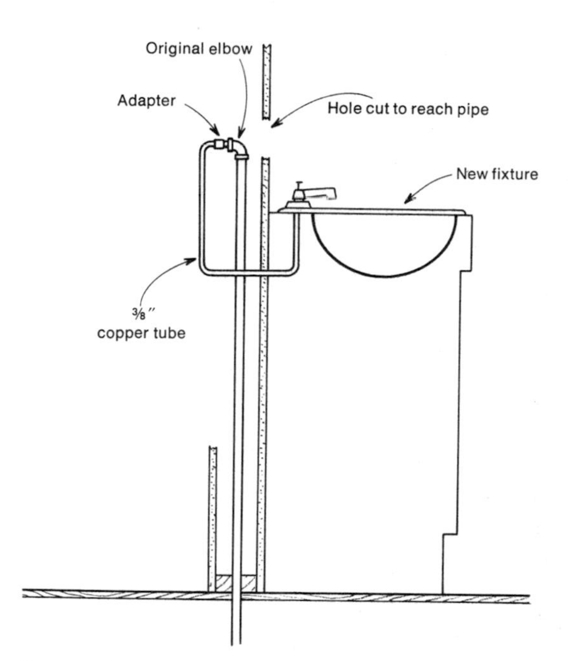

Remove the old sink. Cut a hole in the wall and connect a copper tube to the end of the old pipe. Install a new sink and connect to the tube.

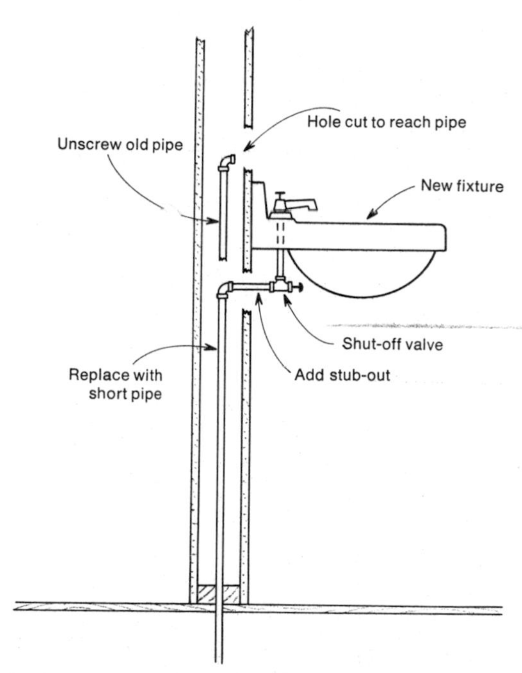

Cut a hole in the wall. Unscrew the old pipe, cut it shorter and rethread. Cut another hole in the wall and connect a stub-out to the short pipe.

unscrew the projecting nipples (stub-outs), enlarge the hole so you can get a wrench inside, then attach a transition fitting to connect some soft copper tube to the ends of the existing pipe. This is illustrated. Use ⅜-inch soft copper for lavatories, ½-inch for kitchen sinks.

Alternately, you can cut a hole farther down the wall, reach in and, with the aid of a wrench, remove the pipe. The pipe in hand, cut it to length and rethread its end. Then replace the pipe and connect a short-turn elbow at the top. From here you can go any way you wish, with any kind of stub-out that's convenient. Stub-outs are discussed in Chapter 12, Roughing-in.

Wall-hung sinks and lavatories. The problem with these is in securely fastening the new bracket from which the new sink or lavatory is to be hung. If the old wall is badly battered you might consider removing enough of it to find the studs and rabbet cross brace in place. If the old wall is not to be touched you have to find the old studs and then drill large holes in the bracket to match the stud spacing (generally 16 inches on center). You can't attach the bracket to the wall with toggle bolts or similar fastenings; they won't hold.

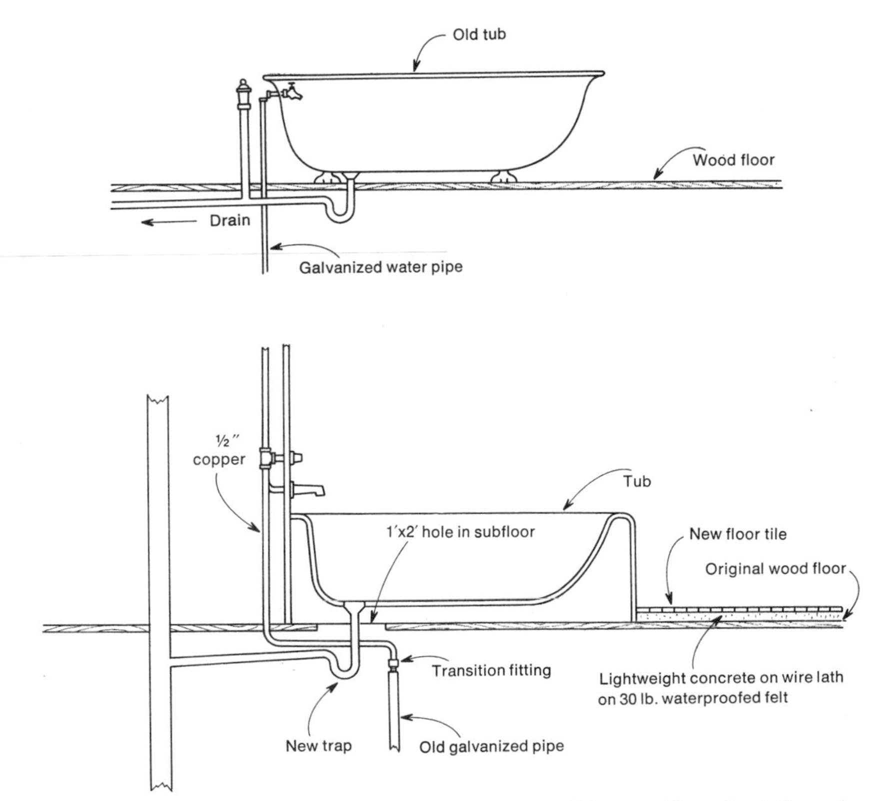

In many old houses, the tub in the bathroom rests on a plain wood floor (top). To replace the tub with a new one, it is not necessary to install an entirely new floor. Install the new tub on the old floor (below), and lay the new floor alongside the tub. Cut old galvanized feed pipes and connect faucets with copper tubing; also, install a new trap.

Self-supporting sinks and lavatories. There is no problem here; the fixture stands on its own support. All you have to do, after you connect it, is fasten it in place.

REPLACING OLD TUBS

If you have an antique tub, and wish to replace it with a modern tub, the job consists of removing the old one, cutting holes in the floor, and running new pipe.

The procedure you follow in preparing to install the new tub will depend on the type and condition of the existing walls and floor.

If the walls and floor are of tile, in good condition, consider a free-standing tub. It is most expensive and needs lots of space, but it is luxurious and most easily installed. You only have to cut a hole for the drainpipe and trap. The new tub is caulked to the tile floor.

Instead of destroying a portion of the tile wall, think about bringing your faucets and spout in from the back side of one of the walls. If you can do it that way it is sometimes possible to save the tile wall and just cut the necessary holes through it.

If the old bathroom has a good wood floor there is no need to remove it to get down to the subfloor. Install the tub as discussed in the previous chapter and lay the tile right on the old floor.

Installing feed pipes. The same methods suggested for replacing old sinks may be used. The old galvanized pipe is cut short and the end rethreaded. A transition fitting is screwed fast and the balance of the run from feed pipe to faucet is made with ½-inch soft copper tubing.

Mounting faucet and shower-head assembly. You can, as stated previously, bring the assembly in from the backside of the tub's wall. It is a little awkward but it saves the side facing the tub. The assembly is first strapped to the wooden braces. Holes are made in the wall. The assembly, attached to the braces, is pushed into position. The braces are then nailed to the studs. This technique is not always practical. When it is not, use the method described in Chapter 13 after you have removed sufficient wall to permit you to get at the studs and mount braces.

BACKING UP FIXTURES

Generally it is easier to install fixtures back to back with existing fixtures than it is to install them independently. In many instances it is easier to remove an existing single-drain Wye, for example, and replace it with a double Wye than it is to install a second single Wye elsewhere on the stack. Then too, the water pipes are usually nearby, so there may be a savings there too. Therefore in planning plumbing additions always consider the possibility of backing up fixtures. You can often save a lot of pipe and labor this way.

Two fixtures can be installed back to back to utilize the same stack. Remove the original single Wye and install a double Wye. The second fixture—in this example, a lavatory—can then be connected to the double Wye.

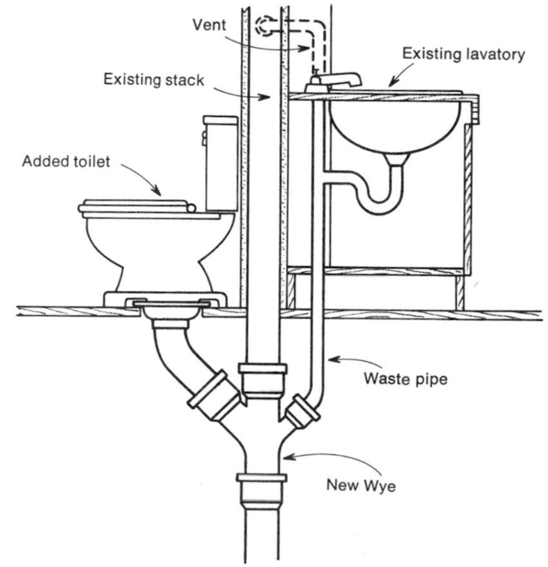

ADDING WASHING MACHINES

So far as plumbing is concerned there is only one major difference between clothes washers and dishwashers. Clothes washers need both hot- and cold-water lines. Dishwashers need only hot. Both need ½-inch feed pipes and both need drains at least as large in diameter as the pipe coming from the machine.

Water connections. Washers made for permanent installation come equipped with flexible water hoses terminating in coarse-thread, female screw couplings. This type of thread is often called water-hose thread because it is similar to that found on garden hoses. To connect a hose one simply makes sure the washer is inside the coupling and screws it fast, most conveniently to a faucet having garden-hose threads on its spout.

To install one of these faucets, and they are comparatively inexpensive unfinished castings, cut the water line at a convenient point, install a Tee and connect the faucet to the Tee by means of a nipple. Or, install as much pipe as you need between the Tee and the faucet, which can be fastened to a nearby wall.

Assembling a washing-machine Tee and drainpipes to connect washer's drain to sink's drain under the counter. Piping is thin-wall brass.

Drain connections. Should you be installing the washer beneath a kitchen-cabinet counter near a sink, you can connect the washer's drain to the sink's drain. Disconnect and remove, or swing to one side, the sink's trap. If necessary disconnect and remove the flexible-ring compression joints and associate fittings attached to the lower end of the sink's tailpipe. Remove the tailpipe (the drainpipe connected to the bottom of the sink). Replace with a washing-machine Tee. This is

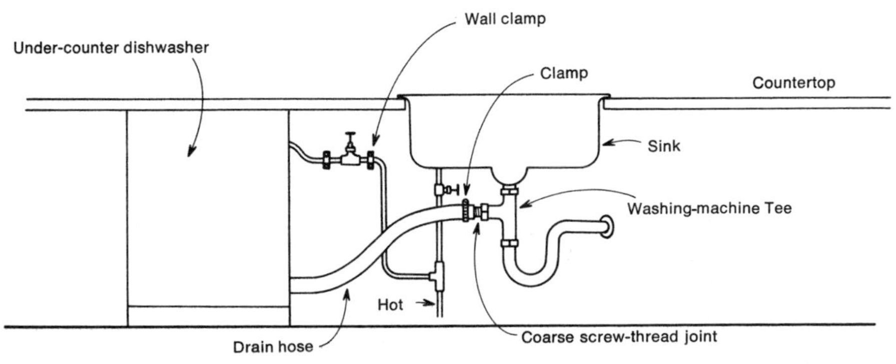

An under-the-sink dishwasher hookup needs only a hot-water line. Machine drain is connected to a washing-machine Tee.

a special thin-wall brass drainpipe having a side connection terminating in male water-hose threads. You may have to cut the special Tee short to fit. Reassemble the drainpipes.

If your washing machine's flexible hose terminates in the proper screw-thread female coupling, simply screw it in place. If it doesn't, use water-hose hardware and fittings to attach a screw coupling to the end of the hose. If you have much more hose than you need, don't cut it. Just make a neat coil that won't kink.

If there is no sink nearby, but a sink or tub in the basement, extend the washer's drain hose to reach the lower-level sink. Use garden hose. Insert a short length of rigid copper tubing into the end of the washer hose and the end of the garden hose. Use automotive hose clamps to clamp the two hose ends fast to the pipe. Clean this pipe out every year or so. Dirt accumulates here.

If you find that the long length of hose siphons the water out of your washer, you have to install an air break in the drain line near the machine. Some suggestions for fabricating an air break are given in an accompanying illustration.

If you cannot connect your washer's drain to a sink's drain or to the sink itself, you will have to install a trap and possibly a vent between the machine's drainpipe and the building's drainage system. When you do this you must follow all the rules outlined in Chapter 11.

Very simply, the washer is considered the fixture and the trap that you must install is considered the fixture trap. You can, if you wish, use a large-diameter trap and merely slip the end of the drain hose into its open end. If, however, your washer is to be connected to the lower reaches of the drain system, it is advisable to make a solid connection between the washer's hose and the trap. If the sewer backs up you don't want muck coming out of this opening.

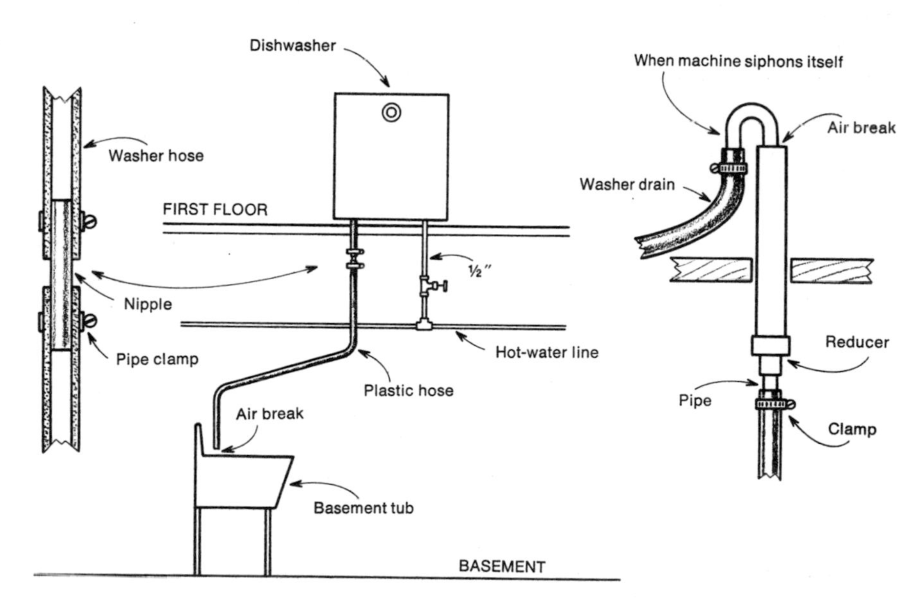

When there is a tub in the basement, washer's drainpipe can be extended to drain the machine into the tub. If the machine siphons itself, install an air break, as shown at right.

Replacing Dual with Single-lever Faucet

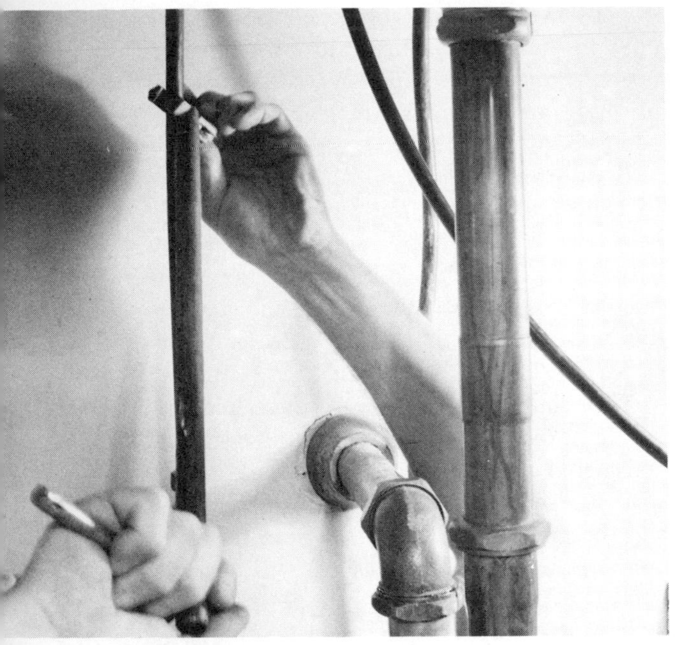

1. Use a basin wrench to remove nuts holding the old faucets in place.

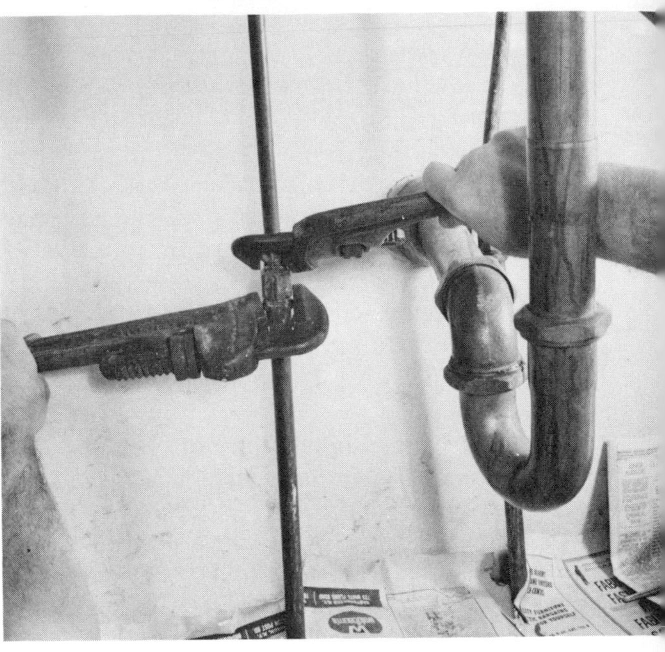

2. Disconnect the feed pipes. Here two stillson wrenches are being used.

3. Remove the old dual faucet, which will come away now that connections are gone.

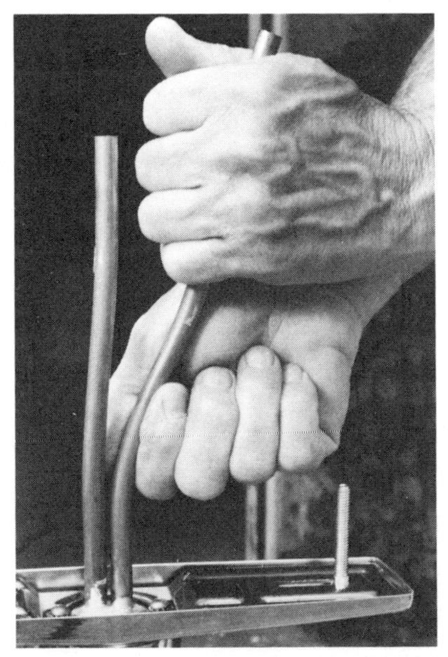

4. Bend apart the feed tubes on the new faucet to give you working room.

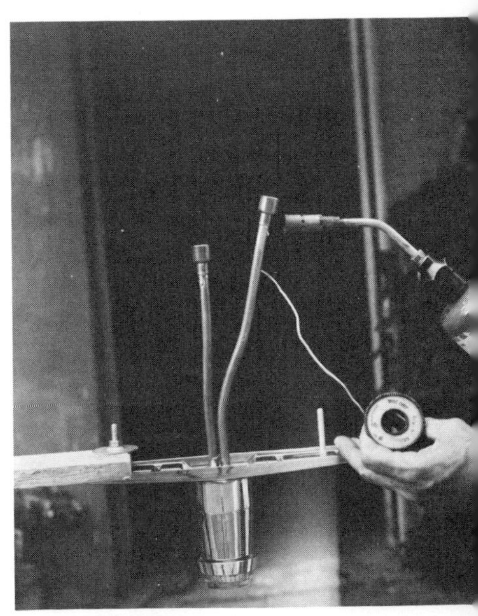

5. Sweat a pair of reducers onto the feed tubes of the new faucet.

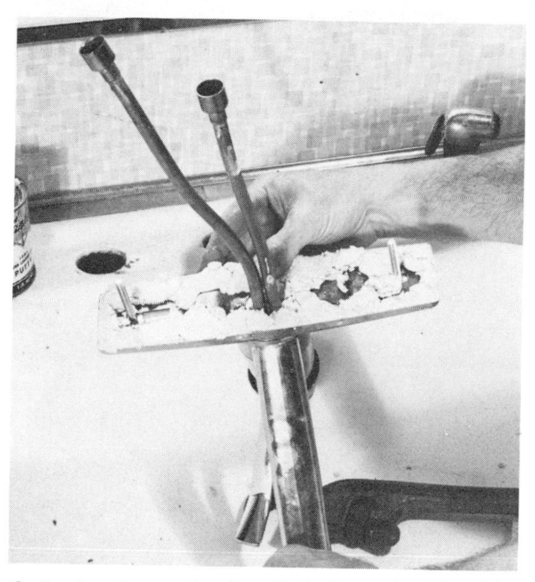

6. Apply a layer of soft putty to the underside of the new faucet.

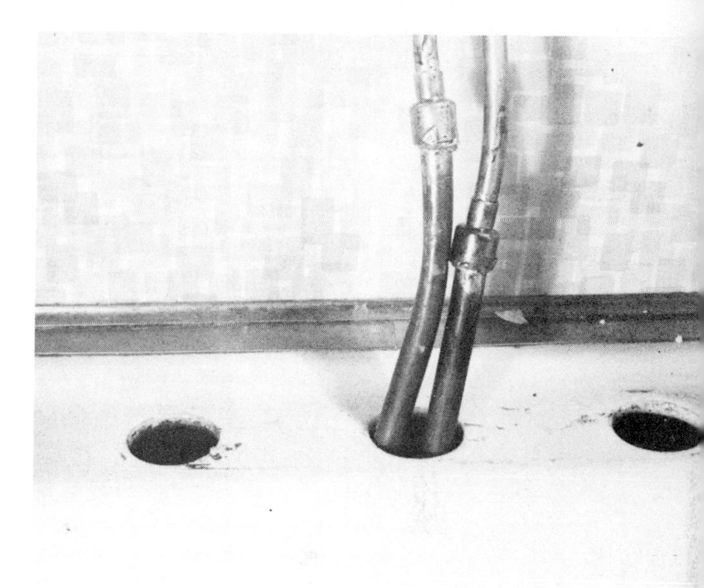

7. Sweat the feed pipes to the reducers. Then lower the faucet and finish the connections.

REPLACING A DUAL FAUCET WITH A SINGLE-LEVER FAUCET

Whether or not the single-lever faucet will operate longer without repairs or attention than the old, conventional dual faucet remains to be seen, claims not withstanding. However, the single-lever faucet is *new* and it can by itself make a sink or lavatory look modern.

The removal and replacement of the old design with the new is straightforward. However, the sink or lavatory or tub must be suitable for the single-lever design you select, or vice versa—you must select a design that can be mounted on your fixture.

Use a flashlight and examine the underside of the old dual faucet and the sink. Most, if not all, single lever faucets require three holes in the sink. The faucet's feed pipes go through the center hole. The mounting bolts go through the outside holes. If the old faucet is on a countertop you may be able to drill a third hole (if there is room). If the old faucets mount on the sink and it is porcelainized steel or cast iron, you will not be able to drill the third hole without damaging the sink. You can drill through stainless steel but it is an awfully difficult job.

If there are three holes, or if you can drill the necessary third hole, measure from the center of one outside hole to the center of the other outside hole. Your new, single-lever faucet must have bolts which fit these holes or which can be adjusted to fit.

To disconnect and remove the old faucet, use a basin wrench to get at the nuts holding the faucet down. Either disconnect the water pipes at the faucet, or, as illustrated, disassemble the feed pipes at the nearest joint, and then separate the water feed pipes from the old faucet.

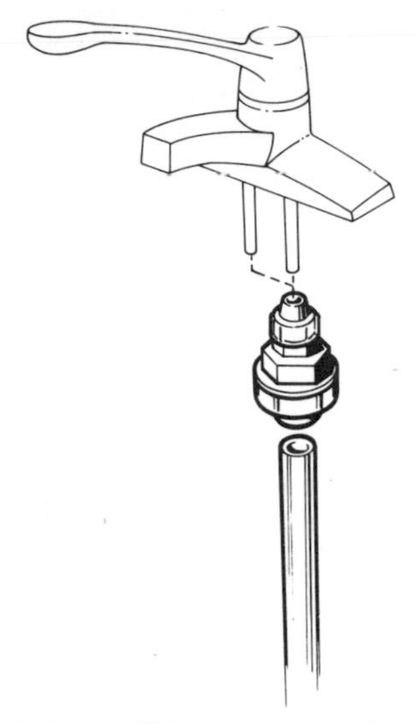

How a Click & Seal fitting can be used to connect a faucet to the water supply of a single-lever faucet.

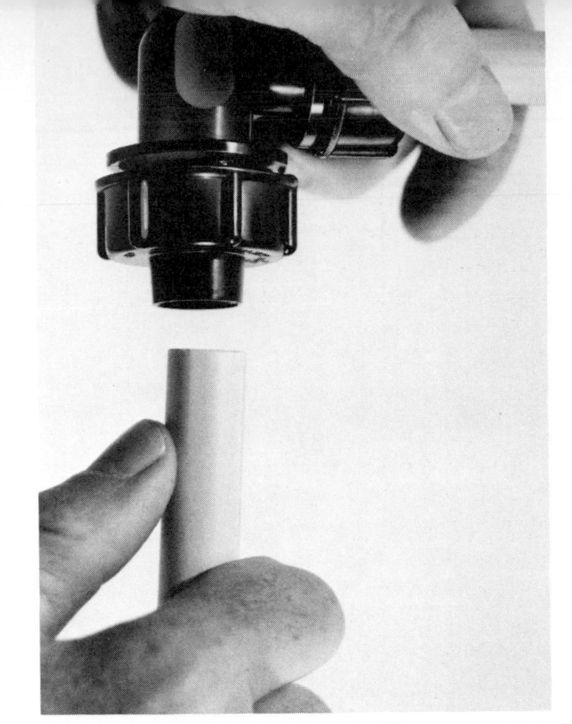

Close up of a Click & Seal angle fitting used to join two lengths of plastic pipe. Note that pipe end is smooth and square. *Courtesy Mobile International.*

Installing the new faucet. Carefully bend apart the two ⅜-inch copper tubes brazed to the bottom of the new faucet. Then sweat a pair of reducing couplings onto the ends of the tubing connected to the faucet.

Apply a generous layer of soft putty to the underside of the new fixture. Bring the old feed lines up through the center hole. Solder the water feed pipes to the couplings on the ⅜-inch tubing. Push the two feed pipes down through the central hole. Press the bottom of the new faucet against the top of the sink. Attach and tighten the two nuts to the bolts holding the faucet in place. Reconnect the feed pipes to the rest of the water system, and you're done.

If you are cramped for working space and can't or don't want to solder the faucet tubing to the feed-pipe tubing, you can use no-solder, Click & Seal fittings. The tube or pipe end is pushed into the fitting and the nut is turned. They work with any type of pipe. Just make certain the pipe ends are clean, smooth and reasonably square.

INSTALLING WATER-CONDITIONING EQUIPMENT

Water that contains a comparatively high percentage of dissolved minerals is called "hard" water. If you have trouble getting soap to lather; if your pots and kettles rapidly acquire an insoluble layer of lime on the inside; if your domestic hot water grows increasingly cooler with each passing month, you are in hard-water country.

If you want to know just how "hard" your water is write to Culligan or any of the other large water-conditioning manufacturers or to Ward or Sears. They all presently offer free water-testing service.

	Grains per Gallon	Parts per Million
Soft	Less than 1.0	Less than 17.1
Slightly hard	1 to 3.5	17.1 to 60
Moderately hard	3.5 to 7.0	60 to 120
Hard	7.0 to 10.5	120 to 180
Very hard	10.5 and over	180 and over

Table 38. Relative hardness of water expressed in calcium-carbonate, by weight. One grain = 17 parts per million.

The operation of a modern water softener is based on the use of zeolite, which in its natural form is a green, sandlike mineral. Synthetic zeolite is white.

Water to be softened is passed through a bed of zeolite. The zeolite absorbs the calcium and magnesium compounds in the water by a process of ion exchange, releasing sodium as it does so, without itself dissolving. In addition, the zeolite will remove iron present in the water.

The white zeolite has some four times the calcium-carbonate removing ability of the natural mineral and can absorb up to 6 ppm (parts per million) of iron while doing so. The natural, green zeolite can handle up to 10 ppm of iron. Both types are manufactured in granules of various sizes.

As zeolite is not dissolved and the magnesium and calcium ions do not become permanently attached, zeolite can be used countless numbers of times. It simply has to be regenerated each time the granules have absorbed their capacity.

Regeneration is accomplished with ordinary salt dissolved in water, that is to say, brine. When the zeolite—natural or synthetic—is immersed in brine the magnesium and calcium leave the zeolite and combine with

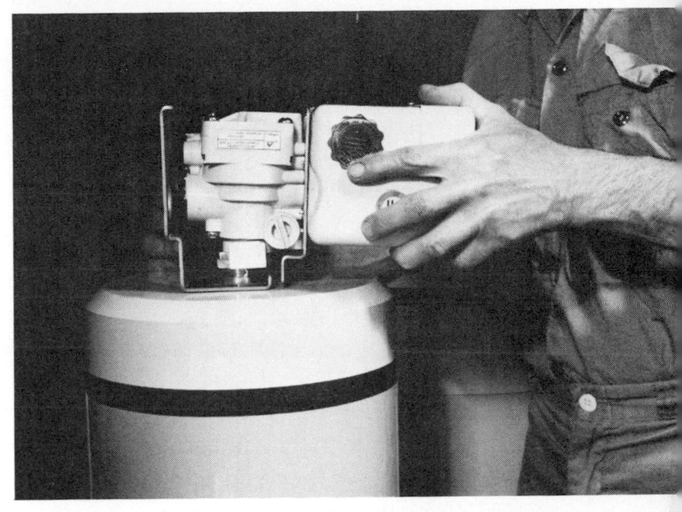

Assembling a Culligan water softener.

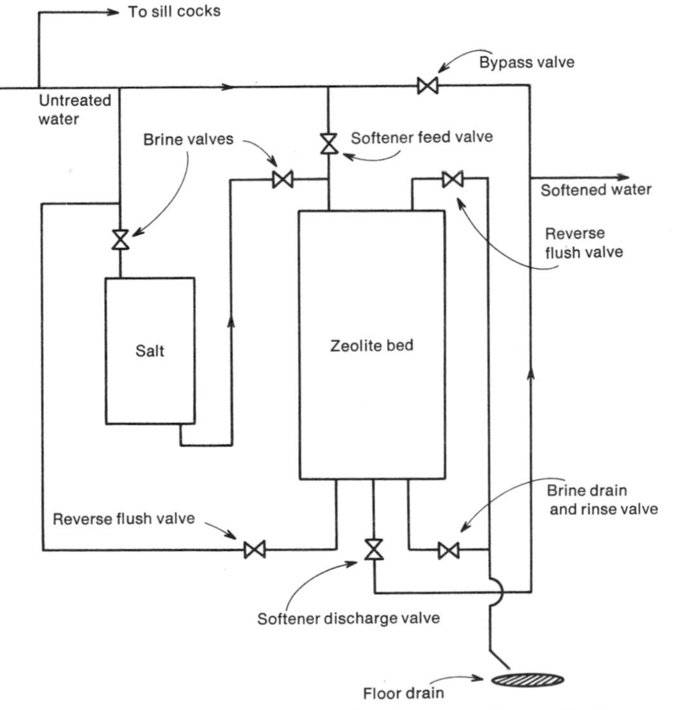

Schematic of a conventional water softener. Modern softener uses ganged valves. Connections are inside.

Setting the timer of a Culligan water softener. Machine automatically opens and closes valves shown in schematic at left.

the salt, and some of the salt's sodium replaces that lost by the zeolite. The zeolite is regenerated and ready to be used again.

Although the piping required by the different designs and makes of water conditioners varies somewhat, their needs are essentially alike. If you are simply having someone come every two or three months and replace the entire unit, you need three gate valves, two pipe-to-conditioner fittings and enough pipe and fittings to span the pipe ends leading to the unit. Use special flexible copper pipe for making the unit connections. These make it easier to attach and remove the unit.

If the unit is to be permanent, you will not need the flexible pipe, nor the couplings—assuming you will sweat the joints. However, you will need a separate pipeline from the cold-water supply ahead of the conditioner, a shutoff valve in that line, and a pipeline to the nearest basement floor drain or washtubs.

CONNECTING A WATER PURIFIER

A water purifier that will remove fine silt and unpleasant tastes and odors is easily connected in the domestic cold-water line, generally in series with the tap of the kitchen sink.

The line is cut and a section removed. A valve on the pressure side of the cold-water line is installed, followed by a union and a threaded fitting. This screws into the filter. Then another threaded fitting is used to continue the pipeline onto the cold-water faucet. The accompanying photo shows how this may be done with copper tubing. Note that the filter is connected by transition fittings; the screw ends entering the filter. To replace the filter unit the glass is unscrewed.

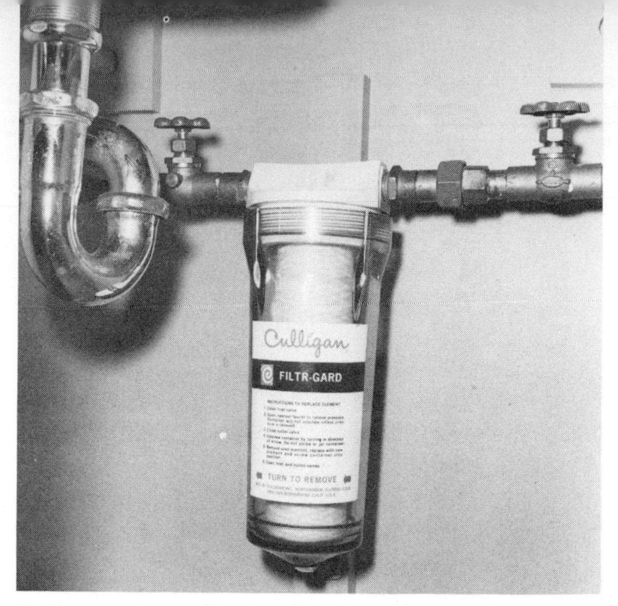

Culligan water purifier installed in cold-water pipeline leading to faucet on kitchen sink. Note two screw-threaded transition fittings on either side of purifier. To replace filter, glass bowl is unscrewed. Two valves will prevent any water from coming out when glass is removed. But one valve will be sufficient if you place pot under glass when you remove it.

INSTALLING A FOOD GRINDER

To install a grinder, remove the strainer basket and following section of drainpipe from beneath the sink. Install the strainer basket that comes with the grinder. This in place, bolt the grinder to its strainer basket. The exact arrangement varies with grinder manufacturer. Some grinders are given a quarter turn and then bolted. Some are just held in place and bolted. Don't forget the gasket.

Now, the fat pipe near the bottom is the new sink drain. Connect this to the old drain. You may make it with a couple of bends; if not, think about a length of flexible auto radiator hose and a couple of clamps. As long as the hose runs downward and is not kinked, all will be well. The small pipe near the top of the grinder accepts a drain hose from a dishwasher.

15

Hot-Water Space Heating

This chapter explains the workings of a hot-water space-heating system, how you can improve your present system, add radiators and make adjustments and repairs.

While the electrical and oil-burner portions of a hot-water heating system are beyond the scope of this book, the water portions of the system are not. You work with the same pipes, fittings and valves, and you use the same tools and methods when working on a hot-water space-heating system as you do on the domestic water system. The major difference is in the direction of water flow, and this is explained. It is not at all complicated.

There are several types of hot-water heating systems. They divide into two major groups: gravity and forced flow (hydronic).

GRAVITY SYSTEM

A gravity system includes a boiler in which the water is heated by oil, gas or coal, pipes leading to and from radiators, and a pipe leading from the boiler to an expansion tank, generally located in an upper floor of the building. A second pipe leads from the expansion tank to a drain or even onto the building's roof. This pipe is always open.

When a gravity-system furnace is fired, the water in the boiler is heated and two things happen simultaneously. One, the water expands. As the system is completely closed except for the expansion tank which has an always-open drain, the additional volume of water produced by the expansion flows through the connecting pipe and into the expansion tank. If the water expands beyond the capacity of the tank, it simply flows out and into the drain or down the roof.

The second thing that happens when the water in a gravity system boiler is heated and expands is that the water in the boiler becomes lighter than the cold or colder water in the radiators on the floors above. The cold water, being heavier, flows down into the boiler, displacing the lighter water, moving it upwards into the radiators. Once in the radiators the hot water gives up its heat, becomes cooler and heavier than the water in the boiler below and moves downward. Thus, water circulation in a gravity system is caused by the difference in weight between water in one part of the system and another.

Gravity systems have several drawbacks. Since the weight difference between hot and cold water is only a matter of ounces, large radiators and pipes must be used to encour-

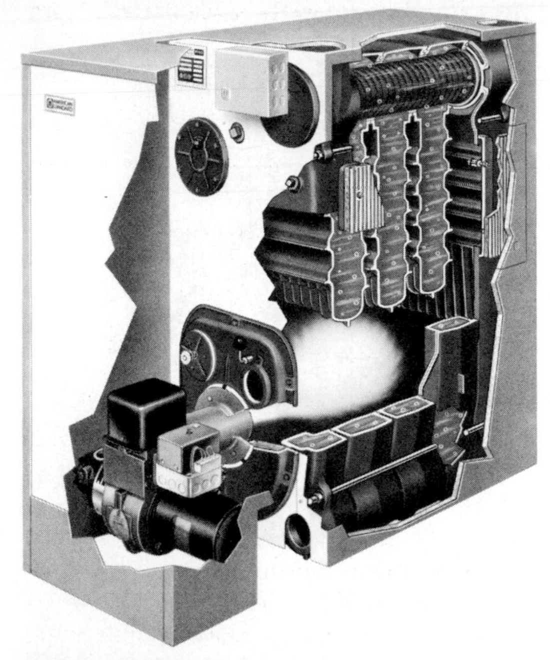

Modern hot-water space-heating furnace cut away to show boiler sections and immersed domestic hot-water heater (ribbed coil at top). *American Standard*

age flow. Even so, the rate of flow is low and heat transfer from the furnace to the radiators is inefficient and slow.

Other problems arise from the always-open expansion tank. Water that leaves the tank is lost and must be replaced. As the water is hot, water loss is also a heat loss. Air entering the system through the open pipe leading to the expansion tank and air mixed with the replacement water entering the system causes the pipes, radiators and boiler to rust.

FORCED-FLOW SYSTEM

The major difference between a gravity and a forced-flow system is the use of a motor-driven pump in the forced-flow system. The pump forces the water to circulate at a rapid rate; thus heat transfer from the boiler to the radiators is rapid and efficient.

Another major difference between the gravity and the forced-flow systems is the use of a closed expansion tank. Because the forced-flow system's expansion tank is closed, no water and no heat is ever lost this way.

Single-pipe system. The single-pipe forced-flow system, sometimes called a loop system because it has one pipe that makes a loop or circuit of the house, is most often used today. The pipe starts at the top of the boiler, makes a circuit of the building and returns to the bottom of the boiler through a centrifugal pump. Each of the radiators is connected to this loop by a pair of Tees, one of which is a special fitting called a directional Tee or a scoop Tee. As the water, forced to travel around the loop by the action of the pump, strikes a scoop Tee, a portion of the water is

Typical layout of a gravity hot-water space-heating system. Arrows indicate water flow. Expansion tank is permanently connected to an overlfow pipe.

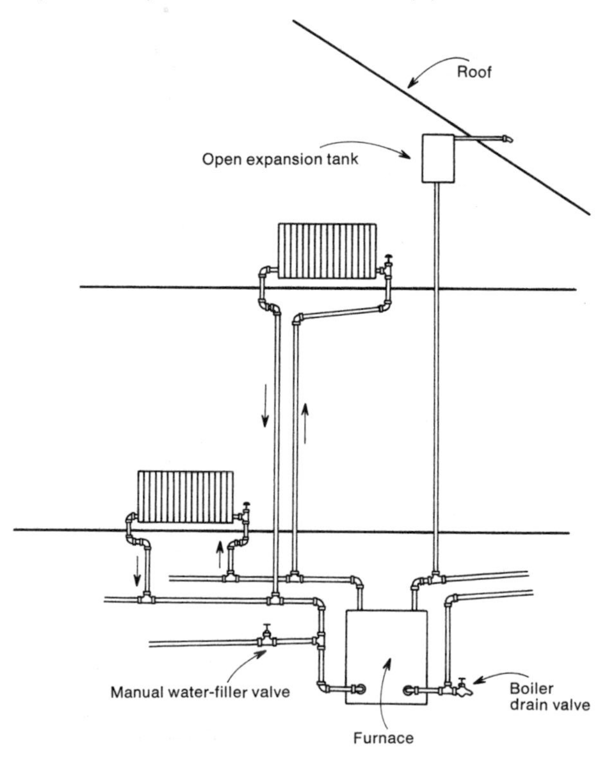

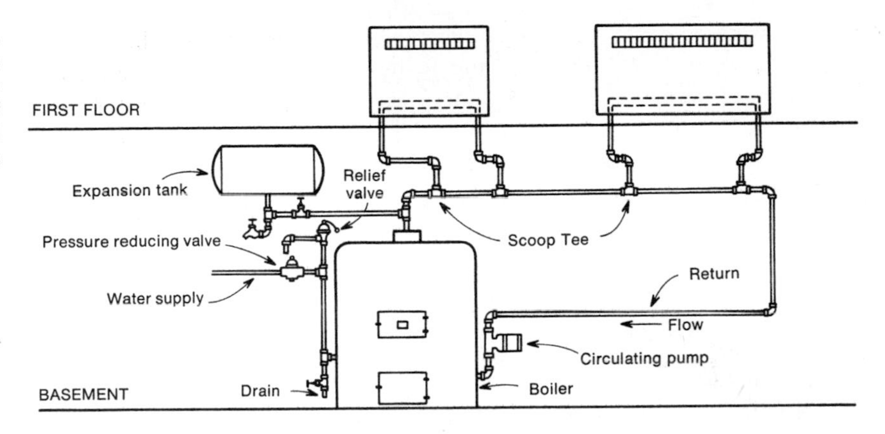

Layout of a single-pipe, forced-flow heating system. Pump drives hot water around loop. Scoop Tees in loop direct a portion of the hot-water flow to individual radiators. As many radiators as needed can be connected.

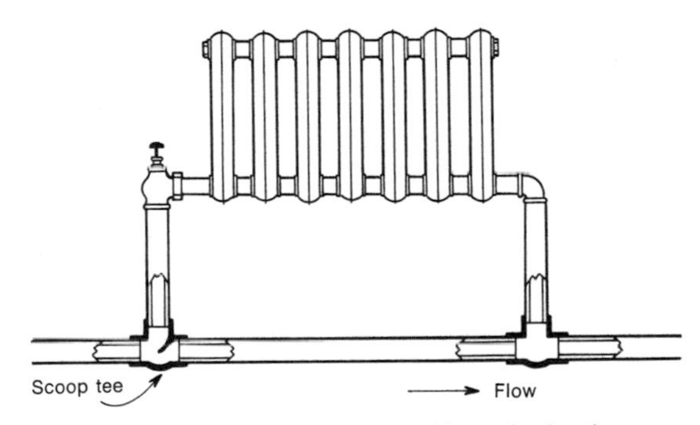

How each radiator is connected in a single-pipe system. Note that the scoop Tee must be connected with its intake end toward the direction of flow and that it always precedes the standard Tee in the direction of flow.

"scooped" up and forced into the radiator. In this way each radiator is supplied with its share of hot water. Shutting one radiator by closing its valve merely leaves a little more hot water for the rest of the radiators connected to the loop.

Two-pipe system. The two-pipe system consists of a main feed pipe which is connected to every radiator by an individual pipe. The other end of every radiator is connected by another individual pipe to a main "return" pipe. The advantage of the two-pipe system is that the heat is usually more evenly distributed among radiators. In the loop system the radiator nearest the boiler gets the

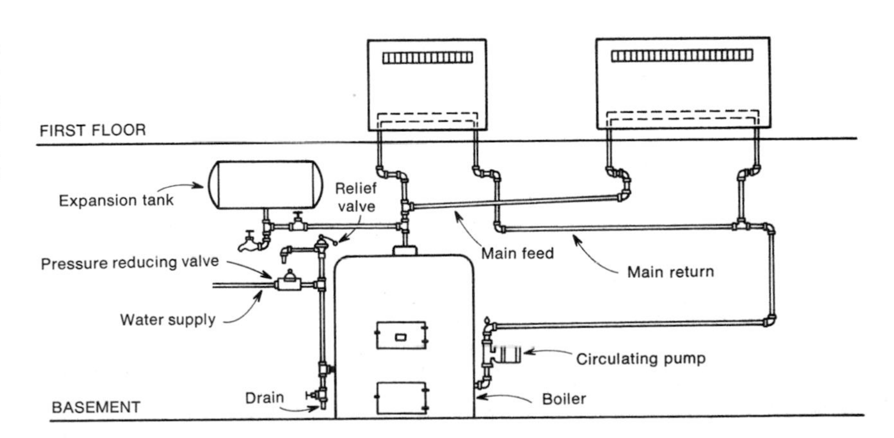

Layout of a two-pipe forced-flow system. Note that each radiator has one pipe connected to the main feed and a second pipe connected to the main return. Additional radiators would be connected the same way.

hottest water. Each radiator following receives successively cooler water and is therefore a little cooler than the radiator preceding it.

The disadvantage of the two-pipe system is that it requires considerably more pipe and labor to install than the single-pipe system.

VALVES, GAUGES AND CONTROLS

Reducing valve. To replace the water lost in the boiler of a gravity system it is necessary to manually open a water valve connecting the domestic cold-water line with the boiler. On the newer single- and two-pipe space-heating systems, the water level is automatically maintained by a water-pressure reducing valve.

The boiler, its pipes, radiators and expansion tank form a closed system. Heating the water in the system produces pressure. However, the pressure is usually lower than the normal pressure in the service line. Thus, if the domestic cold-water line would be connected *directly* to the space-heating hot-water boiler, boiler and system pressure would go higher than we wished. Therefore we use a pressure-reducing valve. To clarify, let us suppose some reasonable figures.

Typically, a space-heating hot-water boiler is pressure tested at 75 psi and red-lined at 30 psi. (The pressure gauge will either have a red line from 30 psi upwards or it will be marked danger.) Again, typically, the boiler will be operated at pressures ranging from 15 psi with boiler water at around 180 degrees F to 19 psi with the water temperature around 200 degrees. (This would be the usual temperature swing of the boiler water with the furnace in operation. Remember, it doesn't run continuously but goes on and off.)

Let us further assume that the feed valve, which is a pressure-regulating valve, has been adjusted to 15 psi. As long as the water in the boiler remains at a constant volume and doesn't drop below 15 psi, boiler pressure is equal to water pressure and no water enters the boiler. Assume that some boiler water is drained from a radiator. Boiler pressure drops. The cold-water pressure is higher than that in the boiler, and water enters until a pressure of 15 psi has been reached, at which point conditions are stable and no more water flows.

To increase the pressure of the incoming cold water, loosen the lock nut at the top of the valve and turn the screw into the valve.

Pressure relief valve. As stated, the modern hot-water space-heating system is closed. As water expands as it gets hot, a pressure-relief valve (safety valve) is installed on every boiler. Generally the valve is set to open at about 30 psi. However, setting accuracy is not very high and it is usual for the valves to actually open 3 to 5 psi higher. This is satisfactory, as the boiler is usually capable of holding 75 psi. When the relief valve opens, boiler water flows out the drainpipe and onto the basement floor or into a

Left to right: A check valve, water-pressure reducing valve, pressure-relief valve with overflow pipe at bottom. Lift handle very gently to release pressure.

tub. Note that it is against the code to connect the safety drainpipe solidly to the house drain. (If the drain plugs up and the boiler runs wild, you can have a serious explosion.)

Gauges. Hot-water space-heating boilers are always furnished with a pressure gauge, temperature gauge and sometimes with an altitude gauge. These may be individual instruments. The pressure gauge indicates the pressure in the system. The temperature gauge indicates boiler-water temperature, which is always a few degrees higher than the temperature of the water that comes out of the immersed, domestic hot-water coil.

The altitude gauge indicates the vertical height of the water in your system in feet. To use, check for the presence of water in your topmost radiator by opening the air-bleed valve. Then note the altitude indication. Should the needle drop below this figure you know the system is short on water. Some indicators have a second hand which you can use to help you remember the altitude when the system is full.

Aquastat. Sometimes called the high-temperature cut-off or safety because it does just this, this control consists of a temperature-sensitive element that projects into the boiler water. The temperature at which it opens the burner circuit and prevents firing under all conditions is determined by setting the little indicating wheel with a screwdriver. This control is a safety; it prevents boiler-water temperature from going too high. However, for proper operation it must always be set at a temperature higher than that of the boiler-water temperature control.

Boiler-water temperature control. Sometimes called the control or relay box because it contains several electrical devices, it also contains a small wheel which sets boiler-water temperature. Unlike the aquastat, which

Boiler gauges. Two pointers on upper half indicate pressure and altitude (note *Danger* at 30 psi). Lower pointer indicates temperature of boiler water.

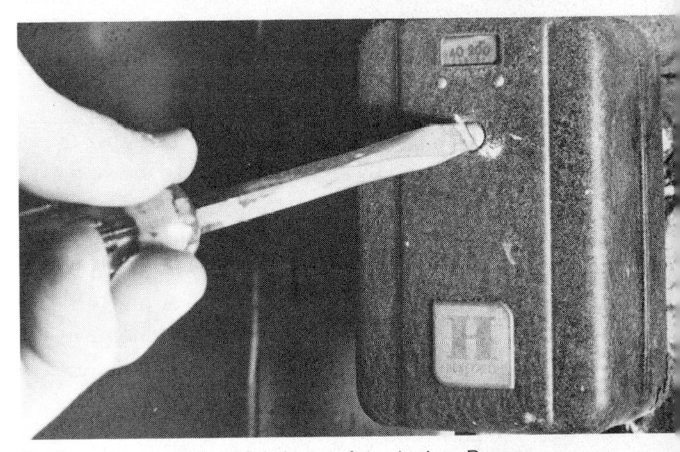

Setting an aquastat, which is a safety device. Burner will be shut off if temperature of boiler water goes above temperature on dial in window.

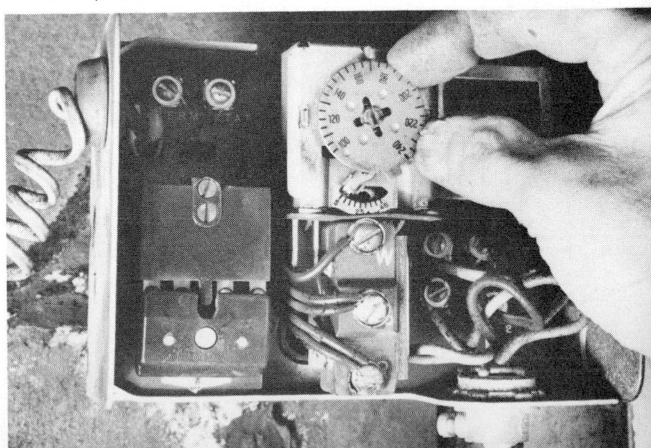

Setting temperature of boiler water. Control will hold temperature to figure on dial under pointer. Control setting must be lower than aquastat setting or furnace will not operate.

merely shuts the burner off, this control turns the burner on when water temperature is too low, and off when water temperature is too high. Generally it is set to around 180–190 degrees in the winter and lower in the summer.

Circulator. The circulator is an electrically driven centrifugal pump in the hot-water heating line. It is turned on and off automatically in response to demands for heat by the room thermostat.

Room thermostat. This is a temperature-sensitive, low-voltage mercury switch that closes when the room's temperature is lower than the temperature to which the "stat" is set. It opens when the room's temperature exceeds the stat's setting.

Chimney stat. This is another temperature-sensitive electrical switch. Its probe is inside the chimney just after the furnace. The chimney stat together with other electrical equipment prevent the burner, which sprays a mist of oil into the furnace, from starting when the furnace is hot. The chimney stat also acts to shut the burner down if there is no ignition.

In normal operation, ignition, a high-voltage electrical arc, is always switched on in the path of the oil mist simultaneously with burner startup. Should there be no arc—ignition—the chimney stat and associate circuits shut off the burner.

The reason for all this is safety. Should the burner start up without ignition and should the chimney and furnace be hot at the time, there would be an explosion. Should the burner go on without ignition and spew oil into the furnace, and should ignition occur following, there would be an explosion.

Check valve. This valve usually has a lever which can be set in one of three positions,

Check valve in hot-water heating loop. Control arm is in "Normal" position, which is correct for winter operation. During the summer, arm is swung to "Closed" position. For fast system drainage, arm is swung to "Open" position.

marked "open," "normal," and "closed." The valve is connected in line with the hot-water heating pipe coming out of the top of the boiler.

When the valve is open, water may pass in either direction. In its normal position, and that is where the lever should be during winter operation, water can only pass through in one direction—from the boiler upwards to the radiators. Operating as a check valve in the "normal" position, the valve prevents water from running backwards through the heating system.

During the summer you can, if you wish, move the lever to the closed position, which closes the valve and prevents water from circulating. In this way you reduce the nuisance heat that is produced by the furnace when it is being used to make domestic hot water.

Thermostat. This is an automatic valve that looks and acts just like the thermostat in your car. It is positioned "in line" with the aforementioned check valve. You'll find it inside the circular housing directly atop the heating pipe emerging from the top of the furnace.

The thermostat remains closed until the heating water temperature rises to about 150 degrees or so. In this way the circulator is prevented from driving cold water through the radiators. When the water's temperature exceeds 150 degrees, the stat opens and the water can pass.

Bypass valve. If you examine your hot-water space-heating system you will find a pipe connecting the hot-water feed pipe on top of the boiler with the hot water return at the bottom of the boiler. There is a gate valve in the middle of this pipe. This is the bypass valve.

When this valve is closed and the circulator kicks in and drives the water through the system, all the water passes through the boiler. On cold days, or after the heating system has been set to a low temperature, the circulating water chills the immersed domestic hot-water coil. Should you be showering at the time, the water may run cold. To prevent this indignity, the bypass valve and pipe have been installed. With the bypass valve open, a portion of the circulating water bypasses the boiler. Thus the domestic hot water isn't chilled every time the room stat calls for heat.

Access to the thermostat in the hot-water heating loop is through the plug on top of the stat's housing. Boiler should be cool and all the water drained out of the system before plug is removed. Pipe leading upwards from the stat is the heating loop. Pipe leading downward goes to the bypass valve.

HOW IT WORKS

Take a moment to study the sequence of events that takes place in a relatively unsophisticated forced-flow heating system. A thorough appreciation of what constitutes a complete cycle may save you hours of useless troubleshooting.

The room stat has been positioned at 60 degrees all night. House temperature may swing from 58 to 62 or so without affecting the room stat. This is normal. Room stats do not respond to small temperature changes. In the morning the room stat is pushed beyond 62 degrees. First the circulator kicks in and drives the water through the system. Minutes later, when the boiler water has cooled below the setting on the boiler-water temperature control, the circulator stops. The burner kicks in and simultaneously a high-voltage ignition spark appears across the electrodes in front of the burner. The oil is ignited. A half minute or so later ignition is turned off. You can hear the change in sound. When the boiler is again relatively hot the circulator starts working. The radiators grow warm. The rooms grow warm. The room stat responds and the burner and circulator are shut off.

In summer, when the room stat is turned way down, the boiler-water temperature control keeps the boiler water within the range established by the setting on the little dial. The circulator doesn't operate and the bypass valve is in the closed position, preventing water circulation by gravity.

MAINTENANCE

A regular maintenance program will extend the life of your heating system and greatly reduce the need for repairs.

- Oil the burner motor twice a year with SAE 30 pure (no detergents) motor oil.
- Oil the circulator pump twice a year with SAE 50 pure motor oil. (Don't use thin oil, it will run out.)
- Oil the circulator pump motor twice a year with SAE 30 pure oil.
- Drain the expansion tank twice a year. Close the valve leading to the tank. Attach a garden hose to the other valve on the tank. Open this valve and let the water gurgle out. Water gets into the tank by virtue of the air present slowly dissolving into the water.
- Move the check-valve arm to "closed" at the start of each summer. Return it to "normal" with the end of summer.
- Bleed each radiator with the start of the winter heating season. Open air-bleed valve until water runs bubble-free.
- Don't let your oil tank run dry—you will pick up all the muck.
- Check your pressure-relief valve once a year. Lift the lever very gently (too high and you will damage the diaphragm), let half a gallon or so of water run out, then release.
- Shut off the furnace, remove the side panels and brush the dirt and soot free once a year. (Some oil companies offer this cleaning service with their oil.)
- Make a practice of looking at the gauges every month or so, just to keep an eye on things. Also watch the overflow tube (pressure-release valve outlet).
- Remove the room stat cover once a year and blow off dust.
- If your fuel line incorporates a filter, remove it, drain off the water that has collected and either wash the element clean or replace it once a year.

Draining the expansion tank. The valve at the other end of the tank (out of sight) is closed tightly. A garden hose is connected to the drain valve and the valve is opened.

Bleeding a radiator. Special small socket wrench or "key" is used to open the air-bleed valve on each radiator. Valve is left open until all air is out and water flows free of bubbles.

Adjusting the zone control valves. In this house, the loop is split into two zones. Partially closing one valve directs more hot water and more heat to other loop, thus making connected radiators hotter.

TROUBLESHOOTING A HEATING SYSTEM

One radiator is cold. There may be air in the radiator; bleed it. Radiator valve may be closed or partially closed. Room may be much colder than others; radiator is chilled. Radiator farthest from boiler, feed pipes may need insulating. Radiator feed pipes smaller than others; getting less water.

Some radiators are cold. Zone valves need to be adjusted. Radiators are in series (one follows another) and one valve is partially closed, affecting all the radiators on the line.

All radiators are cold. Furnace is operating and hot. Circulator appears to be working. (Do not touch to check—just look.) Check valve may be in closed position; check. Thermostat may be in closed position, preventing circulation.

To examine and test the thermostat it is necessary to shut the system down and drain it. To speed drainage, open radiator air-bleed valves. Remove cover from stat housing. Place stat in pan of water and heat. Stat should open long before water boils. If uncertain as to stat's condition, replace or operate system without stat. Refill system with water, drain expansion tank, bleed radiators and resume normal operation.

Circulator doesn't turn. Shut system down. Remove furnace fuse as a safety precaution. Examine coupling between motor and pump. If broken, disassemble. See accompanying photographs. Replace. If not broken, try turning pump with pliers. Pump and its motor should turn easily. If it does, smell motor. If it has burned out it will have a strong, acrid smell. However, these motors are oversize and rarely burn out, so trouble is electrical and beyond scope of this book.

If motor and pump cannot be turned, pump bearing has run dry and pump shaft is "frozen" to bearing. Try soaking in liquid wrench and turning with pliers. If you can loosen it, follow with light oil and resume operation. However, keep close tabs on pump as it may freeze again despite lubrication.

If you cannot loosen pump with pliers, shut system down, remove fuse, drain the system of water and disassemble pump. Either try once again to loosen pump shaft or replace shaft and bearing. (Parts are available. You don't have to replace the entire pump.) Reassemble. Refill system with water, bleed all radiators, empty expansion tank. Resume normal operation.

Circulator drips. This is common to pumps that have run without bearing lubrication. If

Disassembling a Circulator

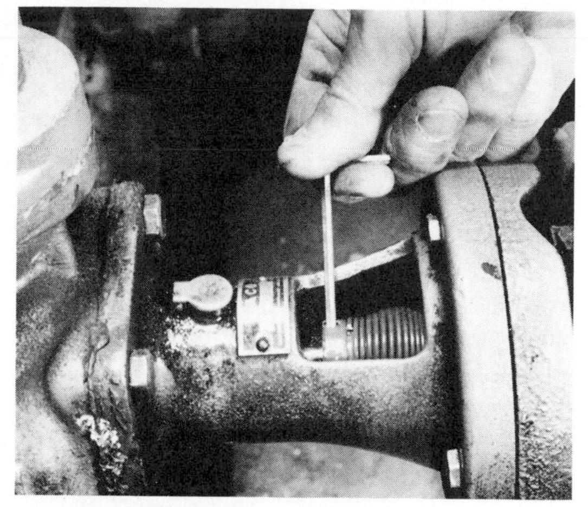

1. Hex wrench is used to loosen the setscrew holding one end of the spring coupling to the pump shaft.

2. The bolts holding the motor to the pump frame are removed with an open-end wrench.

3. The motor is moved backwards and supported. No need to disconnect the wires. If spring coupling is damaged, remove and replace it.

4. If the pump is defective, drain the system, then remove the four corner bolts. Replace faulty parts. Insert gasket between pump bearing mount and housing before reassembling.

drip is slight, collect in pan and otherwise ignore it. If drip is bad, bearing and shaft must be replaced.

Pressure too high. Boiler-water temperature has been set too high; correct. Boiler-water valve (pressure-reducing valve) has been set too high, or is defective. Reduce pressure a little, drain a gallon or so of water from boiler and watch pressure gauge. If that was the trouble, pressure will not go as high as before.

Expansion tank filled with water, drain.

Immersed domestic hot-water coil leaking. To test, drain expansion tank. Close boiler water supply valve. Drain a gallon or two of water from the boiler. Watch pressure gauge. If hot water coil is open, pressure will climb slowly up again.

Relief valve drips. Check pressure. If pressure is above 30 psi, take steps to reduce pressure. If pressure is below 30 psi, at which pressure it should not open, lift little handle very gently. Let water flow out. Seat may be dirty. If leaking continues, lower system operating pressure or replace pressure-relief valve. Units are usually factory set and cannot be easily readjusted.

ADDING RADIATORS

If you can get a pair of Tees into the pipe circuit, you can add a radiator. The amount of heat that the radiator will give off will depend on its size and type and the size and length of pipe used to connect it.

The Tees are connected just far enough apart on the supply pipe to permit the connecting pipes to be attached directly to the radiator. As stated previously, the scoop Tee goes first in the stream or water flow, the standard Tee follows. Check existing pipes to duplicate size and connections. Use black pipe; it is a bit smoother than galvanized, a little cheaper, and good enough for heating.

Types of radiators. There are any number of radiator designs. They all work alike in that they are designed to present a maximum area to the air. Cast-iron units come in sections. Some stand upright and alone, away from the wall. Some are rectangular and square. Some are shallow and are designed to fit into a wall. Baseboard radiators are low and long. Since their heat is closer to the floor they provide somewhat better heat distribution than tall radiators.

Copper radiators, usually tubes with integral fins, are more efficient per foot of length than cast-iron radiators. Copper radiators have more radiating surface per foot of length, and the copper itself conducts heat better, than cast iron. On the other hand, an iron radiator is large and holds more water, providing longer periods of warmth. So if you have the space, I suggest you opt for the cast iron.

Positioning radiators. It is common practice to install a radiator beneath a window, the purpose being to reduce the influx of cold air. If you can do this, fine. If not, don't fret. The room will not be heated as evenly, but then no room is ever heated perfectly evenly.

If the radiator is to be installed within the wall, with one side exposed, be certain to insulate the wall behind the radiator. If your space is limited, at least install a layer of aluminum-covered building paper behind the radiator. And remember, a built-in radiator projects one-third to one-half less heat than a free-standing radiator of equal size.

Judging furnace capacity. A furnace is a go-no-go device. It is either running or shut off. Therefore it is technically impossible to overload a furnace. However, it is possible to add radiators to the point where the temperature of all the rooms is unsatisfactory on very cold days. You can judge this possibility by thinking back to the coldest days of the previous year. If there were days when the heating system never caught up, when it ran continuously, adding radiators will only increase the number of these days. If your oil burner has never had to run continuously because of the cold, your system has heat to spare and you can add radiators without concern.

Radiator connections on a single-pipe system. Scoop Tee is at left, the direction from which the water flows; standard Tee is at right. Tees are spaced so pipes run directly to the radiator.

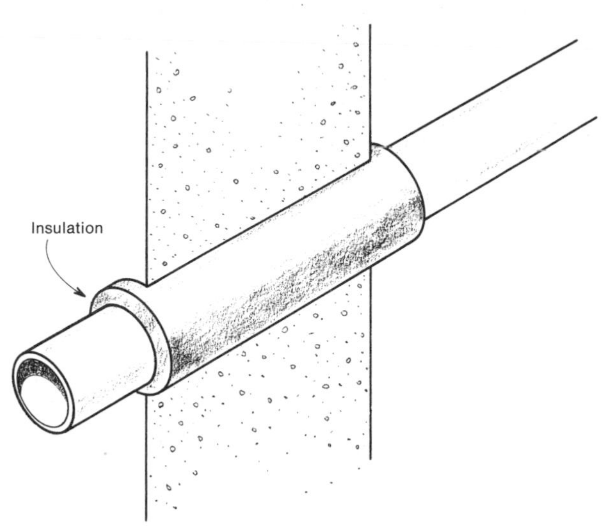

When heating pipe must pass through concrete, insulate it.

Old free-standing cast-iron radiator installed in a trimmed hole in a basement wall to heat two rooms.

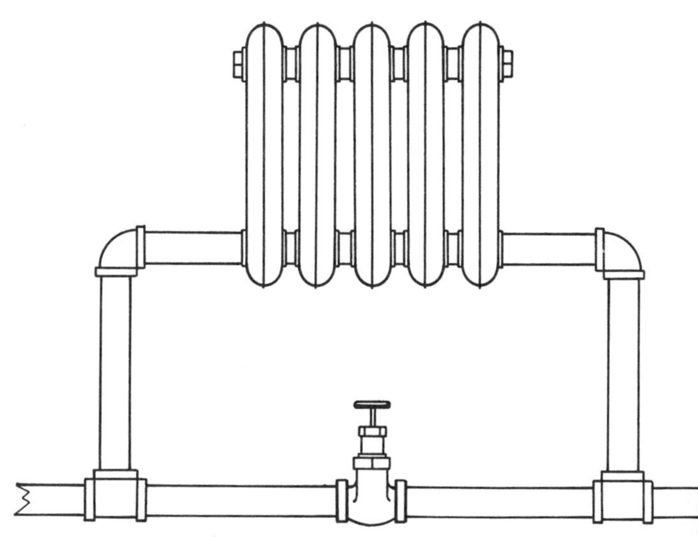

To increase a radiator's share of the heat, install valve in pipeline between two Tees.

GETTING MORE HEAT

If you install a gate valve in the heating loop between the Tees leading to a radiator, and if you fully or partially close that valve, more water is forced to pass through the radiator. The radiator becomes hotter but all the other radiators become proportionately cooler.

If you place a reflecting surface behind a radiator, as for example a shiny sheet of aluminum, more heat will be transferred from that radiator to the room.

If you insulate the heating loop, more heat will emanate from the radiators and less will be lost to the basement and elsewhere.

If you place a fan behind or even in front of a radiator, the moving air will increase the radiator's heat output considerably. With an operating fan a room's temperature can easily be increased 10 or more degrees.

You can increase the heat transferred from a furnace to its radiators by increasing the temperature of the water in the boiler. It is advisable to secure competent advice if you plan to go to 210 degrees or more.

To increase the heat output of the burner, you may be able to have a larger-diameter nozzle installed. Most burners are designed to accept a range of nozzles. If your burner does not have the largest now, a larger can be readily installed. This requires that the air mixture and possibly the flue be adjusted. Again, it's advisable to ask for experienced help.

Note that increasing boiler-water temperature and increasing nozzle size increases the heat output of your furnace. Your radiators will become hotter with higher water temperature and go from cold to hot much more rapidly with the larger nozzle. However, your heating costs will also increase and generally at more than a proportionate rate.

16

Steam Heat

Two types of steam-heat systems are currently used for private homes and small buildings: the single-pipe and the double-pipe systems.

In the single-pipe system, water is brought to a boil in a boiler. The generated steam is led through pipes to radiators. The steam gives up its heat to the radiators, turns back to water and runs down the same pipe and back into the boiler. The movement of steam from the boiler to the radiator is caused by the water expanding more than a thousand times its volume upon turning to steam. The movement of the condensed steam back to the boiler is by gravity alone. The water simply runs downhill.

One difficulty with the single-pipe system is that the steam and water flow in opposite directions. This requires fairly large-diameter piping, which assists the counterflow, but does not eliminate the transfer of heat from the rising steam to the descending water.

To eliminate these drawbacks the two-pipe system was developed. In this system steam is delivered to the radiator by one pipe and the water is drawn off by a second pipe. Smaller-diameter pipe can be used, but there is twice as much pipe and more than twice the labor is needed to install it.

As a result, the single-pipe system is more economical to install and is always selected when the cheapest central-heating system is desired. Both systems, however, are exactly alike in components and operation. The only difference is the second pipe in the two-pipe system.

HOW A STEAM-HEAT SYSTEM WORKS

For the sake of clarity a steam-heat system may be divided into sections and each section studied at a time. While the steam loop described below is a single-pipe system, its parts and operation are identical to a two-pipe system. The only difference would be that the condensed steam would come back to the boiler by a second pipe. All the other components, boiler, valves, controls and gauges are exactly like and operate exactly like those found in a single-pipe system.

The steam loop. By this we mean the pipes that lead from the top of the boiler to the radiators and back to the boiler. As the accompanying drawing shows, it is a loop—but not a simple loop. First, note that the radiators are not in series. Steam does not enter one, pass through and enter a following radiator. Instead, a single pipe leads from the steam-pipe loop to each radiator. Only two radiators are shown, but there can be any number connected, compatible with the capacity of the boiler and the size of the pipes. Note that the steam main rises from the boiler and then slopes downward. This is very important. Should all the radiators be shut off, the condensed steam (water) would still be able to flow back to the boiler.

Note the change in the pipe diameter of the steam loop and where it occurs. The reducer is a good distance below the normal water level in the boiler. And note that the wet return pipe does not connect to the bottom of the boiler but joins the boiler via an arrangement of pipes called a Hartford loop. The Hartford loop balances the steam pressure atop the water in the boiler with steam pressure at the bottom of the water in the boiler. If the loop wasn't there, steam pressure between the top of the water and the top of the boiler would force the water down and out the return line.

When the furnace is fired, the water in the boiler is heated and brought to a boil. The expanding steam forces its way up through the pipes and into the radiators. As the steam front moves forward it pushes the air present in the pipes and radiators out through a number of air-release or air-vent valves positioned at the radiators and along the steam main. The valve on the main is designed to open rapidly and is called a quick-release air valve, or vent. When all the air is out of the lines and the steam strikes the valves, they close. The steam heats the radiators, condenses into water and runs back to the wet return and into the boiler.

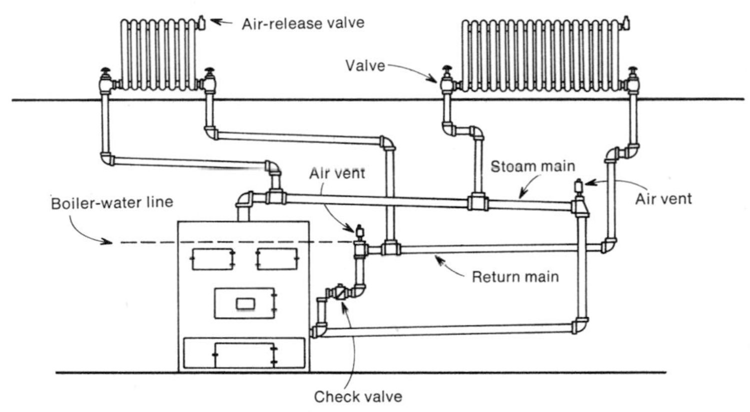

Two-pipe steam system is very similar to the single-pipe system, illustrated on page 11. The difference is that the water draining from the radiators comes down and joins a main return. The returning water does not mix with the up-rushing steam.

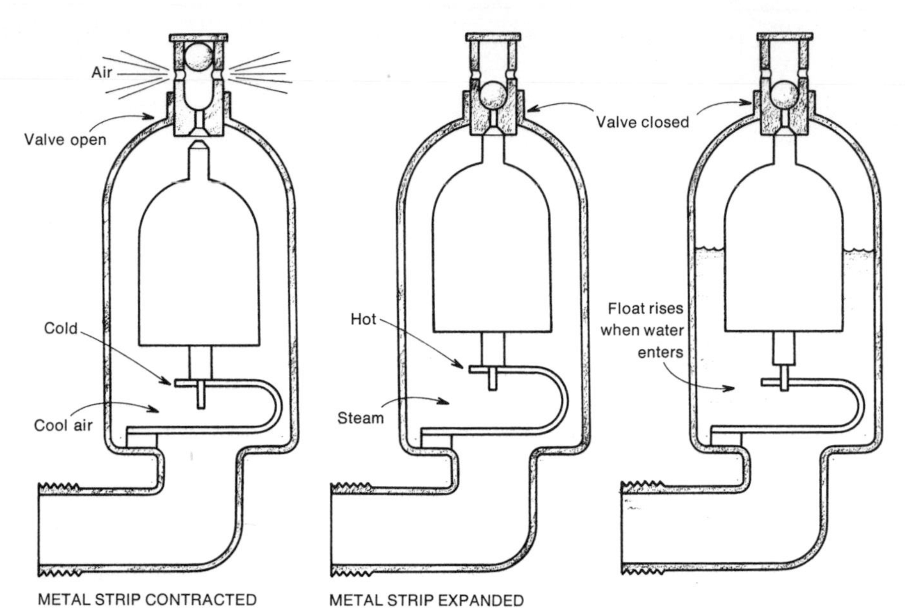

How an air-release valve works. Left: The valve is cold; the metal strip is contracted and the valve is open. Air can escape. Center: Valve is hot; the metal strip expands and closes valve. Steam cannot escape. Right: Water enters valve; float rises and closes valve whether valve is hot or cold.

The process of the steam rising, heating the radiators and turning back to water continues until the furnace cuts out. The radiators grow cold. All the steam turns to water and there is a partial vacuum inside the steam pipes and radiators. When the air-release valves are working correctly, they remain closed and air does not enter the system. However, fresh water entering the system always contains some dissolved air, so even when all the release valves are working perfectly some air gets into the steam loop.

Pressure gauge. All steam boilers are equipped with a steam-pressure gauge which indicates the pressure of the steam above atmosphere. Generally a small domestic boiler is operated at something less than 5 psi.

Water-level gauge. Generally called a sight glass, this is a vertical glass tube connected

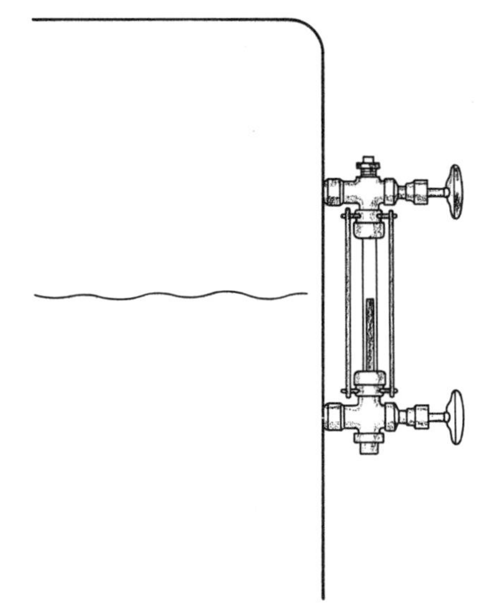

The sight gauge provides a visual indication of the exact level of the water in the boiler. Usually, the gauge is placed so that the correct water level is about the middle, but not always. Check it against level set by automatic water-feed mechanism.

at its top and bottom to the boiler. Thus, when the water level in the boiler is at the same height as the glass you can *see* the water level. Generally the sight glass is positioned in relation to the boiler so that proper water level is in the center of the glass tube.

Try cocks. Some boilers have a sight glass. Some have try cocks, and some have both. Try cocks are small faucets connected directly to the boiler at a point just above and just below proper water level. If water comes out when you open the lower try cock, you know there is water in the boiler to this level. If water comes out when you open the higher level, the water level is equal or higher than the upper try cock. Do not open a cock wide, as outrushing steam will draw water along and give you a false indication.

Make it your habit to always look at the sight glass every time you enter the furnace room. Water level and steam pressure are most important to safe operation.

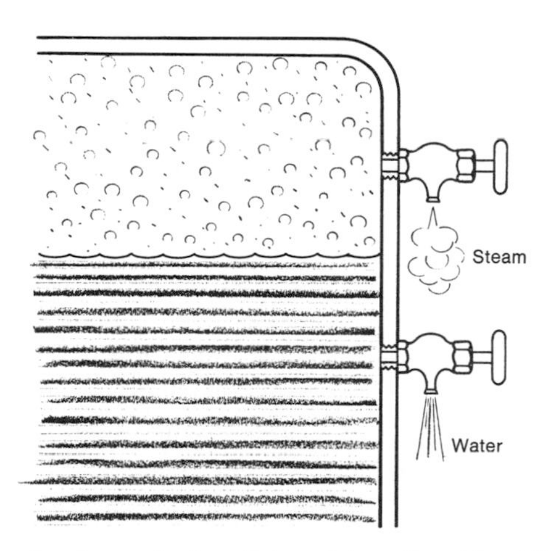

The two try cocks (small faucets) on the side of the boiler are used to determine water level. The lower cock should show water. The upper should show steam. This indicates water level is somewhere between. Open try cock slowly; otherwise steam may blow water out, giving a false indication.

Water-feed controls. Old-time steam boilers were equipped with manual valves alone for supplying water to the boiler. Not only was this a nuisance, it was highly dangerous. If you forgot to let water into the boiler on time, it could explode. For if you heat 30 gallons of water in a closed vessel to 50 psi, which will occur at 297.7 degrees F, you create an explosive force equal to 1 pound of nitroglycerin or 2 million foot-pounds of energy, which is released when the vessel bursts.

Modern boilers have automatic water-feed mechanisms coupled with safety controls. When the water drops below a preset level the device admits water to the boiler. Should water fail to enter the boiler, for any reason whatsoever, and the water level falls dangerously low, the device will shut off the oil or gas burner and prevent its reignition until the boiler receives sufficient water.

Oil burner and controls. The oil burner used with a steam boiler is exactly the same as that used with a space-heating hot-water boiler.

Steam-pressure controls. Generally, steam pressure within a boiler is controlled by a dual, automatic switch mounted in a little metal box that has a pressure-sensitive element within the boiler. When the room thermostat calls for heat (closes) and there is little or no steam pressure within the boiler and system, the pressure-control switch fires the furnace. When steam pressure exceeds a present value, the pressure-control switch shuts off the oil burner or gas flame.

The pressure control has two small levers. One is marked "cut-in." This establishes the low point at which the furnace will be fired (if other controls call for ignition). The other control is marked "differential." The cut-in pressure plus the differential pressure sets the high pressure at which furnace ignition will be cut off. This is a safety and the furnace will be automatically shut off no matter what is happening at the other controls.

During normal winter operation, when heat is called for, the furnace will be fired when the pressure falls below cut-in and shut off when the pressure exceeds cut-in plus differential. So don't be surprised when during a very cold day your furnace starts and stops every five or ten minutes.

The settings on the steam-pressure control are inaccurate. As the pressure gauge is reasonably accurate, watch the pressure gauge as the furnace cuts in and out and correct the pressure-control settings as required. Remember, 5 psi is usually maximum safe operating pressure for a small steam boiler.

Summer water-temperature control. During the summer the room thermostat will not call for heat; therefore the steam pressure control cannot and will not start the furnace. At this time furnace start and stop is controlled by an aquastat. An aquastat is a small box having a temperature-sensitive element in contact with the water in the boiler and a small dial indicating approximately the desired water temperature. When no other control supersedes it, the aquastat will turn the furnace on when the water falls a few degrees below its setting and back off when the water temperature exceeds the setting by a few degrees.

Thus, when the thermostat calls for heat the boiler is fired. If boiler pressure exceeds the pressure control setting, the furnace will be shut off until the pressure drops, whether or not the room thermostat is calling for heat. When there is no call for heat and the boiler's water drops much below the aquastat setting, the aquastat will fire the furnace and turn it off again when water temperature exceeds the aquastat setting by a few degrees. And if, at any time, the water level in the boiler falls dangerously low, the automatic water-feed mechanism will shut the furnace down no matter what position the other controls may be in.

MAINTENANCE

More than any other form of space heating, a steam-heat system requires regular maintenance if maximum equipment life, maximum safety and minimum repairs are to be realized. Any program you arrange should include the following.

Monthly "blow down." Blow down is the term used to denote the process of opening the pet cock at the bottom of the automatic water-feed mechanism and letting some of the water and steam escape. This must be done at least once a month during the heating season and once every two months during the summer.

Open the valve and let the water run until it flows clear. This removes the rust and accumulation from the bottom of the automatic-feed float chamber. If you don't do it, the chamber clogs up and the mechanism will fail to supply water to the boiler and possibly fail to shut off the burner in the event of low water.

Should you drain off more than a bucket full of water without the water running clear, close the valve and after shutting the furnace down, add boiler-cleaning chemicals as per instructions. You can purchase a can at any plumbing supply house.

Check feed action. Each time you drain the boiler watch the sight glass. After you have drawn a few gallons you should hear the fresh water enter and you should see the water level in the glass fall. When you close the blow-down valve you should see the boiler water level return to its proper mid-glass position.

Check boiler-water level. Do this every time you enter the boiler room. Make it a habit. As cannot be repeated too often, low boiler-water level precedes an explosion.

Check low-water shutoff. Do this once a year. Close the manual water-feed valve so that the boiler cannot get more water. Turn the room stat up. Open the blow-off pet cock. Watch the water in the glass go down. Watch the pressure gauge. The automatic water-feed mechanism should shut off the burner just about the time the water disappears in the sight glass. If it doesn't, shut off the burner instantly. There is a switch on the boiler's side. If the pressure goes above 5 psi, shut off the burner. Open the manual water feed valve and let the boiler refill itself. Call in a heating expert; the automatic water feed device is in need of attention. You can continue to operate your system in this condition, but be advised your major safety isn't functioning properly.

Check boiler pressure. Make it your habit to always look at the pressure gauge every time you enter the boiler room. If the pressure is above 5 psi, and your boiler isn't specifically designed for higher pressure, reset the differential on the pressure control (bring it closer to cut-in).

Check the safety valve once a year. Do this by using a long stick and very slowly and gently lifting the little arm. Do not lift it too high or you may damage it. If steam comes out, it is okay. If steam continues to come out after you release it, open it again and hold it open. Let the steam clear the seat of debris. If you cannot lift the arm, if it has rusted solid, replace it. Do not try to repair it.

When you install a new unit, use a little pipe dope near the top of the thread, not the end of the pipe. You don't want any dope to get inside and rest on the seat. Make the unit snug, but not tight. It will tighten of itself when the boiler heats up.

Lubricate the burner motor. Use pure #30 S.A.E. motor oil. Shut the furnace off and look for the oiling points on the motor. Some have two, some have only one. Do this at least four times a year.

Clean the room stat. Remove the cover and blow it free of dust and lint once a year.

Check the fire brick. Once every year or two, shut the furnace down and examine the brick inside. If they have crumbled, as they will in time, replace them. Use fire brick and fire clay. With the fire brick absent the furnace loses heating efficiency.

STEAM-HEAT TROUBLES AND CURES

System on, radiator doesn't get hot. Assuming the valve leading to the radiator is open, the air-release valve is defective (closed). To check, close radiator valve. Remove air-release valve (unscrew it). Open radiator valve. If you hear the rush of air and then see steam, air-release valve is definitely defective. Replace.

Steam comes out of air-release valve. Valve is defective. In normal operation, valve is open when cold, closed when hot. Replace valve.

Noisy radiator. Check radiator tilt. If it doesn't tilt towards single pipe or drainpipe (two-pipe system), water is collecting inside radiator. Incoming steam bangs it about. Or the radiator valve is worn. Incoming steam rattles it. Try closing it tightly.

Squeaks in radiator. Could be caused by loose radiator sections or radiator rubbing against enclosure. Noise is caused by metal expanding and contracting with temperature changes.

Squeaks in pipes. Steam pipe is touching something metallic, another pipe perhaps. Separate pipes with wood block or wire apart.

Banging noise in pipes. Pipe has loosened from support. Valley has formed where water collects. Up-coming steam bangs water about. Find dropped pipe. Lift and fasten in place.

Radiator slow to heat. Defective air-release valve. Partially closed radiator valve. Water in radiator's steam pipe. Radiator distant from boiler. Insulate pipe to keep steam from cooling.

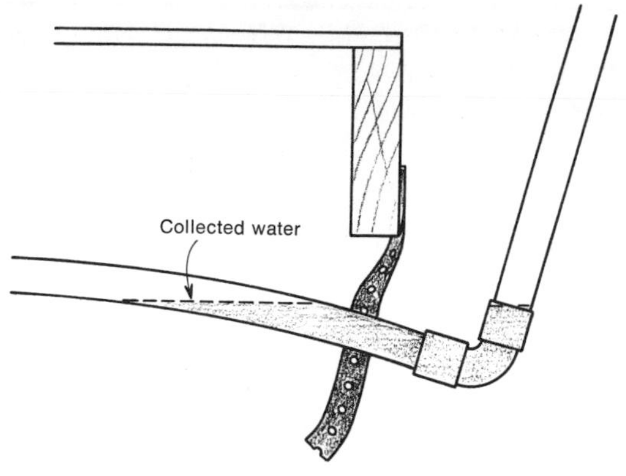

Broken strap lets heat pipe drop and water collect. Noise occurs when steam enters pipe.

Fails to make sufficient hot water in summer. Aquastat set too low. Domestic hot-water coil limed up. Water level too low in boiler.

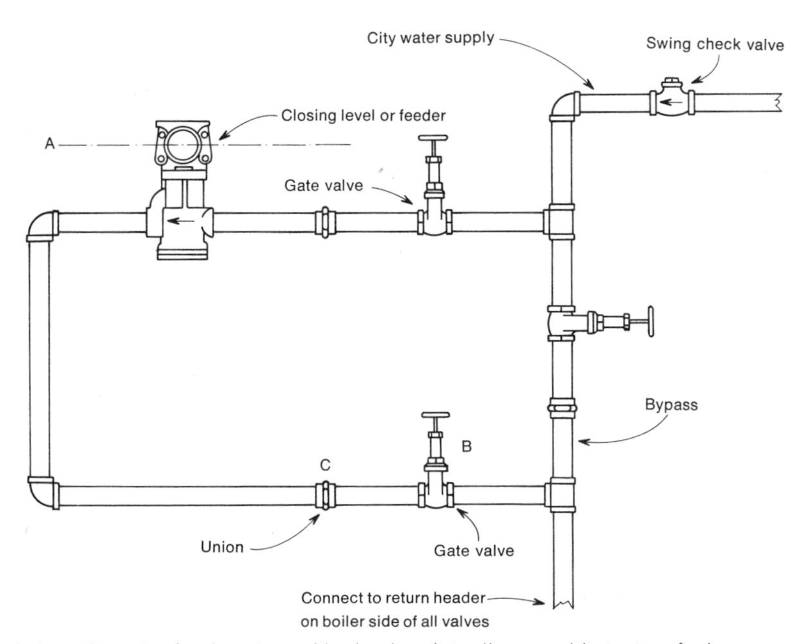

Automatic water-feed system with check points discussed in text on facing page.

Burner goes on and off too fast. Cut-in and offset controls set too close together. Chimney stat old and worn.

Automatic feed problems. Refer to the accompanying drawings of a MacDonnell & Miller automatic water-feed mechanism. Connections and service suggestions are typical of most automatic valves and controls.

Refer to schematic of the feeder and its piping. In normal operation bypass is closed, the two other valves are open. To feed the boiler manually, bypass valve is opened. To replace automatic feed, other two valves are closed.

If boiler-water level is below normal, close valve *B*, open union *C* a crack. If feeder is operating normally, water should flow out of union *C*. If it doesn't and you have water pressure, trouble lies in feed mechanism, which will be discussed below.

If water flows out of open union *C*, trouble lies in plugged water-feed line. Generally plugging occurs at the junction between the cold feed pipe and the hot return header (black pipe on drawing).

If there is excessive water in the boiler, trouble may be due to a defective feed valve. To check, make certain bypass valve is closed. Then open union *C* a crack. If water trickles out, feed valve is not closing.

To check and clean automatic water-feed, close bypass valve. Open union *C* a crack. Use screwdriver and try to lift stem as shown. If valve cannot be easily lifted, float chamber is filled with sediment and float cannot drop. The only cure is to take device apart and clean it out or replace it.

If valve lifts easily, hold it up awhile and let the water flow through. This may clean the valve seat and permit it to close tightly.

Low boiler water. In addition to the automatic-valve difficulty discussed a few paragraphs back, low boiler-water level can also be caused by a defective check valve in the boiler feed line, low water pressure or high steam pressure. If the steam's pressure is higher than that of the feed water, no water will enter the boiler.

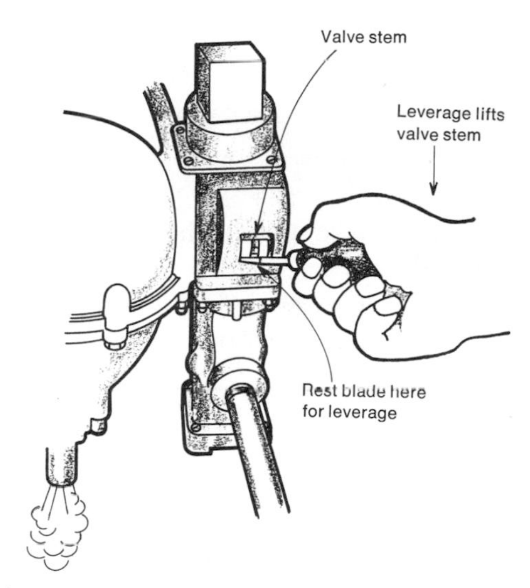

How automatic water-feed valve stem can be lifted with a screwdriver to test and clean valve seat.

High boiler water. Again, in addition to a defective automatic feed valve, excessive water in the boiler may be caused by a leaking bypass valve, a leaking immersed hot-water coil, dirty water which keeps the valve open from time to time and overly high water pressure.

An unusual but still possible cause of too much water in the boiler is an incorrectly mounted feeder valve. If the valve is mounted too high on the boiler's side, water will always be too high because the valve acts to keep the water level at just about its own center.

ADDING STEAM RADIATORS

Low-pressure steam and hot-water radiators are for most purposes, interchangeable. If you have a radiator of suitable size that was drilled for hot water, you can utilize it for steam by plugging one end and adding an air-release valve. If there isn't a hole drilled and tapped for the valve, you can drill one. The valve hole goes about two-thirds of the distance from the bottom to the top of the radiator on the side farthest from the steam feed pipe. Any domestic air-release valve will do. Actually, the best place for the valve is near the bottom, but it might get plugged with collected water, so it is placed high on the side. If you wish, you can even use the hole drilled for the bleeder valve used with hot water to mount the air-release valve. It will work.

If you have a finned radiator you want to use with steam, attach an elbow to one end and connect the air-release valve to the elbow. Then install the radiator with a tilt towards the feed pipe. As long as the air-release valve is above the central tube of the radiator, the arrangement will work.

Adding one or more radiators to a single-pipe or a double-pipe system is straightforward and uncomplicated. You merely break into the steam main where convenient and add a standard Tee. Connect the new radiator to the Tee through a gate valve preferably, but any other valve will do almost as well. If you are hooking onto a double-pipe system you will also have to connect a return pipe to the system's return main. If you connect your new radiator to an existing feed pipe, that is to say a small-diameter pipe leading directly to an existing radiator, your new radiator will more or less share the steam with the old. Thus, your new radiator will probably not get as hot as your old. You can connect your return to any convenient point; there is very little water running down the return.

Radiator tilt. Just be certain if you are installing a single-pipe radiator that it tilts strongly towards it feed pipe. If you are installing a radiator in a double-pipe system the radiator must tilt strongly towards the return or drainpipe.

Pipe size and type. Use black iron pipe. It is cheaper and a mite smoother than galvanized. Use the same diameter pipe that was used in the rest of the system. You can use larger pipe if you have it, but don't use smaller; it will reduce the steam received by the new radiator.

Running steam pipe. There is one major precaution to observe when installing steam pipe: You cannot make a "valley" in the pipe run. For example, should you encounter a girder you cannot pierce, do not run your pipe under it and then back up again. Instead, go above it and then back down to the original level. If you build a valley into the line, water will collect there and make noise and block the passage of steam.

17

Maintaining Septic Tanks

The disposal of human waste has always been a problem. Only a century ago, city folk got rid of waste by dumping it into the street. In the country people dumped their waste into nearby lakes and rivers. Where there were no natural water depositories, an artificial lake was devised. This was a hole in the ground filled with rain and spring water. It was and still is called a cesspool, but fortunately it is no longer permitted by law. Its place has been taken by the septic tank. Today, an estimated ten million Americans use septic tanks for waste disposal. They are practical, safe and accepted by all communities that do not have municipal sewer systems.

The tank itself is a watertight container of metal, vitrified clay, poured concrete or concrete block buried beneath the earth. The house drain is connected to the tank via a sewer line just as it would be connected to the municipal sewer main. Waste and soil flow by gravity from the house into the tank. The inflow connection is positioned just a little higher than the outflow connection. Both are near the top of the tank. Baffles are placed in the path of flow. Thus the waste remains in the tank for a minimum of sixteen hours.

While in the tank the soil and waste are attacked by anaerobic bacteria, which dissolve most of the solids. Matter not dissolved sinks to the bottom of the tank. The effluent thus formed is led by closed piping to a drain field which consists of short, separated lengths of pipe embedded in loose, coarse gravel. The effluent flows out of the pipe into the gravel and into the soil. At this time aerobic bacteria complete the decomposition of the effluent and render it harmless.

A properly designed tank that is not overloaded with grease and excess water needs nothing more than the removal of the accumulated solids every few years or so. The rate of accumulation depends on the nature of the discharged waste and the size of the tank in relation to the number of people using it.

You can clean your septic tank yourself if you wish. It is merely a matter of removing the cover and then removing the muck. It is, however, a dirty job. If you are thinking of using the sediment to fertilize your fields, *don't.* It will be dangerous for at least six months. You are well advised to hire a professional with a giant tank truck, a pump and some place to dump the muck.

The drain field requires no attention until it clogs up. When it does, the only solution is to dig up the field, remove and clean out the pipe, remove and clean the gravel or replace the gravel. However, a well designed and constructed drain field can operate for twenty years without attention.

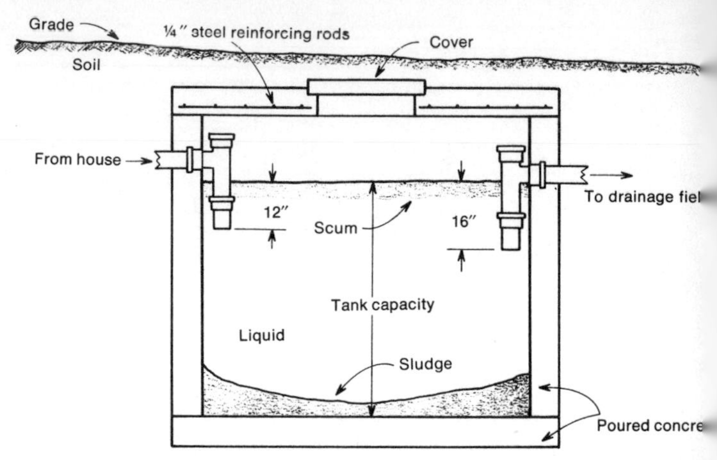

Cross-sectional view of typical septic tank. Soil and waste enter from left and remain in the tank while anaerobic bacteria break down the solids into a liquid, which flows out to the right. Material that is not broken down falls to the bottom of the tank and is removed when the tank is cleaned.

CARE OF SEPTIC TANKS

Limit water inflow. As previously stated, household sewage is decomposed by two types of bacteria. The first are anaerobic, which feed in the absence of air and require time to do their job. If too much water flows into the tank, the waste is not digested by the first group of plants but is carried out into the drain field where aerobic bacteria live. These plants require air, which is available in a shallow field, but as the waste has not been treated by the anaerobic group, the air-loving bacteria find it hard chewing. If there is an odor in your drain field, chances are that too much water has entered the septic tank.

The cause may be one or more roof leaders connected to the house drain. Or possibly a swimming pool has been emptied into the house drain. Reconnect the roof leaders elsewhere and do not continue to drain the pool into the house drainpipe.

Another way to further reduce the quantity of water flowing into the septic tank is to transfer the drain from your clothes washer to a dry well outside the building.

Incidentally, as the bacteria do their best when they are warm, you can sometimes improve the winter operation of your tank by increasing its cover of dirt.

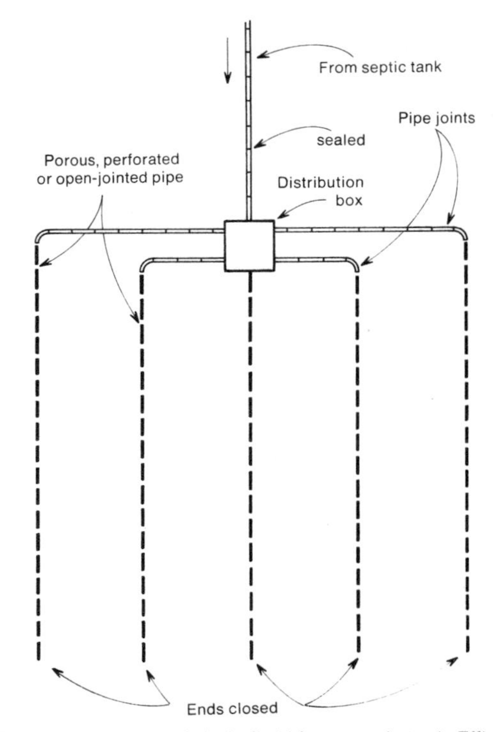

Diagram of a typical drain field for a septic tank. Effluent from septic tank flows through tightly joined pipe to distribution box, and then into a network of drainpipes.

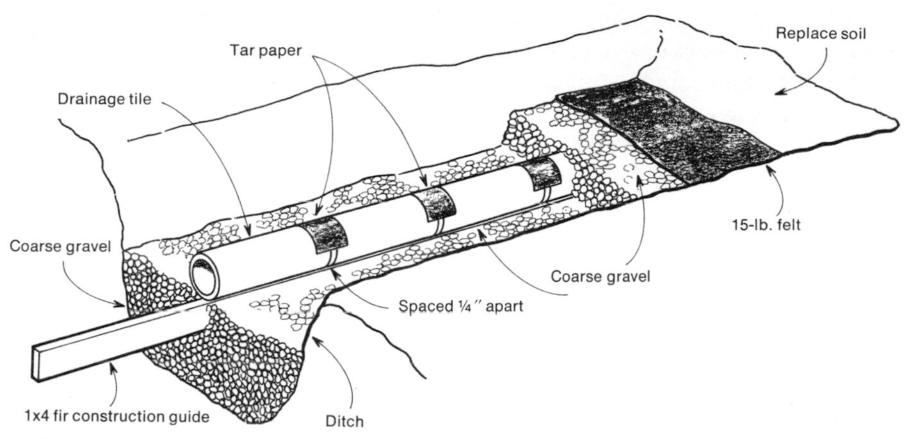

Drain field consists of clay drainpipes laid on a bed of coarse gravel. The open joints are covered with strips of tar paper and covered with more gravel. Liquid is thus exposed to air and can seep into the ground.

Avoid chemicals. When drainpipes that lead to a septic tank clog, do not use chemical drain openers such as Drano and Clobber. Use a snake to clear the stoppage. By the same token, don't dump unused paint thinner, chemical cleaners, photographic chemicals and the like into the house drainpipe. The chemicals wreak havoc with the bacteria, and without the all-important plants the muck in the tank remains a collection of solids and will not flow out into the drain field for further treatment and purification.

If your septic tank is sluggish, meaning the house sewer doesn't drain as rapidly as it should, and yet there is no blockage in the line, the trouble can be lack of bacteria. The only easy cure is to stop emptying chemicals into the pipe and give plants sufficient time to multiply.

Avoid grease. Train your kitchen crew not to dump used cooking fats and oils into the kitchen sink. Fats and oils are not easily digested by bacteria. Fats and oils in the tank tend to flow to the top and form a crust, interfering with soil flow, eventually plugging up the entire tank.

You can use cooking fats and oils as a supplement to your pet's food. You can also use it in your compost heap, and if you have no compost heap, just dump it on the ground in some hidden area. It will seep into the ground and eventually will be digested by oil-eating bacteria.

Avoid grinders. If you are remodeling an existing home, think twice before adding a grinder—food disposer—beneath your new kitchen sink. A grinder will increase the load on a septic tank by approximately 50 percent. If the septic tank cannot handle the increase you are in trouble.

Tank inspection. Should you need to inspect your septic tank, do not remove its cover with a lighted cigarette in your hand. The tank is filled with methane and other explosive gases. Do not enter the tank or even put your head inside for at least a couple of days after you have removed its cover. Septic tanks are filled with poisonous gas.

TROUBLESHOOTING A SEPTIC SYSTEM

Drain field wet. This is probably caused by too much water entering the septic tank. But it can also be caused by an old drain field that has plugged up with normal use. Drain fields consist of open-jointed pipelines laid in a bed of gravel. After twenty years or so the gravel becomes choked with sewage particles and can no longer pass liquid. The only solution is to dig up the field and replace the crushed stone.

Sewer backs up. If the pipeline itself is not plugged, trouble could be due to absence of bacteria caused by chemicals, presence of too much solidified grease and oil, or a filled tank. The last is a process normal to tank use. The bacteria do not liquefy all the solids, some sink to the bottom and fill the tank. Eventually the tank must be cleaned. As stated, this is best left to companies with the proper equipment.

Water Hammer

There is water hammer and there is water noise. The two are somewhat similar, but they are different. They are similar in that both make noise in the pipelines of a house, but the noises are different and so is the effect these noises have on the water system.

Water hammer is, literally, water hammering against the closed end of a pipe or against a closed valve. It happens every time a valve is closed, but it is of consequence only when the water's rate of flow is high and the stoppage is sufficiently abrupt.

Under such conditions the water slams into the obstruction like a solid. When the flow is both sufficiently rapid and stopped sufficiently fast, pressures of up to 1,000 psi can be developed in ordinary domestic water systems. Pressure this high is rare, but pressures sufficient to burst plastic pipe are not uncommon.

The cause of the shock wave produced by suddenly stopping water flow is due to the nature of water itself. Water is heavy, weighing 8.3 pounds per gallon. Water is almost incompressible. Water flowing through a pipe may be likened to a long, friction-free rod. When the flow is suddenly stopped, the entire weight of that heavy rod is stopped. Water, when confined within a pipe, is not the soft, gently spreading liquid we pour out of a glass.

How do you know when water hammer is harmful?

Shock waves strong enough to be damaging to pipes and valves are clearly audible. If you can hear the water make noise, you should take steps to eliminate it.

Assuming you are troubled by water hammer, how can you find where it is occurring? There are several ways. First, the particular type of sound that is made is usually indicative of the device making the trouble. Second, the sound grows louder as you come closer to its source. Third, in some instances it is possible to alter conditions and so localize the trouble.

Some examples:

A single thump is usually caused by a fast-acting valve, most often by an electrically operated valve such as is found in a dishwasher or washing machine. To make certain, partially close the valve in the feed line. With the pressure reduced the noise will be reduced or even eliminated.

Rapid, repeating thumping is usually caused by a worn faucet or valve that is so loose the stem bounces up and down in response to the water pressure. Sometimes you can actually feel them bounce under your hand. Sometimes they will make noise only when partially open.

PREVENTING WATER HAMMER

Since water hammer is caused by rapidly flowing water that is suddenly stopped, there are three approaches to its prevention. We can reduce the speed of flow; we can slow the stopping process; or we can absorb the shock wave.

Pressure-reducing valve. Rapid flow is caused by high water pressure. Water pressure can easily be reduced by use of a pressure-reducing or regulating valve in the water-service line following the water meter. The valve is easily adjusted to reduce the incoming pressure to any value desired.

Unfortunately, pressure-reducing valves are not inexpensive and not always practical. For example, you may need fairly high pressure to supply a third-floor bathroom. If you reduce the incoming pressure to keep a basement washing-machine valve silent, you may reduce top-floor water flow to a dribble when more than one faucet is open.

One occasionally effective alternative to installing a pressure-reducing valve and lowering top-floor pressure is to install an ordinary globe valve in the water line giving trouble. Pressure to that appliance is reduced. This is comparatively easy; however, when all the faucets in the house are open there may be insufficient pressure for the appliance. Most, if not all, are "timed." If the tank isn't full after a predetermined number of minutes, the washing cycle begins anyway. A pressure-reducing valve compensates for pressure changes due to variations in water flow. The fixed valve does not.

Closure speed. While you can easily vary the speed at which you manually close a valve, there is no way of altering an electrically operated water valve. So this method of preventing water hammer isn't practical with electrical valves.

Shock absorption. When alternate methods are impractical, you can install a shock absorber, which you can purchase as a complete unit or assemble yourself.

The commercial shock absorber, sometimes called a water-hammer eliminator or arrestor, consists of a closed chamber more or less divided in half by a flexible membrane. The area above the membrane is sealed. The area below is connected by pipe to the water line just before the troublemaking high-speed valve.

With the shock absorber connected, most of the energy that would be expended by the water slamming into the suddenly closed valve is expended pushing the flexible membrane up against the pocket of air in the shock absorber. In other words, it acts as a kind of pillow.

A homemade shock absorber consists of a length of pipe positioned more or less vertically above the pipeline it is to protect and as close to the troublesome valve as possible. The top end of the shock absorber is tightly capped. The lower end is connected to the pipe by means of a Tee and a shutoff valve. The arm of the Tee is connected to a pet cock.

In operation the pet cock is closed, the connecting valve is open. Energy that would be expended against the suddenly closed valve is expended dashing upwards and compressing the air within the vertical pipe.

For most conditions an air chamber about

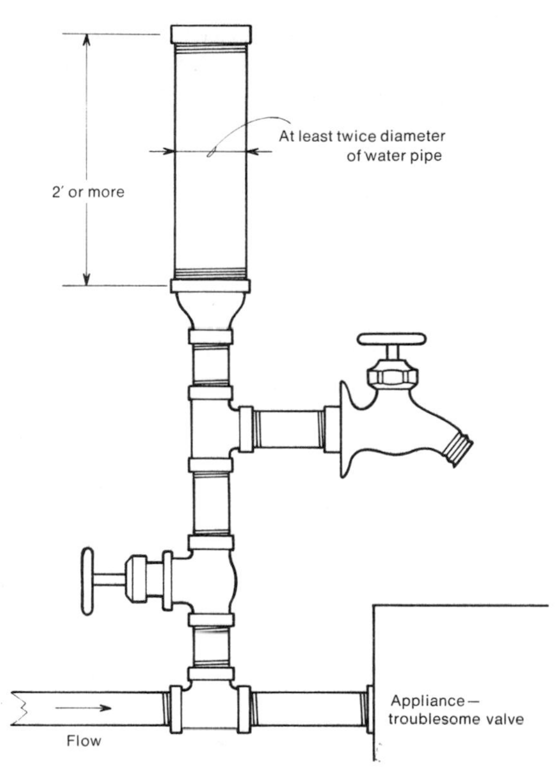

Homemade shock absorber for curing water hammer. In operation, pet cock is closed and valve leading to the water pipe is open. When device becomes filled with water, as evidenced by return of water hammer, air chamber must be emptied by closing valve and opening pet cock.

2 feet long and twice the diameter of the water pipe it serves provides adequate shock absorption. The connection between the air chamber and the water pipe should be made with pipe at least as large as the water pipe. Smaller diameter connecting pipe can be used with somewhat reduced effectiveness.

The membrane shock absorber requires no attention until the membrane breaks, which event can take place a dozen years after installation. The homemade device has to be drained about every six months. Its air dissolves into the water and the air chamber becomes filled with water. Then there is no more cushion and you will once again hear the water hammer.

If you include the pet cock and valve as shown, drainage is simple. If you do not include them you have to drain the entire water system to free the air chamber of water.

WATER NOISE

Water noise is caused by the action of water in a pipe against a moving or vibrating part of a valve. Generally, water noise is usually more of a nuisance than a danger to the pipes and valves. Like water hammer, water noise can be reduced by reducing pressure. Shock absorbers have only a negligible effect. However, it is usually possible to eliminate water noises by repairing or replacing the troublesome valve.

The sound itself will quickly lead you to the source. Once you believe you have discovered the source you can make certain by operating the valve. Opening it all the way or closing it usually stops the sound. When and where you cannot easily operate the valve you can make certain you have the correct valve by shutting off the water leading to the valve. Such would be the case with a squealing or singing ballcock valve. Reducing or stopping the flow of water to the valve would pinpoint it.

When the water pressure is very high many otherwise satisfactory valves become noisy. If several valves in your home "sing," and they are in fairly good condition, the solution is the reduction of water pressure. You can do this with either a pressure-reducing valve in the water-service line or by partially closing the shutoff valves in the fixture-supply lines.

If the noisy valves are old and worn, you can replace the entire valve or try replacing the stem and packing alone. In some instances a new stem will cure the noise.

19

Cold-Weather Precautions

When water is permitted to expand its usual 11 percent, it will freeze at about 32 degrees. When water is not permitted to expand, if it is confined in a closed vessel or pipe, a much lower temperature is necessary to freeze it.

Most flexible plastic pipe is fully capable of expanding 11 percent without permanent damage. However, all flexible plastic-pipe fittings cannot expand equally. So while a length of flexible pipe may escape harm upon freezing, don't assume fittings will not be cracked by very low temperature.

Rigid plastic pipe and both rigid and soft copper tubing can absorb some expansion without permanent harm. However, rigid plastic pipe fittings have very little "give." While piping of these materials can be exposed to below-freezing temperature if there is only a little water inside, they will be damaged if they are filled with water and closed by valves.

Galvanized, black and brass pipe have very little give. They will burst if water in them freezes. However, as they are considerably stronger than the aforementioned pipes, it requires a "harder" frost (lower temperature) to burst them.

If you are closing up a cottage for the winter, here is a checklist to help you prepare your plumbing system for the coming cold spell.

- ☐ Shut off the water at the main.
- ☐ Open all valves; let the water drain out of all pipes.
- ☐ Flush all toilets.
- ☐ Open all cleanout plugs and drain as many of the traps as you can.
- ☐ Drain radiators (if hot water).
- ☐ Drain boiler and hot-water tank.
- ☐ Open small screw-plugs on each of the valves. These drain the inside of the valves.

Add automotive antifreeze (glycerin type) to all the traps you could not drain, including the toilet.

When everything listed has been done there should be nothing in the building that can be damaged by frost. However, there is the water supply itself to take care of.

If you are served by a city main and can shut off the water at the curb stop, do so. Then open the water-service line as close to the basement wall as you can; just uncouple the pipe so that the service pipe is open to the air. If there is a water meter, uncouple it, tilt it and drain it of water.

If you cannot shut off the water at the curb stop, close the main valve and disconnect the pipe from its house side. Then wrap the main valve, and as much of the attached pipe as you can, with insulation. Use wire if, necessary, to hold the insulation in place.

It doesn't matter if your main valve and service pipe is below the frost line. As long as it is exposed to the air (in the basement), its temperature will eventually drop to that of the outside air if it is not insulated.

Should you have your own well system, open the valves and let the water run back into the well. Be certain that the pump is dry. Do not leave any water in the piston or the rotary pump casing, if that's the type you have. If your pump is submerged, let it be. It won't be harmed by frost. If you have a water tank, be certain to drain that too.

Whereas the task of preventing frost damage in an unoccupied home consists of removing all the water from the pipes or adding antifreeze to water that cannot be removed, preventing frost damage in an occupied home consists of finding uninsulated pipes before the frost does.

Look in the attic and beneath added-on rooms where pipe insulation may have been omitted or ripped off by animals. Look for completely closed radiators beneath open windows. Make certain the sill cocks have been opened and drained (after the indoor shutoff valve is closed). Look for bare pipes that pass very close to an uninsulated exterior wall.

Insulate all exposed pipes. Use pipe insulation or wrap some batts of rock wool around the pipe with the aid of galvanized wire. Either close open windows or open the radiators a crack so they won't freeze.

Assuming a deep frost is rushing your way, here are a few things you can do if you cannot insulate.

Let the water run. It is wasteful but it can reduce the chance of the water freezing. Place a turned-on electric light bulb next to the bare pipe. Don't let the bulb's metal shell touch the pipe. Loosely wrap some aluminum foil around the bulb and pipe.

THAWING FROZEN PIPE

The best way to thaw pipe is to wait for spring, and this is not a joke. Thawing frozen pipe can be a time-consuming, thankless task,

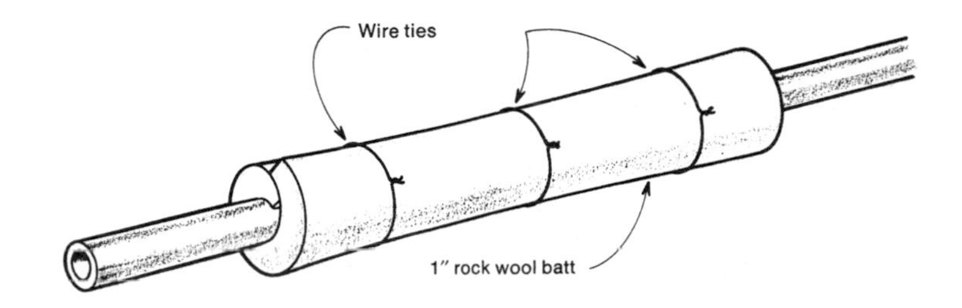

If you can't get regular pipe insulation wrap a bit of house insulation around the pipe and tie it in place with wire.

Thawing Pipes

Place a light bulb against the pipe to be kept warm and wrap some aluminum foil around the bulb and pipe.

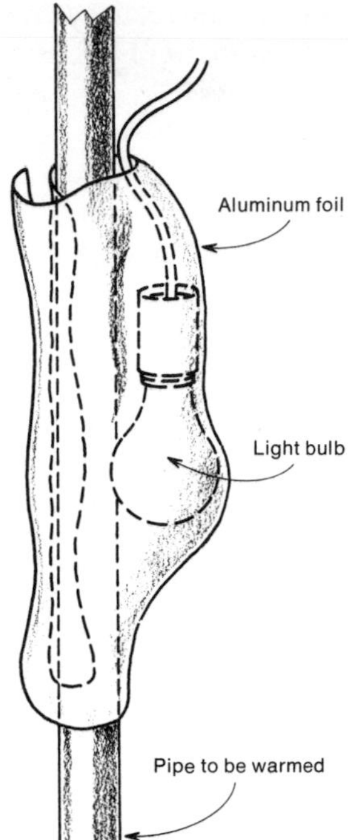

Wrap towels around the pipe and soak with hot water.

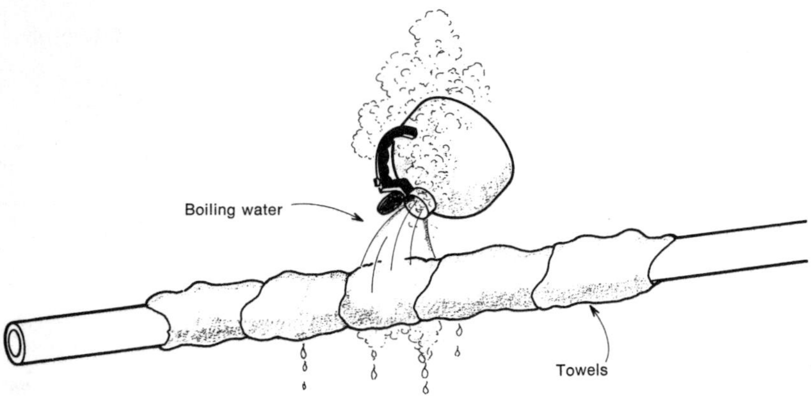

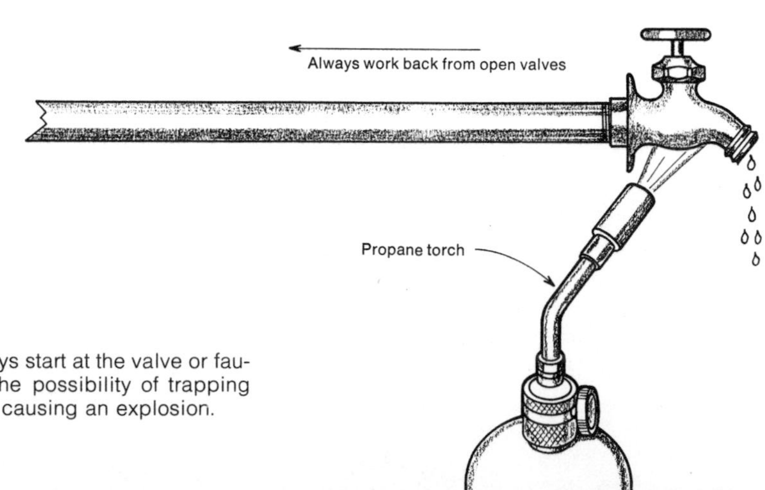

When thawing a pipe, always start at the valve or faucet end. This eliminates the possibility of trapping steam inside the pipe and causing an explosion.

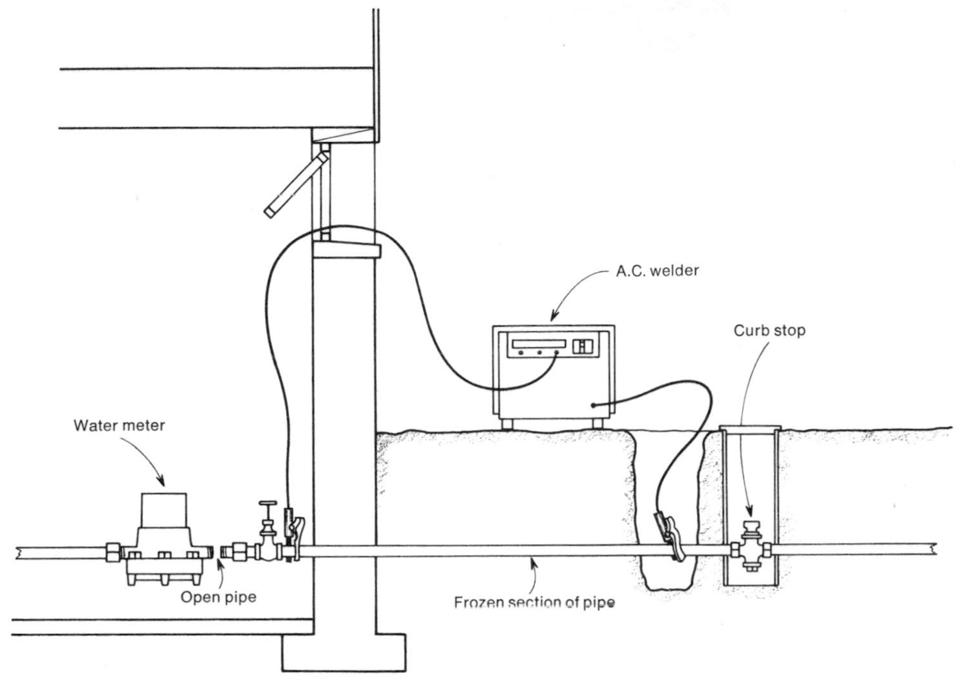

Using a welder to thaw frozen water-service pipe. Note that service pipe has been uncoupled from water meter to let water and steam out.

because if the temperature remains very low you may not be able to apply sufficient heat to unthaw the pipes. But if you can't wait for spring, here are several things you might try.

Place an electric blanket on the pipe. Wrap a towel around the pipe and soak the pipe in boiling hot water. Wire a large soldering iron to the pipe and turn it on. Use a blow torch.

When you do any of these things, and especially when you use the torch, always start at the faucet end of the pipe. For one, you want the ice you melt to run out. For another, applying the torch to the middle of a pipe and bringing the melted ice in the center to a boil can produce an explosion if the remaining ice locks the steam inside.

The best method, and the only one you can use when the pipe is not directly accessible is to use an electric welder. One lead from the generator or transformer is connected to one end of the frozen pipe, and the other lead to the other end of the pipe. Start by keeping the two leads as far apart as possible to make certain you do not overload the welder. Start with the lowest power output, just to be safe, and then gradually increase the power to speed melting. An AC or DC welder may be used. Polarity is unimportant.

This method produces fast results. You have to be certain the pipe ends (faucets and valves) are open so no steam is trapped, and you have to watch soldered copper joints to make certain they are not overheated and melted open.

Once you get more than a trickle of water to flow, you can ease your efforts; the flowing water will clear the rest of the ice, if, as we said before, the temperature isn't too low.

20

How to Run Gas Pipe

If you are planning to install some gas pipe or just hook up a gas stove, you are advised to study this chapter. Gas is fraught with danger. Many explosions in public and private buildings are due to escaped gas. Therefore, be extremely careful in handling gas fixtures, appliances and pipe.

Codes regulating gas pipe are strict and usually enforced. It is therefore advisable not to do slipshod work with whatever material may be at hand, but to determine the applicable local codes before starting and to use only approved pipe, fittings and valves.

APPROVED MATERIALS

For utility gases (gases normally provided by the local gas company), wrought-steel and wrought-iron pipe of standard weight and threading are acceptable. As black pipe is the least expensive, it is the one most often used. Pipe under ⅜-inch in diameter is not permitted.

For LPG (liquefied petroleum gas) standard weight, threaded wrought steel and iron pipe may be used, and in addition, seamless (regular) copper, brass, aluminum and steel tubing rated at a working pressure of no less than 115 psi and having a minimum wall thickness of 0.032 inch may be used. Standard K and L grade copper tubing meet these specifications. The aluminum shall not be in contact with concrete, plaster, the earth nor run along the exterior of a building.

Standard threaded and sweated fittings (check this locally) may be used with the exception of unions and running threads. Unions may be replaced by right- and left-hand couplings. Running threads refer to nipples so short that the thread at one end runs into the thread at the other end and there is no "body" to the pipe.

Standard globe and gate valves cannot be used at all. When the pipe is less than 3 inches in diameter, "gas approved" plug-type stopcocks with cast brass bodies and plugs must be used. When the pipe is larger than 3 inches, "gas approved" globe and gate valves can be used.

PIPE SIZE

It is fairly standard practice to limit pressure drop in any gas line within a building to 0.3 inches of water column (0.108 psi). Commercial gas pressure is very low so, proportionately, the drop is not insignificant.

As we have learned, pressure loss due to pipe friction increases with decreasing pipe diameter. On the other hand, pipe, fitting and valve costs increase rapidly with increasing pipe diameter (while pressure drop decreases). Our task then is to determine the smallest diameter pipe we can use without exceeding the 0.3-inch pressure loss limit dictated by the code.

Pressure loss depends on pipe length as well as diameter and, of course, the viscosity of the gas or fluid that passes through. In our case, the viscosity, measured in specific gravity, varies from 0.45 to 0.65. However, to do things the easy way, we assume whatever gas we are running has a specific gravity of 0.60 and let it go at that.

Pressure drop or loss also varies with the quantity of gas that flows. The more gas that passes through the pipe, the greater the pressure drop.

The first step is to draw a simple sketch showing the relationship of the various appliances you are going to supply with gas and the distance in feet of the pipe that will connect them to the gas meter or your LPG tank. See the accompanying sketch.

Then find the heat that will be generated

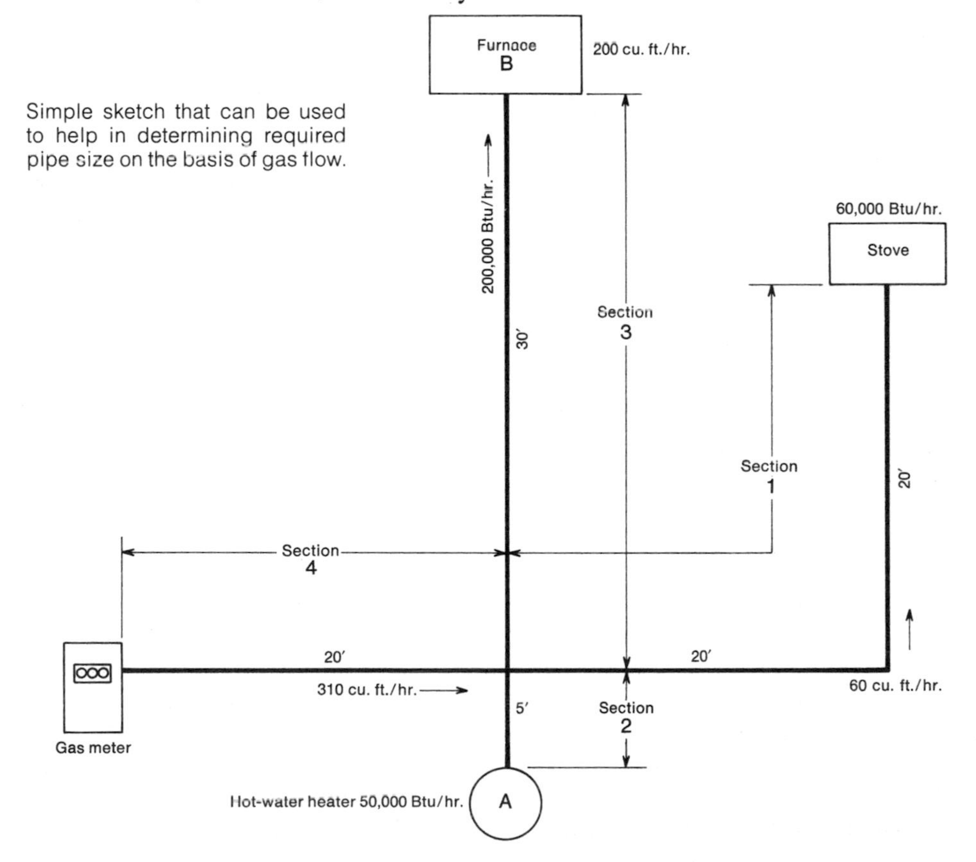

Simple sketch that can be used to help in determining required pipe size on the basis of gas flow.

by each appliance in Btu's per hour. This data should be on the nameplate. If it is not, refer to Table 39 and use the appropriate figure you find there.

Knowing the quantity of heat to be generated per hour, we find the amount of gas that will be burned per hour by dividing the heat generated in Btu by the Btu available per cubic foot of gas per hour.

The Btu content of the various gases used in the home varies, but for our example we will ascribe a figure of 1,000 Btu per cubic foot per hour.

Next we add these figures to our sketch next to the appropriate appliances.

Thus, appliance A is rated at 50,000 Btu/hr

$$\frac{50,000}{1,000} = 50 \text{ cu. ft./hr.}$$

$$\text{Appliance B} = \frac{200,000}{1,000} = 200 \text{ cu. ft./hr.}$$

$$\text{Appliance C} = \frac{60,000}{1,000} = 60 \text{ cu. ft./hr.}$$

Returning to our sketch, we can insert the flow rates in the sections of pipe. Section 1 is 40 feet long and carries 60 cubic feet per hour. Section 2 is only 5 feet long and carries 50 cubic feet per hour. Section 3 is 30 feet long and carries 200 cubic feet per hour. And, section 4 is 20 feet long and feeds all the appliances with a total of 310 cubic feet of gas per hour.

Diversity factor. We now know how much gas will flow through our gas-pipe system when everything is turned on. What we haven't determined is the diversity factor (called demand factor when dealing with water). In other words, we will draw 310 cubic feet per hour when the furnace, hot-water heater and gas stove are operating. But

Table 39

Appliance	Approximate Input Btu per hour
Furnace	100,000 to 250,000
Range	65,000
Wall oven or broiler	25,000
Countertop unit	40,000
50-gal. water heater	50,000
Instantaneous water heater	
2 gal./min.	142,800
4 gal./min.	285,000
6 gal./min.	428,400
Circulating water heater or side-arm	35,000
Refrigerator	3,000
Clothes dryer	35,000

Approximate Btu's per hour produced by various domestic appliances

will they all be operating at the same time? While this is doubtful, there is a good chance that all the appliances will be in simultaneous use for goodly portions of the winter day, so in this particular setup it is advisable to plan on 100 percent flow. In other situations, it may be possible to figure on no more than a percentage of the total demand at any given time and plan accordingly.

Selecting pipe. Refer to Table 40, find the length of the pipe in the column at the left (or use the next longer length), read right until you encounter the flow rate nearest yours (or the next larger quantity). Now read up to find the recommended pipe size.

In our example, pipe section 4 with a flow rate of 310 cu ft/hr and a length of 20 feet requires 1¼-inch pipe. Section 3, with 200 cu ft/hr, would be satisfactory with 1-inch pipe. Section 2, 5 feet long, carries 50 cu ft/hr and can be served with ½-inch pipe. Section 1, 40 feet long, carries 60 cu ft/hr and requires ¾-inch pipe.

Table 40

Length of Pipe	½″	¾″	1″	1¼″	1½″
	Iron-pipe size / Gas flow in cu ft/hr				
15′	76	172	345	750	1220
30′	52	120	241	535	850
45′	42	99	199	435	700
60′	38	86	173	380	610
75′		77	155	345	545
90′		70	141	310	490
105′		65	131	285	450
120′			120	270	420
150′			109	242	380
180′			100	225	350
210′			92	205	320
240′				190	300
270′				178	285
300′				170	270
450′				140	226
600′				119	192

Pipe size guide for use with gas. Suggested pipe size will cause an acceptable pressure drop of 0.3 inches or less when used with gas having a specific gravity of 0.60 or less. To use, find pipe length at left. Read across to gas flow in cubic feet per hour. Column figure appears in is recommended minimum gas pipe size for given flow rate and pipe length.

INSTALLING THE PIPE

Gas piping is installed much like water piping, with a few exceptions. A minimum of pipe dope is used on the threads and is always placed on the male thread, never on the female. Great care is taken cutting thread to make certain all thread is sharp and clean. Gas pipe is never bent. Turns are made with fittings. We are speaking of steel, brass and copper pipe, not tubing.

Branch connections are almost always made from the top and sides of a pipe. If you have to drop a branch, you go sideways at least 6 inches and then drop. All horizontal gas pipes are pitched towards a drip nipple. This is a Tee fitting with a short, capped nipple beneath the Tee, placed there for the express purpose of collecting moisture. Should sufficient water collect there to block gas flow, it can easily be drained.

Pipe smaller than ½ inch is never run lengthwise through the walls or flooring; that

A drop gas branch is always taken from a horizontal section as shown. First run pipe horizontally for at least 6 inches; then use an elbow to drop the pipe.

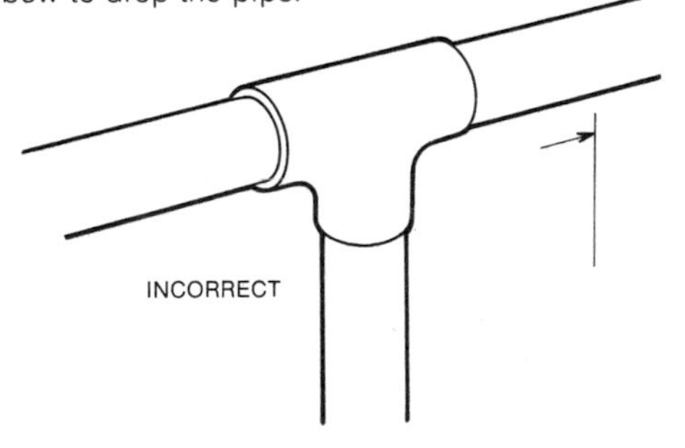

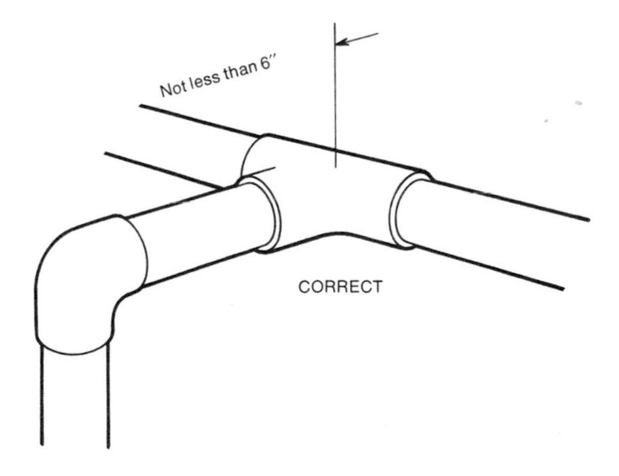

All horizontal gas pipe runs must be pitched. Use spacers under the straps to provide this pitch.

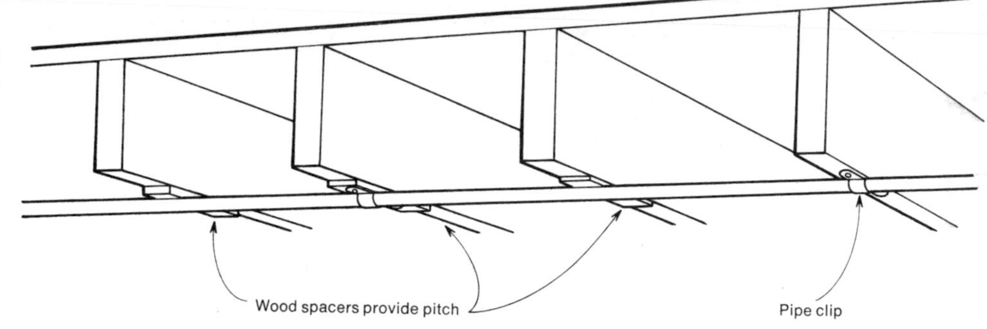

Use a cross-Tee and a drip nipple at the end of a pitched horizontal run and the end of a gas riser. Nipple collects moisture and is easily drained when necessary.

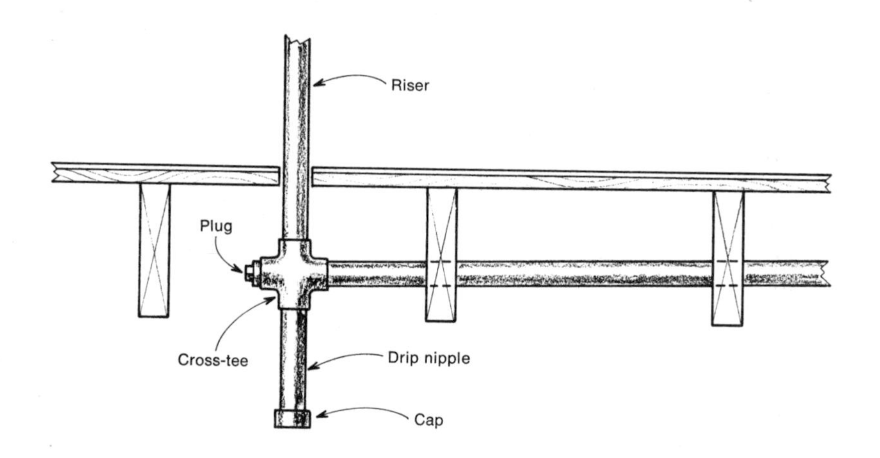

Another arrangement that may be used to produce pitch and drain gas pipe.

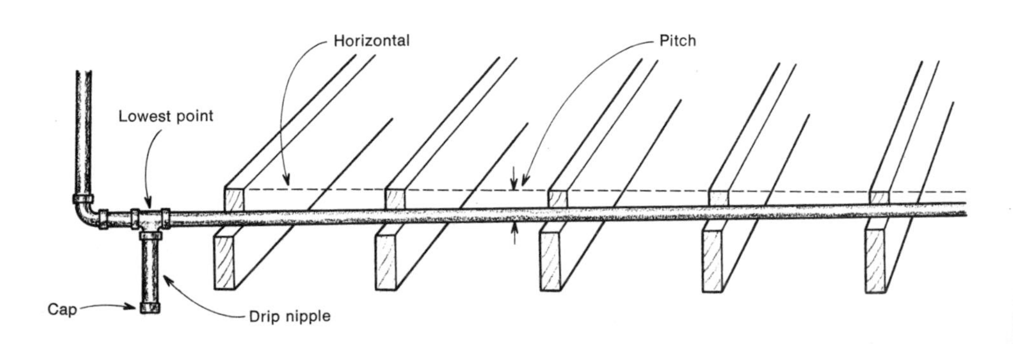

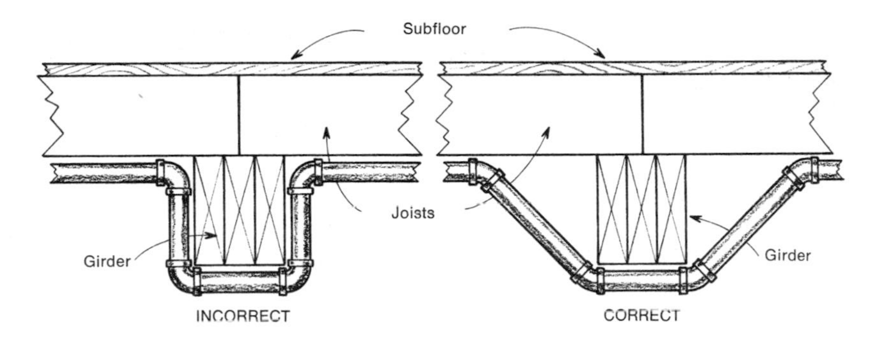

The correct and incorrect way to install gas piping below an obstruction such as a girder.

is to say, it is never concealed. When you run any soft pipe inside a wall, you must take positive steps to protect it from accidental nails and the like.

Gas pipe cannot be laid directly in concrete. If there is no alternative, a channel must be cut and the pipe laid within the channel so it can be reached if necessary.

Gas pressure is limited to 0.5 psi. If the pressure is higher, a pressure regulator should be installed prior to meter. The regulator should be vented to the external air and positioned or protected against tampering.

The usual test is to close all appliance valves and apply at least 3 psi to the gas line and hold it there for no less than 10 minutes. The gauge should be clearly readable to 0.0036 psi. In other words, not a sniff should escape. Some localities demand that the pressure hold 12 hours or more.

Some codes insist that the connection between gas pipe and the appliance be solid. If steel pipe is used, for example, it must be directly connected to the appliance. Others allow a flexible, corrugated copper tube to be used. Check your code, as the tube is far less troublesome to install.

Finding leaks. Never go after a gas leak with a candle or a match. You may find a pocket of gas and go up in glory. It has happened many times. I know that many old-timers use candles and it works. But there are many times when it doesn't.

The proper method is to paint the suscepted joint or valve stem with a soap and water mixture or a special liquid compound sold by plumbing supply houses for the purpose. If bubbles appear, there is a leak.

21

Testing Pipes

When you have gotten this far you may be in no mood to test your pipes. They are in and all the work is fortunately behind you. This may well be your attitude, but all plumbing inspectors insist the systems be tested.

Testing gas pipes. The test is usually made at the junction of your gas line and the company's service line. The pipe is opened at the joint, all appliance valves are closed and 3 pounds of air pressure applied to the pipeline. This is developed with a small pump made for the job and measured with a sensitive gauge. The gauge must be so sensitive it will clearly indicate a pressure drop of 0.0036 psi or less. The standard test allows for no pressure loss for a period of ten minutes. However, some communities require there be no loss of pressure for as long as 12 hours.

Testing water pipes. The usual method is to simply turn the water pressure on and watch for leaks.

Testing vents and drains. The test is made before the fixtures are connected. The drains and vents are either plugged, or in the case of a lead toilet bend, soldered closed. The tops of the stacks are left open and water is poured in until there is 10 feet or more water above the joint to be tested. Usually you have to wait 15 minutes to make certain there are no leaks.

One alternative to the water test is the use of compressed air. A pressure of 5 psi is usually required and must show no drop on a sensitive pressure gauge.

There are also smoke tests and peppermint tests (really) that are used. But the water test is the easiest, and is acceptable to most plumbing departments.

Testing the house sewer. The house sewer must also be similarly tested. Unless the lower end of the sewer has been left open for testing, a special type of plug must be used. It is a kind of balloon that is floated down to the end of the pipe and then inflated to seal the pipe end. Once the pipe end is sealed, any of the aforementioned tests that are acceptable may be used.

Index